Robotics and Automation in Industry 4.0

The book presents the innovative aspects of smart industries and intelligent technologies involving Robotics and Automation. It discusses the challenges in the design of autonomous robots and provides an understanding of how different systems communicate with each other, allowing cooperation with other human systems and operators in real time.

Robotics and Automation in Industry 4.0: Smart Industries and Intelligent Technologies offers research articles, flow charts, algorithms, and examples based on daily life in automation and robotics related to the building of Industry 4.0. It presents disruptive technology applications related to Smart Industries and talks about how robotics is an important Industry 4.0 technology that offers a wide range of capabilities and has improved automation systems by doing repetitive tasks with more accuracy and at a lower cost. The book discusses how frontline healthcare staff can evaluate, monitor, and treat patients from a safe distance by using robotic and telerobotic systems to minimize the risk of infectious disease transmission. Artificial intelligence (AI) and machine learning (ML) are looked at and the book offers a comprehensive overview of the key challenges surrounding the Internet of Things (IoT) and AI synergy, including current and future applications with significant societal value.

An ideal read for scientists, research scholars, entrepreneurs, industrialists, academicians, and various other professionals who are interested in exploring innovations in the applicational areas of AI, IoT, and ML related to Robotics and Automation.

Big Data for Industry 4.0: Challenges and Applications

Series Editors: Sandhya Makkar, K. Martin Sagayam, and Rohail Hassan

Industry 4.0 or the fourth industrial revolution refers to interconnectivity, automation, and real time data exchange between machines and processes. There is tremendous growth in big data from the Internet of Things (IoT) and information services which drives the industry to develop new models and distribute tools to handle big data. Cutting-edge digital technologies are being harnessed to optimize and automate production including upstream supply chain processes, warehouse management systems, automated guided vehicles, drones, etc. The ultimate goal of Industry 4.0 is to drive manufacturing or services in a progressive way to be faster, more effective, and more efficient, which can only be achieved by embedding modern-day technology in machines, components, and parts that will transmit real-time data to networked IT systems. These, in turn, apply advanced soft computing paradigms such as machine-learning algorithms to run the process automatically without any manual operations.

The new book series will provide readers with an overview of the state of the art in the field of Industry 4.0 and related research advancements. The respective books will identify and discuss new dimensions of both risk factors and success factors, along with performance metrics that can be employed in future research work. The series will also discuss a number of real-time issues, problems, and applications with corresponding solutions and suggestions. Sharing new theoretical findings, tools, and techniques for Industry 4.0, and covering both theoretical and application-oriented approaches. The book series will offer a valuable asset for newcomers to the field and practicing professionals alike. The focus is to collate the recent advances in the field, so that undergraduate and postgraduate students, researchers, academicians, and industry professionals can easily understand the implications and applications of the field.

Industry 4.0 Interoperability, Analytics, Security, and Case Studies
Edited by G. Rajesh, X. Mercilin Raajini, and Hien Dang

Big Data and Artificial Intelligence for Healthcare Applications
Edited by Ankur Saxena, Nicolas Brault, and Shazia Rashid

Machine Learning and Deep Learning Techniques in Wireless and Mobile Networking Systems
Edited by K. Suganthi, R. Karthik, G. Rajesh, and Ho Chiung Ching

Big Data for Entrepreneurship and Sustainable Development
Edited by Mohammed el Amine Abdelli, Wissem Ajili Ben Youssef, Ugur Ozgoker, and Imen Ben Slimene

Entrepreneurship and Big Data
The Digital Revolution
Edited by Meghna Chhabra, Rohail Hassan, and Amjad Shamim

Microgrids
Design, Challenges, and Prospects
Edited by Ghous Bakhsh, Biswa Ranjan Acharya, Ranjit Singh Sarban Singh, and Fatma Newagy

Handbook of Artificial Intelligence for Smart City Development
Management Systems and Technology Challenges
Edited by Sandhya Makkar, Gobinath Ravindran, Ripon Kumar Chakrabortty, and Arindam Pal

For more information on this series, please visit: https://www.routledge.com/Big-Data-for-Industry-4.0-Challenges-and-Applications/book-series/CRCBDICA

Robotics and Automation in Industry 4.0

Smart Industries and Intelligent Technologies

Edited by
Nidhi Sindhwani, Rohit Anand, A. Shaji George,
and Digvijay Pandey

CRC Press
Taylor & Francis Group
Boca Raton London New York

CRC Press is an imprint of the
Taylor & Francis Group, an **informa** business

Cover credit: Shutterstock

First edition published 2024
by CRC Press
2385 NW Executive Center Drive, Suite 320, Boca Raton FL 33431

and by CRC Press
4 Park Square, Milton Park, Abingdon, Oxon, OX14 4RN

CRC Press is an imprint of Taylor & Francis Group, LLC

© 2024 selection and editorial matter, Nidhi Sindhwani, Rohit Anand, A. Shaji George and Digvijay Pandey; individual chapters, the contributors

Reasonable efforts have been made to publish reliable data and information, but the authors and publishers cannot assume responsibility for the validity of all materials or the consequences of their use. The authors and publishers have attempted to trace the copyright holders of all material reproduced in this publication and apologize to copyright holders if permission to publish in this form has not been obtained. If any copyright material has not been acknowledged please write and let us know so we may rectify in any future reprint.

Except as permitted under U.S. Copyright Law, no part of this book may be reprinted, reproduced, transmitted, or utilized in any form by any electronic, mechanical, or other means, now known or hereafter invented, including photocopying, microfilming, and recording, or in any information storage or retrieval system, without written permission from the publishers.

For permission to photocopy or use material electronically from this work, access www.copyright.com or contact the Copyright Clearance Center, Inc. (CCC), 222 Rosewood Drive, Danvers, MA 01923, 978-750-8400. For works that are not available on CCC please contact mpkbookspermissions@tandf.co.uk

Trademark notice: Product or corporate names may be trademarks or registered trademarks and are used only for identification and explanation without intent to infringe.

ISBN: 9781032329437 (hbk)
ISBN: 9781032329444 (pbk)
ISBN: 9781003317456 (ebk)

DOI: 10.1201/9781003317456

Typeset in Times
by Deanta Global Publishing Services, Chennai, India

Contents

Editors ix
Contributors xi
Preface xvii

Chapter 1 Integration of Nature-Inspired Mechanisms to Machine Learning in Real Time Sensors, Controllers, and Actuators for Industrial Automation 1

Jay Kumar Pandey, Jayasri Kotti, Parimita, Dharmesh Dhabliya, Vipin Sharma, Sagar Choudhary, and Rohit Anand

Chapter 2 The Implications of Cloud Computing, IoT, and Wearable Robotics for Smart Healthcare and Agriculture Solutions 26

Jay Kumar Pandey, Santanu Das, Prashant Vats, Dharmesh Dhabliya, Sushma Jaiswal, K. Manikandan, and Digvijay Pandey

Chapter 3 Role of Human-Robot Interaction in Building Automated Mixed Palletization Based on Industrial Applications 46

K. Hamela, Vivek Veeraiah, Parimita, Dharmesh Dhabliya, V. Vidya Chellam, S. Praveenkumar, and Nidhi Sindhwani

Chapter 4 Emerging Soft Computing and Machine Learning Techniques Applicable in Industrial and Automated Applications 65

Jay Kumar Pandey, Vivek Veeraiah, M. Ranjith Kumar, P. Lalitha Kumari, Jayasri Kotti, M.K. Dharani, and Ankur Gupta

Chapter 5 PSO-Based Nature-Inspired Mechanisms for Robots during Smart Decision-Making for Industry 4.0 89

Harinder Singh, Veera Talukdar, Huma Khan, Dharmesh Dhabliya, Rohit Anand, Abhra Pratip Ray, and Sanjiv Kumar Jain

Chapter 6 Application of AI-Based Learning in Automated Applications and Soft Computing Mechanisms Applicable in Industries 110

B. Mahendra Kumar, Veera Talukdar, Huma Khan, Suryansh Bhaskar Talukdar, Ashok Koujalagi, R. Ganesh Kumar, and Ankur Gupta

Chapter 7 Investigating Scope and Applications for the Internet of
Robotics in Industrial Automation 132

*Harinder Singh, Vivek Veeraiah, Huma Khan, Dharmendra
Kumar Singh, Veera Talukdar, Rohit Anand, and Nidhi Sindhwani*

Chapter 8 Role of Artificial Intelligence in Making Wearable Robotics
Smarter 152

*Vivek Veeraiah, B. Karthiga, Chinnahajisagari Mohammad
Akram, Ashok Koujalagi, S. Nanthakumar, Vipin Sharma, and
Digvijay Pandey*

Chapter 9 Role of Machine Learning Approaches in Optimization of
AI-Based Industry 4.0 Healthcare Management Systems 175

Amogh Shukla, Gautam Chettiar, Vinit Juneja, and Sonakshi Singh

Chapter 10 Leveraging Financial Data to Optimize Automation: An
Industry 4.0 approach 198

*Kamakshi Mehta, Nitin Kulshrestha, Dharini Raje Sisodia,
Venkata Harshavardhan Reddy Dornadula, P. Krishna Priya,
and Ravi Ranjan*

Chapter 11 Assessing Cybersecurity Risks in the Age of Robotics and
Automation: Frameworks and Strategies for Risk Management 215

*Venkateswararao Podile, P.M. Rameshkumar, Vinay,
Suprateeka, Bhuvaneswari, and Sai Divya*

Chapter 12 The Role of Education in Addressing the Workforce Challenges
of Automation: A Global Perspective 229

*Deepti Sharma, P. Krishna Priya, Sachin Tripathi,
G. Bhuvaneswari, M. Kavitha, and V. Lakshmi Prasanna*

Chapter 13 The Intersection of Human Resource Management and
Automation: Opportunities and Challenges for HR Professionals .. 244

*Ajay Sidana, Juliet Gladies Jayasuria, Luigi Pio Leonardo
Cavaliere, S. Ramesh Babu, M. Kavitha, and P. Balaji*

Contents

Chapter 14 Data-Driven Insights for Agricultural Management: Leveraging Industry 4.0 Technologies for Improved Crop Yields and Resource Optimization 260

Ashok Kumar Koshariya, P.M. Rameshkumar, P. Balaji, Luigi Pio Leonardo Cavaliere, Venkata Harshavardhan Reddy Dornadula, and Barinderjit Singh

Chapter 15 Data Analytics and ML for Optimized Performance in Industry 4.0 275

K.V. Daya Sagar, K.K. Ramachandran, Purnendu Bikash Acharjee, Pratibha Singh, Harish Satyala, and Barinderjit Singh

Chapter 16 Business Intelligence in Action: Way of Successful Implementation of Automated Systems 288

Venkata Naga Siva Kumar Challa, K.K. Ramachandran, P.M. Rameshkumar, Luigi Pio Leonardo Cavaliere, Purnendu Bikash Acharjee, and Venkata Harshavardhan Reddy Dornadula

Chapter 17 A Review of Dielectric Resonator Antennas (DRA)-Based RFID Technology for Industry 4.0 303

Manvinder Sharma, Rajneesh Talwar, Digvijay Pandey, Vinay Kumar Nassa, Binay Kumar Pandey, and Pankaj Dadheech

Chapter 18 Leveraging Blockchain for Improved Supply Chain Management and Traceability in Industry 4.0 325

P. Balaji, Luigi Pio Leonardo Cavaliere, B. Nagarjuna, S. Ramesh Babu, M. Kavitha, and Barinderjit Singh

Chapter 19 Securing Automated Systems with BT: Opportunities and Challenges 337

Luigi Pio Leonardo Cavaliere, Swati Rawat, Neeru Sidana, Purnendu Bikash Acharjee, Latika Kharb, and Venkateswararao Podile

Chapter 20 Transforming Healthcare with Industry 4.0: The Impact of Social Media on Mental Health 349

Shashidhar Sonnad, Luigi Pio Leonardo Cavaliere, H.V. Vinay, Sonali Vyas, Lims Thomas, and S. Durga

Chapter 21 Multiband Antenna Design for Internet of Things (IoT) Applications 362

Aikjot Kaur Narula and Amandeep Singh Sappal

Chapter 22 A Counter-Propagation Based Neuro Solution Model for Categorization and Fee Fixation of Engineering Institutions 380

Krishna Kumar Nirala, Nikhil Kumar Singh, and Vinay Shivshanker Purani

Chapter 23 Industry 4.0 in the Nutraceutical Industry and Public Health Nutrition 396

Swapan Banerjee, Damanjeet Kaur, Digvijay Pandey, Sulagna Ray Pal, Ahamefula Anselm Ahuchaogu, Muhammad Omer Iqbal, and Binay Kumar Pandey

Index 423

Editors

Dr. Nidhi Sindhwani is currently working as an Assistant Professor at Amity Institute of Information Technology, Amity University, Noida, India. She earned her PhD (ECE) from Punjabi University, Patiala, Punjab, India, and has teaching experience of more than 15 years, is a Life Member of the Indian Society for Technical Education (ISTE), and a member of Institute of Electrical and Electronics Engineers (IEEE). She has published three book chapters, ten papers in Scopus/SCIE Indexed journals, and four patents. Dr. Sindhwani has presented various research papers at national and international conferences and has also chaired a session at two international conferences. Her research areas include wireless communication, image processing, optimization, machine learning, and Internet of Things (IoT).

Dr. Rohit Anand is currently working as an Assistant Professor in the Department of Electronics and Communication Engineering at DSEU, GB Pant Okhla-1 Campus, New Delhi, India. He earned his PhD (ECE) from IKG Punjab Technical University, Kapurthala, Punjab, India. Dr. Anand has teaching experience of more than 21 years and is a Life Member of the Indian Society for Technical Education (ISTE). He has published 16 book chapters, 18 papers in Scopus/SCIE Indexed journals, and 4 patents. He has presented more than 20 research papers at national and international conferences. He has chaired a session at fourteen international conferences. He has been the recipient of awards such as the Indian & Asian Record Holder, Integral Humanism Award, and Best Teacher Award. His research areas include wireless communication, electromagnetic field theory, antenna theory and design, image processing, optimization, optical fiber communication, and Io.

Dr. A. Shaji George has served the education and ICT industry for over 25 years. He is a member of the board of trustees and professor of Crown University International and cooperates with the advisory board of trustees as a special assistant to Pro-Chancellor at Chartered Int'l Da Vinci University Inc., Nigeria, West Africa. Dr. George is a trainer and counselor of the National Human Rights and Humanitarian Federation – NHRF, Kerala India. He is a recognized technical expert in ICT Systems, Networks amp; Telecommunication, designed and deployed large-scale ICT projects, and has published more than 60 research papers in national and international journals. His research interests include wireless, networking, cloud computing, Big Data, IoT, image processing, and industrial automation systems.

Dr. Digvijay Pandey is currently working as an Senior Lecture and Ex Acting Head of the Department in the Department of Technical Education, Kanpur, Government of Uttar Pradesh, India. He is also a faculty member at IERT Allahabad. He has teaching and industry experience of more than 10 years and has written 16 book chapters and 60 papers that have been published in Scopus Indexed Journals. Dr. Pandey has presented several research papers at national and international conferences and has chaired a session at IEEE International Conference on Advanced Trends in Multidisciplinary Research and Innovation (ICATMRI-2020). He has four patents that have been published in the Patent Office Journal and two that are currently being processed in the Australian Patent Office Journal. He serves as a reviewer for several prestigious journals. His research interests include medical image processing, image processing, text extraction, information security, and other related fields.

Contributors

Purnendu Bikash Acharjee
Christ University
Pune, India

Ahamefula Anselm Ahuchaogu
Department of Pure and Industrial
    Chemistry
Abia State University
Nigeria

**Chinnahajisagari Mohammad
    Akram**
Department of Mechanical Engineering
Acharya Nagarjuna University College
    of Engineering & Technology
Guntur, India

Rohit Anand
Department of Electronics and
    Communication Engineering
GB Pant DSEU Okhla-I Campus
    (Formerly GB Pant Engineering
    College)
New Delhi, India

S. Ramesh Babu
Department of MBA
Koneru Lakshmaiah Education
    Foundation
Vaddeswaram, India

P. Balaji
PG & Research Department of
    Commerce
Guru Nanak College (Autonomous)
Chennai, India

Swapan Banerjee
Department of Nutrition
Seacom Skills University
Birbhum, India

Bhuvaneswari
Koneru Lakshmaiah Education
    Foundation
Vaddeswaram, India

G. Bhuvaneswari
Vellore Institute of Technology
Chennai, India

Luigi Pio Leonardo Cavaliere
Department of Economics
University of Foggia
Foggia, Italy

Venkata Naga Siva Kumar Challa
Department of BBA
KL Business School, Koneru
    Lakshmaiah Education Foundation
Green fields, Vaddeswaram, India

V. Vidya Chellam
Department of Management Studies
Directorate of Distance Education,
    Madurai Kamaraj University
Madurai, India

Gautam Chettiar
Vellore Institute of Technology
Vellore, India

Sagar Choudhary
Department of Computer Science and
    Engineering
Sardar Patel University
Balaghat, India

Pankaj Dadheech
Computer Science and Engineering
Swami Keshvanand Institute of
    Technology, Management, and
    Gramothan
Jaipur, India

Santanu Das
Department of Biotechnology
Seshadripuram First Grade College
Bangalore, India

Dharmesh Dhabliya
Department of Information Technology
Vishwakarma Institute of Information Technology
Pune, India

M.K. Dharani
Department of Computer Science and Engineering
Kongu Engineering College
Erode, India

Sai Divya
Koneru Lakshmaiah Education Foundation
Vaddeswaram, India

Venkata Harshavardhan Reddy Dornadula
Startups and IIC, Chairman Office,
Sree Venkateswara College of Engineering
Nellore, India

S. Durga
Koneru Lakshmaiah Education Foundation
Vaddeswaram, India

Ankur Gupta
Department of Computer Science and Engineering
Vaish College of Engineering
Rohtak, India

K. Hamela
Department of Computer Science
Government First Grade College
Malur, India

Muhammad Omer Iqbal
Laboratory of Glycoscience and Glycoengineering
School of Medicine and Pharmacy,
Ocean University of China
Qingdao, China

Sanjiv Kumar Jain
Department of Electrical Engineering
Medi-Caps University
Indore, India

Sushma Jaiswal
Department of Computer Science and Information Technology
Guru Ghasidas Vishwavidyalaya
Bilaspur, India

Juliet Gladies Jayasuria
College of Business Management
University of Doha for Science and Technology
Doha, Qatar

Vinit Juneja
Vellore Institute of Technology
Vellore, India

B. Karthiga
Department of Electronics and Communication Engineering
Dhanalakshmi Srinivasan Engineering College
Perambalur, India

Damanjeet Kaur
Department of Health
Deakin University
Melbourne, Australia

M. Kavitha
PG and Research Department of Commerce
Guru Nanak College (Autonomous)
Chennai, India

Contributors

Huma Khan
Department of Computer Science and Engineering
Rungta College of Engineering and Technology
Bhilai, India

Latika Kharb
Department of IT
Jagan Institute of Management Studies
Delhi, India

Ashok Kumar Koshariya
Department of Plant Pathology
School of Agriculture, Lovely Professional University
Jalandhar, India

Jayasri Kotti
Department of Computer Science and Engineering
GMR Institute of Technology
Vizianagaram, India

Ashok Koujalagi
Department of Computer Science and Engineering
Godavari Institute of Engineering and Technology (Autonomous)
Rajamahendravaram, India

Nitin Kulshrestha
Christ Deemed to be University
Bangalore, India

B. Mahendra Kumar
Department of MCA
Dayananda Sagar College of Engineering
Bengaluru, India

M. Ranjith Kumar
Amrita School of Engineering
Amrita Vishwa Vidyapeetham
Chennai, India

R. Ganesh Kumar
Department of Computer Science and Engineering
CHRIST (Deemed to be University), School of Engineering and Technology, Kengeri Campus
Bangalore, India

P. Lalitha Kumari
Department of Computer Science and Engineering
Malla Reddy Institute of Technology
Secunderabad, India

K. Manikandan
School of Computer Science and Engineering
Vellore Institute of Technology (VIT)
Vellore, India

Kamakshi Mehta
TAPMI School of Business
Manipal University
Jaipur, India

B. Nagarjuna
School of Commerce and Management, MB University
Tirupati, India

S. Nanthakumar
Department of Mechanical Engineering
PSG Institute of Technology and Applied Research
Coimbatore, India

Aikjot Kaur Narula
Department of Computer Engineering
Punjabi University
Patiala, India

Vinay Kumar Nassa
Rajarambapu Institute of Technology
Sangli, India

Krishna Kumar Nirala
Gujarat Technological University
Ahmedabad, India

Sulagna Ray Pal
Department of Nutrition
Seacom Skills University
Birbhum, India

Binay Kumar Pandey
Department of Information Technology
College of Technology, Govind Ballabh Pant University of Agriculture and Technology
Pantnagar, India

Digvijay Pandey
Department of Electonics Engg.,
Department of Technical Education (Government of U.P)
Uttar Pradesh, India

Jai Kumar Pandey
Department of Electronics and Communication Engineering
Shri Ramswaroop Memorial University
Barabanki, India

Parimita
Warner College of Dairy Technology
Sam Higginbottom University of Agriculture, Technology and Sciences
Allahabad, India

Venkateswararao Podile
KL Business School
Koneru Lakshmaiah Education Foundation
Vaddeswaram, India

V. Lakshmi Prasanna
Humanities and Sciences
Gokaraju Rangaraju Institute of Engineering and Technology
Hyderabad, India

S. Praveenkumar
Centre for Tourism and Hotel Management
Madurai Kamaraj University
Madurai, India

P. Krishna Priya
KL Business School
Koneru Lakshmaiah Education Foundation
Guntur, India

Vinay Shivshanker Purani
Government Engineering College
Valsad, India

K.K. Ramachandran
Department of Management/Commerce/International Business
Dr. G R D College of Science
Coimbatore, India

P.M. Rameshkumar
Department of Corporate Secretaryship
Dwaraka Doss Goverdhan Doss Vaishnav College (Autonomous)
Chennai, India

Ravi Ranjan
School of Business and Commerce
Manipal University
Jaipur, India

Swati Rawat
MM Institute of Computer Technology and Business Management (MCA)
Maharishi Markandeshwar (Christ Deemed to be University)
Haryana, India

Abhra Pratip Ray
Sancheti English Medium School and Junior College
Pune, India

Contributors

K.V. Daya Sagar
Department of Electronics and Computer Engineering
Koneru Lakshmaiah Education Foundation
Guntur, India

Amandeep Singh Sappal
Department of Electronics and Communication Engineering
Punjabi University
Patiala, India

Harish Satyala
Department of Operation Management
Indian Institute of Management
Ranchi, India

Deepti Sharma
Department of Management
Uttaranchal University
Dehradun, India

Manvinder Sharma
Department of ECE
Malla Reddy Engineering College and Management Sciences
Hyderabad, India

Vipin Sharma
Department of Mechanical Engineering
Sagar Institute of Research and Technology
Bhopal, India

Amogh Shukla
Vellore Institute of Technology
Vellore, India

Ajay Sidana
Amity International Business School
Amity University
Noida, India

Neeru Sidana
Amity School of Economics
Amity University
Noida, India

Nidhi Sindhwani
AIIT, Amity University
Noida, India

Barinderjit Singh
Department of Food Science and Technology
IK Gujral Punjab Technical University
\Punjab, India

Dharmendra Kumar Singh
Department of Electronics and Communication Engineering
Dr CV Raman University
Bihar, India

Harinder Singh
Department of CS and IT
Sant Baba Attar Singh Khalsa College
Sandaur, India

Nikhil Kumar Singh
Government Engineering College
Gandhinagar, India

Pratibha Singh
Department of CSE
Guru Ghasidas Vishwavidyalaya
Bilaspur, India

Sonakshi Singh
Vellore Institute of Technology
Vellore, India

Dharini Raje Sisodia
Department of Management
Army Institute of Management and Technology
Greater Noida, India

Shashidhar Sonnad
Department of Electronics and
 Communication Engineering
 (Co-Education)
Sharnbasva University
Kalaburagi, India

Suprateeka
Koneru Lakshmaiah Education
 Foundation
Vaddeswaram, India

Rajneesh Talwar
Department of DICE
Chitkara University
Punjab, India

Suryansh Bhaskar Talukdar
School of Computer Science and
 Engineering
VIT Bhopal
Bhopal, India

Veera Talukdar
Kaziranga University
Jorhat, India

Lims Thomas
Department of Social Work
Vimala College (Autonomous)
Thrissur, India

Sachin Tripathi
Symbiosis Law School
Nagpur Symbiosis International
 (Deemed) University (SIU)
Nagpur, India

Prashant Vats
Department of Computer Science and
 Engineering
Faculty of Engineering and Technology,
 SGT University
Gurugram, India

Vivek Veeraiah
Department of R and D Computer
 Science
Adichunchanagiri University
Mandya, India

Vinay
Koneru Lakshmaiah Education
 Foundation
Vaddeswaram, India

H.V. Vinay
Department of MBA
BMSIT & M
Bengaluru, India

Sonali Vyas
School of Computer Science
UPES University
Dehradun, India

Preface

Manufacturers from all over the world are optimistic about the opportunities that Industry 4.0 will present to them in their respective industries. In the aftermath of the pandemic, it will be absolutely necessary for manufacturing businesses to combine artificial intelligence and hybrid cloud computing in order to improve their responsiveness, degree of automation, and level of intelligence across all of their essential jobs. This will be accomplished by increasing the level at which artificial intelligence and hybrid cloud computing are used. It's feasible that the implementation will be on the most fundamental level possible, but it might also be on an intermediate or even an advanced level. The efforts of the World Economic Forum (WEF) are apparently making the manufacturing business more competitive in comparison to past years. The change that has taken place in these industries has introduced some new problems that need to be conquered in order to conform to the standards. This has resulted in the creation of new opportunities, but it has also resulted in the emergence of some new difficulties. The manufacturing sectors of the country are currently being challenged by a number of difficult conditions in certain sections of the country.

The term "Industry 4.0," also known as the "fourth industrial revolution," refers to the integration of information technology into manufacturing facilities. This is the interpretation of what "Industry 4.0" means. It is most commonly used to refer to the process of digitally converting an industry; however, it can also pertain to the use of cyber-physical systems, the Internet of Things (IoT), cloud computing, and machine learning. In most cases, it is used to refer to the digital transformation of an industry. The construction of what is currently being referred to as a "Smart Factory" is going to be the result of the acceptance of the ideas that are defined in Industry 4.0. Many different computer systems are now able to communicate in real time with one another as well as with other human-computer systems and operators, all thanks to the Internet of Things (IoT). When applied within the context of Industry 4.0, machine learning makes it feasible to discover and extract patterns that are already present in the data. This makes it possible to automate the previously manual processes. The current state of human civilization is fast progressing in the direction of the widespread deployment of intelligent robots and other forms of mechanized automation. This generalization refers to a broader category of mechanized automation. The development of a wide variety of technologies, such as sensing, planning, perception, control, machine intelligence, and cloud computing, will, in the not-too-distant future, lead to the construction of robots that are more capable of performing a diverse range of occupations across a variety of industries and in everyday life. These robots will be able to perform these occupations in a manner that is more in line with human capabilities. The robots will have the ability to carry out these tasks in a manner that is more comparable to that of a human. As a consequence of this, it is of the utmost importance to stay informed of the quickly emerging ideas and technologies that are significant to the general advancement of the area of robotics, in addition to mastering the foundations of robotics.

Control system research, in particular insofar as it is relevant to robotics and automation, is generally regarded as one of the most essential areas of academic inquiry in the modern era. This is particularly the case in light of the recent developments in these fields. Robots have firmly established themselves as valuable instruments in a broad variety of industries, including manufacturing, health, the armed forces, and even some domestic activities. This includes the ability of robots to perform some tasks that were previously performed by humans. In addition, as improvements are made in intelligent control systems, robots are becoming "smarter" and more productive all the time. It is anticipated that this pattern will carry on well into the foreseeable future. The advancement of the robot's intelligence was made possible, in large part, by Artificial Intelligence (AI), in addition to modern and self-regulating controllers. Both of these factors worked in tandem to make the advancement possible. Robotics is a multidisciplinary topic that entails the design, construction, operation, and use of robots for a wide variety of automated tasks. Robots can be used in a variety of settings, including manufacturing, medicine, and the military.

The field of robotics encompasses not only the creation of robots but also their operation, maintenance, and application. The development of this field has not only led to a reduction in the amount of work that is done by people, but it has also led to the introduction of new standards of precision and accuracy in the many fields that it is used in. These new standards have been brought about as a result of the advancements that have been made in this field. It is impossible to foresee where advancements in robotics and automation will lead us in the future, but one possibility is that these technologies could lead to a significant advance in the progression of humankind. The idea of wearing robots increases the power that a human possesses physically; allowing them to lift and carry greater weights as well as meet other challenges and expectations in real life. The concept of wearable robots makes this kind of development conceivable. The researchers will gain a greater understanding of the numerous sensors and other components that are employed in wearable solutions as a direct result of their participation in this project. This comprehension will be directly applicable to the researchers' future work. The researchers will have a greater understanding of the multiplicity of concerns that arise in real-life circumstances and how best to address them, such as the most effective approach to assist workers who are hauling enormous products, the most effective way to assist elderly people, and other challenges that are encountered in the real world.

1 Integration of Nature-Inspired Mechanisms to Machine Learning in Real Time Sensors, Controllers, and Actuators for Industrial Automation

Jay Kumar Pandey, Jayasri Kotti, Parimita,
Dharmesh Dhabliya, Vipin Sharma,
Sagar Choudhary, and Rohit Anand

1.1 INTRODUCTION

Teaching a computer to correctly predict the future is called machine learning (ML). Resources such as storage, programs, servers, and development tools can be accessed from remote data centers. In order to better understand their customers, automate production processes, and generate better products, firms might use machine learning and analytic techniques [1]. Because of this, a company's profit margins will be higher than those of its competitors. This notion is the basis for machine learning. It is possible to alter the technological architecture using this paradigm. Using cloud computing and machine learning to create an "intelligent cloud" is the best way to go.

1.1.1 Machine Learning

Using data, computers can be taught to make better predictions, why do we use ML, and. If a person crosses the road in front of a self-driving car, it may be predicted if the phrase "book" refers to paperbacks or hotel bookings. It can also be predicted if an email is spam. Apples and bananas may now be distinguished from each other by computers for the first time in their history.

DOI: 10.1201/9781003317456-1

1.1.2 A Look at AI, ML, and Deep Learning

ML is a subset of Artificial Intelligence (AI), as depicted in Figure 1.1, while AI is a subset of ML [1]. Many businesses stand to benefit from artificial intelligence's productivity-boosting abilities.

Machine learning will become increasingly important as computing power continues to rise. It's possible that computers will one day be able to outperform us mentally. Detecting malignancies using computer algorithms is more accurate than using radiologists. But radiology is a new field that has barely begun. Millions of jobs could be displaced if artificial intelligence and automation become more widespread. This refers to any type of computer program that can "learn" on its own without the participation of a human. In the context of big data analytics and data mining, the phrase "machine learning" is often used to refer to a broad variety of software. Most predictive programmers, such as spam filters, product recommendation engines, and fraud detection systems, rely on machine learning algorithms as their "brains." Deep learning (DL) is a subset of machine learning [1–5].

1.1.3 IoT

The Internet of Things is abbreviated as IoT. In IoT, things (items) are synchronized over the internet [2]. This idea is possible with a little human help. Figure 1.2 illustrates this point.

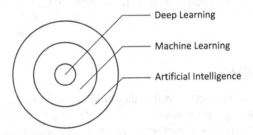

FIGURE 1.1 Interrelationship among AI, ML and deep learning.

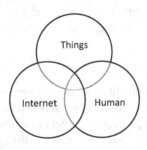

FIGURE 1.2 Tri-sectional intersection among things, internet, and human.

The IoT is the outcome of three different variables interacting with each other: things, the internet, and people. IoT adoption does not happen overnight. Rather, it was a slow process. There is a good likelihood that Artificial Intelligence (AI) will be included in IoT devices and apps in the future [6–10].

1.1.4 A Comparison between Traditional Internet and IoT

Table 1.1 sheds light on the key changes IoT has brought to the traditional internet.

1.1.5 Applications of IoT

The IoT can have a positive impact on a wide range of industries and marketplaces. People who want to save money on utility bills, as well as large organizations that want to streamline their operations, are all part of the user base. It's not just useful, but nearly critical in many areas, as technology advances and we move toward sophisticated automation [11–13].

i. When it comes to manufacturing, marketing, service delivery, or public safety, IoT can help. A wide range of processes may now be monitored in greater detail than ever before thanks to IoT. It is possible to respond more quickly and effectively to opportunities, such as clear client requests and non-conforming items and malfunctioning equipment, thanks to the IoT. When it comes to manufacturing industrial machinery shields, Joan runs a successful business. Shield composition and function are automatically updated by robots when regulations change, and engineers are alerted when the new criteria have been accepted.

TABLE 1.1
Comparison in Traditional Internet and IoT

Topic	Traditional Internet	IOT
Content creation	By human	By machines
Mechanism of combining content	By explicitly defined links	By explicitly defined operations
Value	Answers the query	Action and timely info
Connection type	Point-on-point and multipoint	Multipoint
Digital data	Readily provided	Not generated until augmented or manipulated
Data formats	Homogeneous	Heterogeneous
Composition	PC, server and smart phones	RFID and WSN nodes
Content Consumption	By request	By pushing info and triggering actions

ii. Using IoT in public administration and public safety can help enhance law enforcement, defense, municipal planning, and economic management Using new technology, many of the previous flaws in these programs have been addressed. There are several ways that city planners and governments might benefit from using IoT. For example, Joan lives in a small town. Her area has recently seen a rise in crime, and she's afraid to return home at night because of this. As a result of the system flags, local law enforcement has expanded their presence in the new "hot" region. In an effort to keep one step ahead of crooks, law enforcement has followed up on leads identified by area monitoring technology [14–17].

iii. IoT provides a personalized experience in our daily lives, from our homes and workplaces to the companies with which we do business. Our lives are better because of it, and we're better protected from harm as a result. Using IoT, we may create a workspace that is tailored to our individual needs. As an illustration, consider Joan's position as an advertiser. The technology recognizes her as soon as she enters her office. To her liking, it allows her to control the room's temperature and brightness. It reactivates her gadgets and opens applications to get her back to where she was before. Her office door had previously detected and recognized numerous visits from a coworker before she arrived. Joan's computer system automatically reads this visitor's messages.

iv. IoT advances our picture of the future of medicine, one in which a complex network of medical equipment is firmly intertwined, Physiology and Pharmacology, for example. IoT might change healthcare in a number of different ways. Integrating all aspects of medical research and organizations can lead to greater precision, more attention to detail, quicker reactions to events, and continuous growth and progress. As an example, Joan is a nurse who works in an emergency room. There has been an accident, and a man is in need of help. The computer recognizes the patient and retrieves the patient's medical information. As soon as a paramedic arrives on the scene, their equipment instantly captures and transmits important information to people in the hospital. The system analyzes both fresh and old data in order to provide suggestions. Patients' conditions are automatically reflected in the system as the travel progresses. By accepting the system's efforts for the distribution of medicine and manufacturing of medical equipment, Joan is encouraged by the system [18–22].

v. IoT acts in a comparable and deeper manner than existing technologies, analytics, and big data, promoting and disseminating information. It is possible to generate measurements and trends over time with this data, but it is often lacking in depth and precision. One way to make it better is by utilizing the IoT. More information and specificity mean more precise measurements and patterns. A company's customers' preferences and requirements can be better understood and met with customer profiling software. In both the business and consumer worlds, giving just relevant material and solutions improves both efficiency and strategy [23–27].

vi. In the current condition of advertising, the overflow of messages and the sloppy selection of audience members are causing problems for the advertising industry. Even with today's advanced data, advertisements continue to fail. As a result of the IoT, personalized advertising is possible. IoT enables consumers to participate in advertising rather than merely passively receiving it, which is a major benefit. To individuals looking for answers in the market or unclear of whether those answers exist, this increases the utility of advertising.

vii. A few of the numerous potential applications for IoT in environmental monitoring include environmental protection, monitoring of extreme weather conditions, the safety of bodies of water, conservation of endangered species, and commercial farming. The sensors in these setups keep tabs on and capture every detail of the surroundings. For example, IoT enables for more frequent sampling, a wider range of samples to be collected, and on-site testing that can be far more advanced than is now possible. This allows us to avert environmental catastrophes such as huge pollution. Commercial farming's lack of precision makes reliable weather monitoring difficult, necessitating human labor. A little level of automation is also included. Reduced human involvement in system operation, agricultural analysis, and monitoring are all made feasible by IoT. Systems are used to keep tabs on anything from crops to soil to the environment. They use a lot of data in order to improve their daily routines. Preventing health concerns like E. coli is another benefit of these devices [28–31].

viii. As in content distribution, IoT allows greater real-time insights in manufacturing. Because of this, there is no need to conduct significant market research before, during, and after a product goes on sale. IoT minimizes the risks of delivering new or updated products to the market since it provides more accurate and complete data. Consequently, it has a higher level of trustworthiness and reliability than information obtained from a range of dubious sources.

ix. With a more efficient use of resources and less waste at a lower cost, IoT instruments and sensors are utilized to do maintenance inspections and testing that would traditionally need human labor, saving time and money for manufacturers and the supply chain as a whole. For example, it is also possible to reduce many types of waste such as energy and materials with the use of IoT, which enhances operational analytics. As a result, it examines the entire process, not just a component of it at a particular location, so that any improvements can have more impact. Essentially, it reduces waste throughout the entire network and evenly distributes the savings that are gained from doing so.

x. Product Use and Protection: Even a sophisticated system cannot prevent faults, nonconforming items, and other dangers from reaching the market. Sometimes these incidents are caused by arguments that have nothing to do with the actual production process. The usage of IoT in manufacturing can help prevent product recalls and dangerous product distribution. There are fewer problems to deal with because it has more control, visibility, and integration [32, 33].

xi. We can automate a wide range of chores and needs in both residential and commercial settings by employing IoT in the context of buildings and other structures. Costs are lowered, safety is enhanced, productivity is increased, and the quality of life is improved, as can be seen in manufacturing and energy-related application areas.

1.1.6 Real-Time Sensors, Controller, and Actuators

Things + Controller + Sensors + Actuators + Networks make up an IoT gadget (Internet).

1.1.6.1 Real-Time Sensors

In IoT, the job of a sensor is to gather and convey information. In fact, sensors are one of the most significant ways to capture data in the IoT era. A sensor is a device that converts electrical impulses from physical occurrences or characteristics. This is a piece of hardware that converts environmental input into something useful for the system [34–39].

Physical characteristics such as temperature can then be converted into electrical impulses via a thermometer, for example. Real-time sensors provide data continuously using WSN and the cloud under IoT [40–43].

1.1.6.2 Controller

They utilize this information to make judgments. By and large, sensors don't give any thought to what they perceive. Data analysis requires controllers who can sift through and make sense of it all.

1.1.6.3 Actuators

When a mechanism or system is moved or controlled by an actuator, it is referred to as an actuator. The environment is sensed by the device's sensors, which create control signals for the actuators based on the tasks that need to be performed.

Servo motors are one kind of actuator. Actuators may be either linear or rotatory, allowing them to move to a certain angle or distance. Servo motors may be used in IoT applications to rotate a motor by a specified amount, such as 90 degrees, 180 degrees, etc. The actuator is the part of the control system that makes contact with the outside world. Both power and a command signal are required for operation. When a signal is sent to the energy source, it is converted into mechanical motion.

The IoT has proven to be a game-changer in the industrial automation sector. Implementing IoT technology might provide significant benefits for businesses operating in industrial automation industry [44–46]. With the use of IoT, innovative solutions may be created to current problems, resulting in smoother operations and higher output. The Internet of Things (IoT) is the networked linking of individually identifiable electronic devices using internet "data plumbing" technologies including

IP, CC, and WS. Tablets, smartphones, virtualized systems, cloud storage, and more are all essential for IoT's influence on industrial automation.

1.1.7 Cloud Computing

The term CC refers to the practice of paying a monthly fee or usage-based charges to a cloud service provider for the use of the provider's resources, such as software, development tools, and networking capabilities. The cloud service provider levies a monthly fee or usage-based fees for the use of these resources [34]. Depending on your choice of cloud services and how they compare to traditional IT, you may be able to:

- Save money on IT expenses by hosting some or your entire infrastructure in the cloud instead of on-site
- Reduce the time it takes to implement a service or product by moving it to cloud. Rather than waiting weeks or months for IT to reply to a request, acquire and set up suitable hardware, and install software, your business may start utilizing corporate applications in minutes. Developers and data scientists may benefit greatly from the cloud's self-service provisioning options
- Benefit greatly from the cloud's self-service provisioning options, in the case of developers and data scientists; make the most of your cloud service provider's worldwide infrastructure to expose your apps to as many people as possible

Figure 1.2 shows the primer of cloud computing.

By using on-premises cloud technologies, businesses may optimize utilization and save money while still offering customers the same degree of self-service and agility that they would receive from conventional IT infrastructure. You probably use some kind of cloud service on a regular basis, whether it's email or entertainment [34].

The only other way to get this kind of return was to make a significant investment in machine learning. An investment in infrastructure, as well as highly qualified programmers and data analysts, was prohibitively expensive at a time when there was little data to work with, making machine learning impossible. Compared to large corporations, small and medium-sized enterprises were concerned about this. These days it's a lot simpler thanks to the emergence of cloud computing. Once upon a time, businesses had to pay a hefty sum to a third party in order to get access to ML algorithms and technologies, customize them to their requirements, and finally start reaping the rewards. Because of this, cloud computing is crucial for machine learning. Companies of all sizes are hesitant to invest in the research, prototyping, testing, and production phases of building their own ML algorithms. These companies don't have to train their staff to be specialists in machine learning to enjoy its advantages. Increasing profits while decreasing investment risk is a win-win scenario for all parties involved.

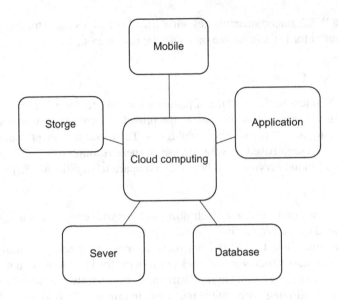

FIGURE 1.3 Primer of cloud computing.

1.1.8 Cloud Computing Platforms for Machine Learning

AWS, Google Cloud, and IBM Cloud are all well-liked cloud computing systems for machine learning (as shown in Figure 1.4). In this context, a brief explanation will be given as follows:

1.1.8.1 *AWS (Amazon Cloud Platform)*

AWS is a subsidiary of Amazon that provides a cloud computing service. This cloud computing platform for machine learning, which was initially presented in 2006, is currently one of the most popular. Machine learning may be accomplished using a number of AWS offerings, including:

- Amazon SageMaker allows you to develop and train ML models with ease.
- Amazon's Augmented AI's main function is to provide human oversight to automated prediction models.

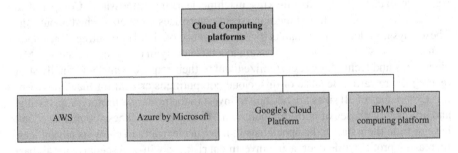

FIGURE 1.4 Cloud computing platform.

Integration of Nature-Inspired Mechanisms to Machine Learning

- Amazon Forecast – accuracy in these forecasts is enhanced with the application of machine learning.
- Amazon Translate translates words and phrases across languages by using ML and natural language processing (NLP).
- Using ML, Amazon's "Personalize" tool makes specific suggestions for users.
- When developing DL solutions, Amazon Machine Images (AMIs) are often used on AWS.
- Amazon Polly is a service that can convert text into speech with a human-like tone.

1.1.8.2 Azure by Microsoft

The firm is responsible for the creation of Microsoft Azure, a cloud computing platform. Since its introduction in 2010, it has seen widespread application in the fields of ML data analytics. For machine learning, Microsoft Azure offers:

- It offers intelligent cognitive services for Azure apps to make use of.
- Microsoft Azure's Azure Data Bricks provide analytics based on Apache Spark.
- Cloud-based service that offers scalable and intelligent bot solutions; developed by Microsoft.
- Microsoft Azure Cognitive Search is a solution for both offline and online applications that uses ML.
- Microsoft Azure ML allows users to build and deploy ML models on the cloud.

1.1.8.3 Google's Cloud Platform

The Google Cloud Platform is a cloud service offered by Google. It debuted in 2008 and uses the same backend systems that power Google's own offerings. The Google Cloud Platform has a wide selection of ML resources, such as:

- An ML model may be trained and refined with the help of Google Cloud Auto-ML.
- With Google Cloud AI Platform, you can develop, train, and control AI models.
- Google Cloud Voice-to-Text is, as the name says, a speech recognition technology that works with 120 different languages.
- It is used in the creation of cloud-based ML models that perform tasks such as text recognition.
- Google Cloud Text-to-Speech is the go-to service for voice-overs and other text-to-speech projects. Through the use of Google Cloud Natural Language, you can do NLP-based text analysis and classification.

1.1.8.4 IBM's cloud computing platform.

One such public cloud service is IBM's Cloud Platform. It supports public, private, and hybrid cloud deployments. The IBM Cloud has several machine learning options, such as:

In order to construct machine learning and artificial intelligence models, and to prepare and analyze data, IBM Watson Studio is used. This is a speech-to-text technology that can interpret spoken words and sounds. IBM Watson's Text-to-Speech technology can convert written text to spoken speech with a more human sound. IBM Watson NLU is what they use for processing language automatically. Machine learning is used by IBM Watson Visual Recognition to find and classify visual pictures. Virtual assistants can be created and managed using IBM Watson Assistant.

1.1.9 NIC

What we call Nursing Interventions Classification (NIC) is based on ideas from the natural world. The way nature works is very remarkable. At times, the unseen reasons that lie behind the observable events are many. Throughout the years, philosophers and scientists have studied these occurrences in the wild in an effort to explain them, adapt them to human needs, and recreate them in laboratory settings. The intricacy of the interplay between the living and nonliving worlds, and the many agents and forces within them, is far beyond the scope of human understanding. These forces shape the natural world and keep the balance, beauty, and vitality of life in check by working together and sometimes at odds with one another. The idea of evolution in the natural world is where we find this dialectics of nature. The natural world displays a recognizable pattern in the progression of complexity. Even in the absence of a single overseer, natural systems are capable of carrying out information processing in a decentralized, self-organized, and optimum fashion. Mechanical, physical, chemical, biological, and social forms are arranged from least complicated to most complex. Structure and history are used to illustrate the interdependence and interrelation of these elements. Alterations are made to the schedule because of the new conditions. Attempts to understand the underlying principles and mechanisms of natural, physical, chemical, and biological organisms that perform complex tasks in a befitting manner with limited resources and capability, as well as problem-solving techniques inspired by nature, are emerging as new fields of science, technology, and computing. Over the course of millennia, scientists have engaged in a conversation with nature that has resulted in the development of new ideas, techniques, and instruments, as well as clearly defined fields of study. Since then, people have been using their ingenuity to better observe and study the natural world. To create new computational tools like algorithms, hardware, synthesis of patterns, behaviors, and organisms, the field of NIC draws on a wide range of scientific disciplines, including but not limited to physics, chemistry, biology, mathematics, and engineering.

1.2 CONCEPTS RELATING TO NIC

As a computer paradigm, one might draw inspiration from nature in a pretty broad sense. There is still a wide variety of problems to be solved, phenomena to be synthesized, and questions to be answered, despite the fact that science and engineering have advanced over many hundreds of years and many sophisticated instruments and

Integration of Nature-Inspired Mechanisms to Machine Learning

techniques are available for their solution. In general, you should think about using natural computing techniques when:

- Many factors and solutions are at play, and the challenge has more than one goal to achieve.
- Complex pattern recognition and classification tasks are insufficient for modeling the situation at hand.
- Traditional methods of finding the best answer are impractical, time-consuming, or unreliable, but there is a quality metric that can be used to evaluate and compare different answers.
- In this case, having a variety of potential answers to the issue is either necessary or desirable.

The term NIC is used to describe a group of meta-heuristic algorithms that are modeled by, or otherwise draw inspiration from, natural events. All meta-heuristic algorithms that take their inspiration from the natural world have a similar characteristic: they use a mixture of rules and chance to simulate various events in the natural world. Several new computer paradigms that take cues from nature have emerged in recent years. The three main categories pseudobulbar affect, congenital bronchial atresia, and bronchial asthma.

1.2.1 Autonomous Entity

The principles of self-organization and complex biological systems provide the basis of a typical NIC software system. The decision-making process of a population of independent creatures in their natural surroundings may be recreated in the software. There are two types of autonomous entities in NIC software systems: effectors and detectors. There might be a plethora of sensors collecting data about the surrounding agents and their surroundings. It's possible for there to be a number of effectors, each of which may act in a certain way, undergo internal changes, and ultimately affect their surroundings. Using effectors, autonomous entities are able to more easily share data with one another.

Critical to the functioning of any autonomous entity are the rules of conduct implemented inside NIC software systems. When a group of sensors gathers and shares data about a local stimulus, the data is utilized to make a decision on how the group as a whole should respond to that stimulus. Learning is possible for autonomous entities because they may adapt to new environments by revising their behavioral norms over time.

Robotics, motion tracking, medical picture analysis, intelligent heat solutions, and aerodynamic design optimization are just a few examples of the fascinating new subject of Nature-inspired computing.

1.3 LITERATURE REVIEW

This section focuses on the information gathered by the researcher regarding prior studies, relating to machine language, IOT, cloud computing, and NIC. All these studies acted as the foundation of our research. A brief description is provided ahead:

Aderemi O. Adewumi, et al. (2017) developed and developing nations accept credit cards as a form of payment for online purchases. There were several advantages to using credit cards for online purchases. However, thieves found new ways to commit fraud, which led to a rise in the rate of fraud. Millions of dollars were lost throughout the world due to credit card fraud, which was worrying. Cybercriminals were always developing new and more sophisticated strategies; hence it was imperative that technology be enhanced and dynamic in order to keep pace with the continuously changing fraudulent tendencies. This task was very challenging because of the ever-changing nature of fraud and the lack of data available to academics. In their research, they looked at several methods for detecting credit card fraud. New ML and Nature-inspired based methods in the field of credit card fraud detection were the primary focus of their investigation. Apparently, the number of cases of credit card fraud that have been uncovered has been increasing lately. It also highlighted some of the limitations and contributions of existing credit card fraud detection algorithms while providing essential context for future research in this area. Their analysis served as a starting point and reference for further research by financial institutions and individual customers interested in discovering more advanced strategies for detecting credit card fraud [1].

Afsheen Ahmed, et al. (2018) explained that building elements for a smart cyber-physical global environment were already available thanks to the Internet of Things (IoT). There were many uses for the IoT in everyday life. Healthcare businesses, in particular, profited from the widespread availability of health monitoring, emergency response services, electronic medical invoicing, and other services. It was impossible for IoT devices to effectively deliver e-health services or analyze and store vast amounts of acquired data because of their low storage capacity. To overcome the limitations of IoTs, Cloud Computing technology was combined with IoTs in multi-cloud form, which provided a secure and on-demand shared pool of resources that may be used to supply effective and well-organized e-health facilities. When it comes to improving patient care, the use of IoT and multi-cloud platforms may help, but they can also put patients' privacy and dependability at risk. Identifying and evaluating the most significant security threats, as well as the current security approaches used to address them, were the primary goals of this comprehensive literature study [2].

Saru Kumari, et al. (2018) discussed that the term "Internet of Things" has become a catchphrase for all Internet-connected gadgets, including embedded devices, sensors, and other everyday things. Many embedded devices were already commonplace in our everyday routines because of the rapid advancement of this technology. However, in order for these embedded devices to share resources and communicate with one other, they must be connected to a vast pool of resources like a cloud. For government and commercial sectors, the promised IoT applications may be realized by combining cloud servers with these embedded devices. When information was shared between devices, there were security concerns, such as the protection of data and the authentication of devices. The recent publication of an ECC authentication solution for IoT and cloud servers boasted that it satisfied all security standards and was immune to a wide range of attacks. However, the method developed by Kalra and Sood was shown to be susceptible to insider attacks, password guessing from outside the network, device anonymity, SKA, and mutual authentication. Due to the backlash against Kalra and Sood's approach, they introduced a new authentication scheme based on ECC that can be used with IoT devices and cloud servers. Their paper suggests using the Automated Validation of Internet Security Protocols and Applications tool, which is widely recognized as the gold standard for formal security research. The suggested technique was more powerful, efficient, and secure than other similar schemes when compared against different known assaults, according to security and performance studies [3].

Debajyoti Misra, et al. (2018) created a dynamic waste management system which necessitates the use of clever solid waste bins. It was shown in their research how a solid waste management system may be automated using a unique integrated sensor system. An ultrasonic-level sensor and a variety of gas sensors were used to automatically detect dangerous gases and the maximum amount of garbage in the proposed smart bin. It was a new technique that employed cloud and mobile app-based monitoring to keep an eye on everything. In addition to checking the bin's maximum trash level, it also tested for a variety of foul gases. The information must also be conveyed to the appropriate authorities, which was the second element of the job. Because of the benefits it offers in terms of usability, accessibility, and disaster recovery, this novel strategy relied on a cloud server. It was possible to attach the data to the municipality's web server and take action right away. People can find the location of each trash can by its unique serial number. When using the eccentric method, everyone can get all the details they need about a bin's physical state to anyone needs to know about it. The data was linked by means of cloud-based web-information technologies on the host server [4].

Sanjay Sareen, et al. (2018) informed that ebola is a dangerous virus that spreads swiftly from person to person through direct contact, and it may lead to death in certain cases. Preventing the Ebola virus illness from spreading requires constant identification and remote surveillance of affected persons (EVD). Healthcare services built on IoT and CC are evolving into a more proactive and efficient choice for delivering remote patient monitoring. A novel architecture based on radio-frequency identification (RFID),, infrastructure was proposed for the diagnosis and monitoring of patients with Ebola. Their strategy was developed to halt the transmission of the disease in the first phases of an epidemic. The J48 decision tree was used to rank the severity of an illness depending on the user's reported symptoms. Users may be able to identify one another in close proximity using RFID. Temporal network analysis was used to characterize the current outbreak and track its progress (TNA). Amazon EC2 cloud was used to evaluate their proposed model's speed and accuracy using simulated data from two million users. Their suggested model showed a classification accuracy of 94 % and a resource usage efficiency of 92 % [5].

Chinu Singla, et al. (2018) presented IoT as a multi-tiered, interconnected network based on cutting-edge computational intelligence. Distributed entities and multimedia smart devices made a large-scale IoT ecosystem. Mobile Cloud Computing (MCC) recent growth in popularity may be attributed to the proliferation of high-powered smartphones. Since there are several clouds offering the same services, users have a hard time deciding which cloud to use when migrating computationally intensive applications. Therefore, selecting the most efficient cloud from a set of alternatives was an multiple-criteria decision analysis issue. By using triangular fuzzy elements and linguistic values to parameterize uncertainty and subjectivity, their idea offered an evaluation model based on an fuzzy analytic hierarchy process and an fuzzy technique for order preference by similarities to ideal solution. An improved understanding of the assessment process and a more effective decision support tool were some of the benefits of the computational intelligence decision modeling approach that was offered [6].

Nam Yong Kim, et al. (2018) cited that IoT in the cloud has become more popular as more and more network data is handled. In the dispersed processing of networks, for handling large amounts of data fog computing played a significant role. Fog computing provided the benefit of dispersed processing, but a new architecture was needed to assure excellent performance and easy network administration. The IoT network helps future IT research by allowing for efficient and easy administration of the system [7].

A.A. Zaidan, et al. (2018) discussed IoT technologies, which were innovative and disruptive, smart houses now include communication components that are widely scattered and bound. For their study to be successful, they needed to be aware of the methodologies that were used and the limits that existed at that time. The authors conducted an exhaustive literature review of literature pertaining to "smart homes," "IoT," and related applications, and used this information to create a taxonomy. Research publications on IoT-based smart home technologies were located through a search of Science Direct, IEEE Xplore, and the Web of Science. Using the rubric "Communication components elements," a total of 82 articles were selected and organized into four distinct subcategories. In the first category, internet devices were modeled after the needs of each development stage; in the second, analytical studies were conducted to track any potential shifts in the variables used in a case study; in the third, evaluations and comparisons were made to determine the devices' relative value, and in the fourth, surveys were conducted to gain a comprehensive understanding of the field. The following step included analyzing the literature's findings about motivation, application barriers, and the creation and usage of smart homes to inform an analysis of IoT-based technologies in this context. In addition to the 82 articles that were evaluated before, their research also included IoT solutions, communication protocols, the IoT stack protocol, and quality of service for IoT-based smart home technologies [8].

Laizhong Cui, et al. (2018) portrayed how the IoT has become a major network paradigm and how it links a plethora of intelligent gadgets together. The explosion of IoT apps and services was in large part due to the vast amounts of data being produced by connected devices. Machine learning has been successfully used in a variety of domains, including CV, CG, NLP, speech recognition, decision-making, and intelligent control. The study of networks was also a part of it. Researchers were interested in applying machine learning to a variety of networking challenges, including routing, traffic engineering, RA, and security. The use of machine learning to improve IoT applications and to provide IoT services including network and traffic engineering, security, and quality of service optimization has increased recently. In their review paper, they looked at how machine learning can be used in the Internet of Things. Significant developments in ML for the Internet of Things addressed a wide range of use cases. Because of ML for IoT, we can now create intelligent IoT apps that provide deep insights. The authors wrote this piece to both keep up with recent developments in ML for IoT and set themselves apart from previously published surveys. Their study covers a wide range of topics related to machine learning for IoT, including traffic profiling, device identification, network management, and common IoT use cases. Yet they also brought to light problems and topics that still need to be addressed in the field of study [9].

Mario W. L. Moreira, et al. (2018) reported that researchers in the field of natural computing have access to an unlimited supply of computational models and paradigms that can be derived from nature's limitless variety. It's an innovative technique to achieving efficient and generalized artificial neural networks (ANNs) by using biologically inspired algorithms to optimize the architectures and weights simultaneously. Their approach combined a simple structural model with low rates of training error. Complexity and a high incidence of errors are now plaguing clinical decision support systems (CDSSs) for prenatal care. Particle swarm optimization (PSO), which takes its inspiration from nature, was designed to cut down on the computational overhead of multi-layer perceptron (MLP), an ANN-based approach, without sacrificing its accuracy. The results indicated that the PSO technique greatly improved the performance of the computational model, with lower validation error rates. This technique has the potential to speed up the MLP algorithm's training process. The performance and accuracy of the CDSSs are improved by the proposed algorithm inspired by natural processes and its approach to adjusting parameters. Their approach may be useful in e-health systems as a means of clarifying information for high-risk pregnancies. Alternatively stated: on average, the new approach was 26.4% more accurate than the old one, 14.9% faster in terms of total physical response, and 35.4% less false positive rate than the old one. It surpassed the MLP algorithm by 2.3% in accuracy and by 10.2% in area under the receiver operating characteristic curve when serving pregnant women [10].

Nadia Boukhelifa, et al. (2018) have noted that evaluating interactive ML systems is still difficult. The human learns from these systems as much as the systems learn from and adapt to humans. It's difficult to get a handle on the nuances of cooperative behavior and adaptability. Throughout this chapter, they described their efforts to develop and evaluate interactive ML applications in a variety of fields. Algorithm-centered analysis and human-centered assessment should be combined in order to analyze the system's computational behavior as well as its usefulness and efficacy for end-users. An example of their work was a guided search visual analytics tool constructed utilizing an interactive evolutionary method. As a result, they believed that human-centered design and assessment may complement algorithmic analysis, and play a significant role in resolving the "black-box" impact of machine learning. Finally, they addressed research possibilities that need the use of human–computer interaction approaches to support both the visible and hidden roles that people play in interactive machine learning [11].

1.4 PROPOSED WORK

1.4.1 PROBLEM STATEMENT

There have been a number of studies conducted in the field of industrial automation that have resulted in important advances in the field. However, it has been noticed that the work of implementing an integrated system for industrial automation that should be supported by IoT is a tough undertaking. For Internet of Things-based industrial automation systems, there is an urgent need to develop a model that is both scalable and efficient [47].

1.4.2 PROPOSED METHODOLOGY

Figure 1.5 shows the proposed research methodology and Figure 1.6 shows the flow chart of the proposed work. Methods from IoT have been used in a variety of recently conducted industrial-related research projects. In the field of mobile computing, research has often focused on the problem-solving aspects of the field. In spite of this, adopting an IoT strategy in industrial automation presents a number of obstacles, one of which is the need to include an optimization mechanism in order to ensure the integrity of an industrial automation system while operating in an Internet of Things environment. In addition to this, traditional techniques of research need to have a greater capacity for scalability. There is an immediate need to provide a different approach for the industrial automation system to improve levels of accuracy and handle issues with performance.

Flowchart of proposed work

1.5 RESULTS AND DISCUSSION

1.5.1 CONFUSION MATRIX IN THE CASE OF A CONVENTIONAL MODEL

Accuracy for the typical model is shown below. A confusion matrix from traditional model testing is shown in Table 1.2, and an accuracy table derived from Table 1.2 is shown in Table 1.3.

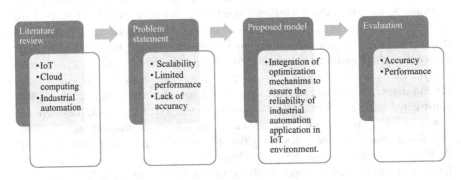

FIGURE 1.5 Research methodology.

Integration of Nature-Inspired Mechanisms to Machine Learning

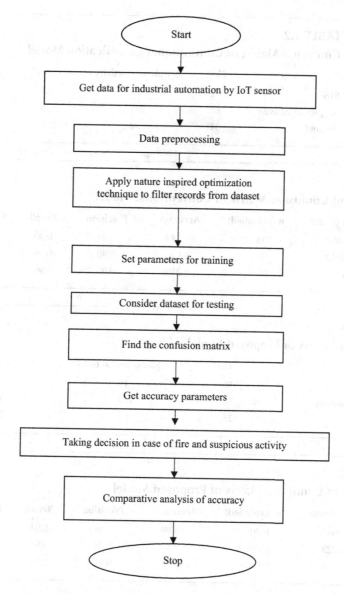

FIGURE 1.6 Flow chart of proposed work.

1.5.1.1 Results
TP: 2692

Overall Accuracy: 89.73%

Table 1.3 displays the retrieved accuracy metrics after applying accuracy, recall, precision, and F1-score to Table 1.2.

TABLE 1.2
Confusion Matrix of Conventional Classification Model

	Fire	Suspicious activity	Normal
Fire	892	41	67
Suspicious activity	64	901	35
Normal	31	70	899

TABLE 1.3
Accuracy of Confusion Matrix of Conventional Model

Class	n (truth)	n (classified)	Accuracy	Precision	Recall	F1 Score
1	987	1000	93.23%	0.89	0.90	0.90
2	1012	1000	93%	0.90	0.89	0.90
3	1001	1000	93.23%	0.90	0.90	0.90

TABLE 1.4
Confusion Matrix of Proposed Model

	Fire	Suspicious activity	Normal
Fire	956	19	25
Suspicious activity	32	950	18
Normal	25	54	921

TABLE 1.5
Accuracy of Confusion Matrix of Proposed Model

Class	n (truth)	n (classified)	Accuracy	Precision	Recall	F1 Score
1	1013	1000	96.63%	0.96	0.94	0.95
2	1023	1000	95.9%	0.95	0.93	0.94
3	964	1000	95.93%	0.92	0.96	0.94

1.5.2 CONFUSION MATRIX OF PROPOSED WORK

The precision of the suggested model is shown here. The results of testing the suggested model are shown in Table 1.4, and an accuracy table is shown in Table 1.5.

1.5.2.1 Results
TP: 2827
   Overall Accuracy: 94.23%
   Table 1.5 displays the derived accuracy metrics after applying accuracy, recall, precision, and F1-score to Table 1.4.

Integration of Nature-Inspired Mechanisms to Machine Learning

TABLE 1.6
Comparison Analysis of Accuracy

Class	Conventional model	Proposed model
1	93.23%	96.63%
2	93%	95.9%
3	93.23%	95.93%

1.5.3 Comparative Analysis

1.5.3.1 Accuracy

Results from verifying completed and scheduled tasks for Classes 1, 2, 3, and 4 are shown in Table 1.6. After comparing the proposed work to the gold standard, it was judged to be correct.

Taking into account the data in Table 1.6, we can compare the precision of the proposed model to that of the standard model in Figure 1.7.

1.5.3.2 Precision

Accuracy of previous and predicted work is compared for all three classes in Table 1.7. It has been pointed out that the proposed model is more exact than the prevalent paradigm.

Taking into account the data in Table 1.7, we can now display the superior accuracy of the proposed model over the baseline one in Figure 1.8.

1.5.3.3 Recall Value

As shown in Table 1.8, recall values for Classes 1, 2, and 3 were calculated from previous and future work. It is shown that the recall value is greater while using the proposed model as opposed to the baseline model.

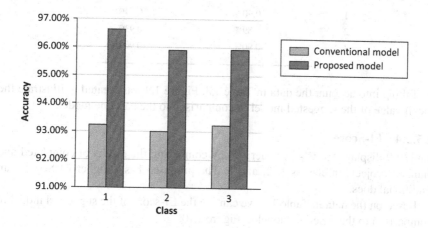

FIGURE 1.7 Comparison analysis of accuracy.

TABLE 1.7
Comparison Analysis of Precision

Class	Conventional model	Proposed model
1	0.89%	0.96%
2	0.90%	0.95%
3	0.90%	0.92%

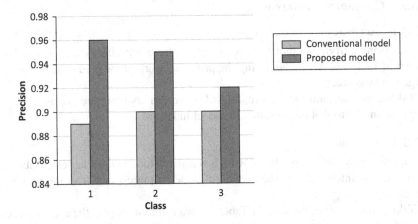

FIGURE 1.8 Comparison analysis of precision.

TABLE 1.8
Comparison Analysis of Recall Value

Class	Conventional model	Proposed model
1	0.90%	0.94%
2	0.89%	0.93%
3	0.90%	0.96%

Taking into account the data in Table 1.8, Figure 1.9 was created to illustrate the recall value of the suggested model in comparison to the baseline model.

1.5.3.4 F1-Score

Table 1.9 displays the F1-score derived by comparing the results of completed and planned projects in classes 1, 2, and 3. The proposed has a higher F1-Score than traditional does.

Based on the data in Table 1.9, we can see the F1-score of the suggested model in comparison to the baseline model in Figure 1.10.

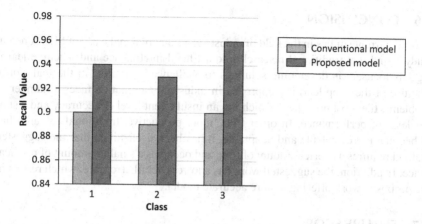

FIGURE 1.9 Comparison analysis of recall value.

TABLE 1.9
Comparison Analysis of F1-Score

Class	Conventional model	Proposed model
1	0.90%	0.95%
2	0.90%	0.94%
3	0.90%	0.94%

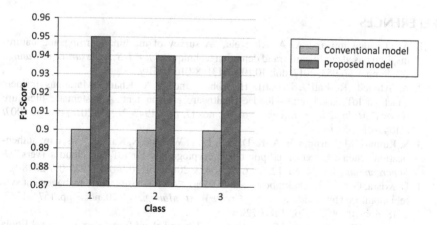

FIGURE 1.10 Comparison analysis of F1-score.

1.6 CONCLUSION

Several current studies in the field of industrial automation make use of ML. A recent study found that traditional research into artificial intelligence and machine learning has focused on developing solutions to challenges that arise in the real world. The use of the deep learning approach in industrial automation faces a number of problems, the most notable of which are an insufficient level of accuracy and a sub-par level of performance. In order to be more productive, traditional research has to become more scalable and adaptable. It has been determined that the suggested method requires a shorter amount of time and occupies a smaller amount of physical space. In addition, the suggested work has a lower overall error rate, which results in the proposed work offering a more accurate model.

1.7 FUTURE SCOPE

Although artificial intelligence is still in its infancy in India, it is steadily making its way into almost every major industry, including industrial automation, agriculture, healthcare, education, infrastructure, transportation, cyber security, banking, manufacturing, and the hospitality industry. Artificial intelligence is still in its infancy in the United States as well. In the fields of industrial automation and healthcare, artificial intelligence has a broad variety of applications, some of which include machine learning (ML) algorithms and other cognitive technologies [48]. To put it another way, artificial intelligence, or AI, is the capability of computers and other technologies to mimic human cognition in terms of learning, thinking, and carrying out actions.

REFERENCES

1. A. O. Adewumi and A. A. Akinyelu, "A survey of machine-learning and nature-inspired based credit card fraud detection techniques," *Int. J. Syst. Assur. Eng. Manag.*, vol. 8, pp. 937–953, 2017, doi: 10.1007/s13198-016-0551-y.
2. A. Ahmed, R. Latif, S. Latif, H. Abbas, and F. A. Khan, "Malicious insiders attack in IoT based multi-cloud e-Healthcare environment: A systematic literature review," *Multimed. Tool. Appl.*, vol. 77, no. 17, pp. 21947–21965, 2018, doi: 10.1007/s11042-017-5540-x.
3. S. Kumari, M. Karuppiah, A. K. Das, X. Li, F. Wu, and N. Kumar, "A secure authentication scheme based on elliptic curve cryptography for IoT and cloud servers," *J. Supercomput.*, vol. 74, no. 12, pp. 6428–6453, 2018, doi: 10.1007/s11227-017-2048-0.
4. D. Misra, G. Das, T. Chakrabortty, and D. Das, "An IoT-based waste management system monitored by cloud," *J. Mater. Cycles Waste Manag.*, vol. 20, no. 3, pp. 1574–1582, 2018, doi: 10.1007/s10163-018-0720-y.
5. S. Sareen, S. K. Sood, and S. K. Gupta, "IoT-based cloud framework to control Ebola virus outbreak," *J. Ambient Intell. Humaniz. Comput.*, vol. 9, no. 3, pp. 459–476, 2018, doi: 10.1007/s12652-016-0427-7.
6. C. Singla, N. Mahajan, S. Kaushal, A. Verma, and A. K. Sangaiah, "Modelling and analysis of multi-objective service selection scheme in IoT-cloud environment," *Lect. Notes Data Eng. Commun. Technol.*, vol. 14, pp. 63–77, 2018, doi: 10.1007/978-3-319-70688-7_3.

7. N. Y. Kim, J. H. Ryu, B. W. Kwon, Y. Pan, and J. H. Park, "CF-CloudOrch: container fog node-based cloud orchestration for IoT networks," *J. Supercomput.*, vol. 74, no. 12, pp. 7024–7045, 2018, doi: 10.1007/s11227-018-2493-4.
8. A. A. Zaidan et al., "A survey on communication components for IoT-based technologies in smart homes," *Telecommun. Syst.*, vol. 69, no. 1, pp. 1–25, 2018, doi: 10.1007/s11235-018-0430-8.
9. L. Cui, S. Yang, F. Chen, Z. Ming, N. Lu, and J. Qin, "A survey on application of machine learning for Internet of Things," *Int. J. Mach. Learn. Cybern.*, vol. 9, no. 8, pp. 1399–1417, 2018, doi: 10.1007/s13042-018-0834-5.
10. M. W. L. Moreira, J. J. P. C. Rodrigues, N. Kumar, J. Al-Muhtadi, and V. Korotaev, "Nature-inspired algorithm for training multilayer perceptron networks in e-health environments for high-risk pregnancy care," *J. Med. Syst.*, vol. 42, no. 3, 2018, doi: 10.1007/s10916-017-0887-0.
11. N. Boukhelifa, A. Bezerianos, and E. Lutton, *Evaluation of Interactive Machine Learning Systems*. Springer International Publishing, 2018.
12. M. Mahmud, M. S. Kaiser, M. M. Rahman, M. A. Rahman, A. Shabut, S. Al-Mamun, A. Hussain, "A brain-inspired trust management model to assure security in a cloud based IoT framework for neuroscience applications," *Cognit. Comput.*, vol. 10, no. 5, pp. 864–873, 2018, doi: 10.1007/s12559-018-9543-3.
13. S. S. Gill, R. C. Arya, G. S. Wander, and R. Buyya, *Fog-Based Smart Healthcare as a Big Data and Cloud Service for Heart Patients Using IoT*, vol. 26. Springer International Publishing, 2019.
14. P. Kaur and M. Sharma, "Diagnosis of human psychological disorders using supervised learning and nature-inspired computing techniques: A meta-analysis," *J. Med. Syst.*, vol. 43, no. 7, 2019, doi: 10.1007/s10916-019-1341-2.
15. F. Mehmood, I. Ullah, S. Ahmad, and D. H. Kim, "Object detection mechanism based on deep learning algorithm using embedded IoT devices for smart home appliances control in CoT," *J. Ambient Intell. Humaniz. Comput.*, 2019, doi: 10.1007/s12652-019-01272-8.
16. IEEE Control Systems Society. Chapter Malaysia and Institute of Electrical and Electronics Engineers, "Proceedings, 2020 16th IEEE International Colloquium on Signal Processing & Its Application (CSPA 2020): 28th–29th February 2020: conference venue, Hotel Langkawi, Lot 1852 JalanPenarak, Kuah 07000 Langkawi, Kedah, Malaysia," 2020 16th IEEE Int. Colloq. Signal Process. Its Appl., no. Cspa, pp. 219–224, 2020.
17. S. Naveen and M. R. Kounte. *In Search of the Future Technologies: Fusion of Machine Learning, Fog and Edge Computing in the Internet of Things*, vol. 31. Springer International Publishing, 2020.
18. A. Shakarami, M. Ghobaei-Arani, and A. Shahidinejad, "A survey on the computation offloading approaches in mobile edge computing: A machine learning-based perspective," *J. Grid Comput.*, vol. 182, 2020.
19. R. Hou, Y. Q. Kong, B. Cai, and H. Liu, "Unstructured big data analysis algorithm and simulation of Internet of Things based on machine learning," *Neural Comput. Appl.*, vol. 32, no. 10, pp. 5399–5407, 2020, doi: 10.1007/s00521-019-04682-z.
20. S. Pande, A. Khamparia, D. Gupta, and D. N. H. Thanh, *DDOS Detection Using Machine Learning Technique*, vol. 921. Springer, 2021.
21. S. Anupam and A. K. Kar, "Phishing website detection using support vector machines and nature-inspired optimization algorithms," *Telecommun. Syst.*, vol. 76, no. 1, pp. 17–32, 2021, doi: 10.1007/s11235-020-00739-w.
22. B. Pourghebleh, A. AghaeiAnvigh, A. R. Ramtin, and B. Mohammadi, "The importance of nature-inspired meta-heuristic algorithms for solving virtual machine consolidation problem in cloud environments," *Clust. Comput.*, vol. 24, no. 3, pp. 2673–2696, 2021, doi: 10.1007/s10586-021-03294-4.

23. M. Etemadi, M. Ghobaei-Arani, and A. Shahidinejad, "A cost-efficient auto-scaling mechanism for IoT applications in fog computing environment: A deep learning-based approach," *Clust. Comput.*, vol. 24, no. 4, pp. 3277–3292, 2021, doi: 10.1007/s10586-021-03307-2.
24. R. S. Gupta, V. K. Nassa, R. Bansal, P. Sharma, and K. Koti, "Investigating application and challenges of big data analytics with clustering," *2021 International Conference on Advancements in Electrical, Electronics, Communication, Computing and Automation (ICAECA)*, pp. 1–6, 2021, doi: 10.1109/ICAECA52838.2021.9675483.
25. V. Veeraiah, H. Khan, A. Kumar, S. Ahamad, A. Mahajan, and A. Gupta, "Integration of PSO and deep learning for trend analysis of meta-verse," *2022 2nd International Conference on Advance Computing and Innovative Technologies in Engineering (ICACITE)*, pp. 713–718, 2022, doi: 10.1109/ICACITE53722.2022.9823883.
26. B. K. Pandey, D. Pandey, V. K. Nassa, S. George, B. Aremu, P. Dadeech, A. Gupta, "Effective and secure transmission of health information using advanced morphological component analysis and image hiding," in Gupta, M., Ghatak, S., Gupta, A., Mukherjee, A.L. (eds) *Artificial Intelligence on Medical Data: Lecture Notes in Computational Vision and Biomechanics*, vol. 37. Springer, 2023, doi: 10.1007/978-981-19-0151-5_19.
27. V. Veeraiah, K. R. Kumar, P. LalithaKumari, S. Ahamad, R. Bansal, and A. Gupta, "Application of biometric system to enhance the security in virtual world," *2022 2nd International Conference on Advance Computing and Innovative Technologies in Engineering (ICACITE)*, pp. 719–723, 2022, doi: 10.1109/ICACITE53722.2022.9823850.
28. R. Bansal, A. Gupta, R. Singh, and V. K. Nassa, "Role and impact of digital technologies in e-learning amidst COVID-19 Pandemic," *2021 Fourth International Conference on Computational Intelligence and Communication Technologies (CCICT)*, pp. 194–202, 2021, doi: 10.1109/CCICT53244.2021.00046.
29. A. Shukla, S. Ahamad, G. N. Rao, A. J. Al-Asadi, A. Gupta, and M. Kumbhkar, "Artificial intelligence assisted IoT data intrusion detection," *2021 4th International Conference on Computing and Communications Technologies (ICCCT)*, pp. 330–335, 2021, doi: 10.1109/ICCCT53315.2021.9711795.
30. V. Pathania, S. Z. D. Babu, S. Ahamad, P. Thilakavathy, A. Gupta, M. B. Alazzam, D. Pandey, "A database application of monitoring COVID-19 in India," in Gupta, M., Ghatak, S., Gupta, A., Mukherjee, A.L. (eds) *Artificial Intelligence on Medical Data. Lecture Notes in Computational Vision and Biomechanics*, vol. 37. Springer, 2023, doi: 10.1007/978-981-19-0151-5_23.
31. K. Dushyant, G. Muskan, Annu, A. Gupta, S. Pramanik, "Utilizing machine learning and deep learning in cybesecurity: An innovative approach," in *Cyber Security and Digital Forensics: Challenges and Future Trends*. Wiley, pp. 271–293, 2022, doi: 10.1002/9781119795667.ch12.
32. S. Z. D. Babu, D. Pandey, G. T. Naidu, S. Sumathi, A. Gupta, M. Bader Alazzam, B. K. Pandey, "Analysation of big data in smart healthcare," in Gupta, M., Ghatak, S., Gupta, A., Mukherjee, A.L. (eds) *Artificial Intelligence on Medical Data. Lecture Notes in Computational Vision and Biomechanics*, vol. 37. Springer, 2023, doi: 10.1007/978-981-19-0151-5_21.
33. B. Bansal, V. Nisha Jenipher, R. Jain, R. Dilip, M. Kumbhkar, S. Pramanik, S. Roy, A. Gupta, "Big data architecture for network security," in *Cyber Security and Network Security*. Wiley, pp. 233–267, 2022, doi: 10.1002/9781119812555.ch11.
34. A. Gupta, D. Kaushik, M. Garg, and A. Verma, "Machine learning model for breast cancer prediction," *2020 Fourth International Conference on I-SMAC (IoT in Social, Mobile, Analytics and Cloud) (I-SMAC)*, pp. 472–477, 2020, doi: 10.1109/I-SMAC49090.2020.9243323.

35. N. Sreekanth et al., "Evaluation of estimation in software development using deep learning-modified neural network," *Appl. Nanosci.*, 2022, doi: 10.1007/s13204-021-02204-9.
36. V. Veeraiah, N. B. Rajaboina, G. N. Rao, S. Ahamad, A. Gupta, and C. S. Suri, "Securing online web application for IoT management," *2022 2nd International Conference on Advance Computing and Innovative Technologies in Engineering (ICACITE)*, pp. 1499–1504, 2022, doi: 10.1109/ICACITE53722.2022.9823733.
37. V. Veeraiah, G. P., S. Ahamad, S. B. Talukdar, A. Gupta, and V. Talukdar, "Enhancement of meta verse capabilities by IoT integration," *2022 2nd International Conference on Advance Computing and Innovative Technologies in Engineering (ICACITE)*, pp. 1493–1498, 2022, doi: 10.1109/ICACITE53722.2022.9823766.
38. N. Gupta, S. Janani, R. Dilip, R. Hosur, A. Chaturvedi, and A. Gupta, "Wearable sensors for evaluation over smart home using sequential minimization optimization-based random forest," *Int. J. Commun. Netw. Inf. Sec. (IJCNIS)*, 14(2), pp. 179–188, 2022, doi: 10.17762/ijcnis.v14i2.5499.
39. H. Keserwani, H. Rastogi, A. Z. Kurniullah, S. K. Janardan, R. Raman, V. M. Rathod, and A. Gupta, "Security enhancement by identifying attacks using machine learning for 5G network," *Int. J. Commun. Netw. Inf. Sec. (IJCNIS)*, 14(2), pp. 124–141, 2022, doi: 10.17762/ijcnis.v14i2.5494.
40. N. Sindhwani, R. Anand, M. Niranjanamurthy, D. C. Verma, and E. B. Valentina (Eds.), *IoT based smart applications*. Springer Nature, 2022.
41. J. Kaur, N. Sindhwani, R. Anand, and D. Pandey, "Implementation of IoT in various domains," in *IoT Based Smart Applications*. Springer International Publishing, pp. 165–178, 2022.
42. A. Tripathi, N. Sindhwani, R. Anand, and A. Dahiya, "Role of IoT in smart homes and smart cities: Challenges, benefits, and applications," in *IoT Based Smart Applications*. Springer International Publishing, pp. 199–217, 2022.
43. B. Kumar Pandey, D. Pandey, V. K. Nassa, T. Ahmad, C. Singh, A. S. George, and M. A. Wakchaure, "Encryption and steganography-based text extraction in IoT using the EWCTS optimizer," *Imaging Sci. J.*, vol. 69, pp. 1–19, 2022.
44. A. Gupta, A. Asad, L. Meena, and R. Anand, "IoT and RFID-based smart card system integrated with health care, electricity, QR and banking sectors," in *Artificial Intelligence on Medical Data, Proceedings of International Symposium, ISCMM 2021*. Springer Nature Singapore, pp. 253–265, 2022 July.
45. D. Pandey, B. K. Pandey, and S. Wariya, "An approach to text extraction from complex degraded scene," *IJCBS*, vol. 1, no. 2, pp. 4–10, 2020.
46. D. Pandey, S. Wairya, M. Sharma, A. K. Gupta, R. Kakkar, and B. K. Pandey, "An approach for object tracking, categorization, and autopilot guidance for passive homing missiles," *Aerosp. Syst.*, pp. 1–14, 2022.
47. S. Meivel et al., "Mask detection and social distance identification using internet of things and faster R-CNN algorithm," *Comp. Intell. Neurosci.*, pp. 1–5, 2022.
48. R. Anand, N. Sindhwani, and S. Juneja, "Cognitive Internet of things, its applications, and its challenges: A survey," in *Harnessing the Internet of Things (IoT) for a Hyperconnected Smart World*. Apple Academic Press, pp. 91–113, 2022.

2 The Implications of Cloud Computing, IoT, and Wearable Robotics for Smart Healthcare and Agriculture Solutions

Jay Kumar Pandey, Santanu Das, Prashant Vats, Dharmesh Dhabliya, Sushma Jaiswal, K. Manikandan, and Digvijay Pandey

2.1 INTRODUCTION

People from all around the world are eager to learn about new and cutting-edge tools and technology that will simplify their daily lives. Researchers and inventors from every field are always working to the best of their abilities to deliver the most recent advancements to people, despite the fact that the process is limitless and involves many different known and unknown variables. The consumer market and businesses began using the Internet professionally by the 1990s, although there were certain limitations since the cost to connect to the network at that time was so cheap. But by the turn of the millennium, because of the rapid advancements made in connection speed by the Internet, it had been already established as a strong foothold as a reliable source of information in a variety of commercial and industrial fields. Despite this, the effective integration of the Internet into a variety of industries and sectors, including education, agriculture, healthcare, and business, is still not entirely possible due to a number of obstacles, including security, deployment, transparency, and other issues.

2.1.1 INTERNET OF THINGS (IoT)

By connecting everyday objects with the Internet and making them intelligent, technology has made it possible for humans to do any task quickly, simply, and safely. This has sparked new forms of communication between people and objects as well as among objects themselves. Consequently, it is simple to state that IoT's primary motto is A6 [1, 2].

- Anyone
- Anywhere
- Anytime
- Anything
- Any-path
- Any-service

Figure 2.1 provides the explanation of the A and C pairings that shows how people and things may be connected to Anyone, at any location, and every time via Anything on Any-Path as well as Any-Service. Therefore, key elements could be easily incorporated like Communication, Connection, Content, Convergence, Computation, and Collection with the A's of Internet of Things whether discussing connectivity between things and people or among things themselves. As a result, the C and A components are strongly connected.

The phrase "Internet of Things" refers to a networked system where each particular computer device, piece of physical or digital equipment, object, animal, and even person is given a unique identification and capability to transmit information over the network. All this is possible without any involvement of humans. Users might interact with software, websites/computers at the user interface. The best user interfaces give the intended outcomes with the least amount of user interaction as feasible. The user interface, or UX, is essentially the interactive user-side interface offered by various apps and domains where IoT is implemented. Moreover, it interacts with people. Knowing the potential of IoT, it is reasonable to claim that

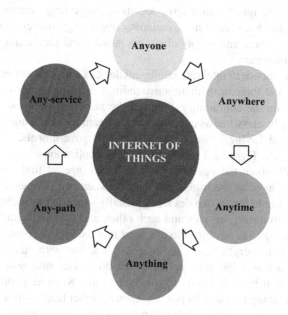

FIGURE 2.1 Main Motto of IoT.

INTERNET OF THINGS
- Smart Medicine
- Smart Agriculture
- Smart Powerhouse
- Smart City
- Smart Infrastructure
- Smart Didactics

FIGURE 2.2 Latest examples of IoT.

the possibilities for creating apps based on it are limitless. Sensors, connection of internet, semiconductors, microcontrollers, IDs, hardware, protocols, databases, and many more technologies are all used in these applications. Figure 2.2 illustrates a few of most recent softwares of Internet.

Discussions of the ambience domain take into account a wide variety of applications concerned with the protection of natural resources. These include product recycling, water safety, energy conservation, animal security, irrigation and agriculture, monitoring the weather, and the management of environmental services. One of the most important ways in which environmental changes are tracked and assessed is via the use of sensors.

Reimagining warehouses, increasing manufacturing efficiency while cutting costs and improving quality, managing logistics, integrating wearables and smart tools, Industrial IoT has several uses, including improving production transparency, expediting smart robots and supply chain operations, and facilitating the development of autonomous cars.

It is common practice to integrate intelligent devices and sensors that are installed in public areas and their immediate surroundings into applications that have far-reaching effects on the community. Some of the newest applications in the social domain that may be classified as cyber-physical system include "smart" workplaces, "Ambient Assisted Living," "smart towns," "smart governments," "smart grids," "smart parking," etc. Figure 2.3 shows IoT domain applications.

When we talk about the beginning of the digital age, at first, communication was simply thought of as the transfer of data between two devices. As technology progressed, sharing between two devices gradually expanded to include infrastructure and the environment, devices and each other, and eventually, humans as well. However, technology has advanced to the point that the Internet is accessible everywhere and can link everything. In addition to interacting with other humans, non-sentient items such as furniture, appliances, clothing, etc., may now be spoken to and interacted with by individuals in their free time. Keeping people and things connected and making the internet pervasive, on the other hand, will require a robust and tightly integrated mix of multiple technologies such as pervasive computing, ES, sensors, AI, IP, and Converged Technology, among others.

Impact of Technology on Smart Healthcare and Agricultural Solutions 29

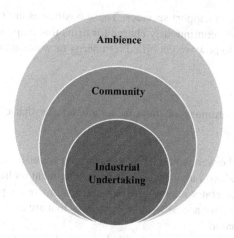

FIGURE 2.3 IoT domain application.

Actuators, sensors, embedded systems, ubiquitous computing, and IP are all shown in Figure 2.4 as the foundation of the IoT. Users may establish direct connections and transfer data by using sensors, actuators, and embedded systems, which together make up the physical IoT components. All of the Internet of Things' intricate user interactions are taken care of by ubiquitous computing since it adheres to Internet standards and establishes transmission between the devices.

2.1.2 Cloud Computing

A technique known as "cloud computing" [1] offers a variety of services that make it easier for us to transport data or other sorts of information. The delivery of computer services is what it is. Servers, warehousing, databases, programming, networks, and network-based inquiry are among the many services. Several companies provide

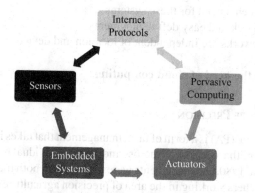

FIGURE 2.4 Simple structure of IoT.

these kinds of technical support services. These businesses are referred to as cloud specialist co-ops. They continuously charge for firms that employ appropriated processing. It is similar to how we get billed for energy or water at home.

2.1.3 Advantages

The following is a summary of the various advantages that cloud computing [3] offers.

1. Operators on the web would have access to remote utilities.
2. Using the cloud, you may access online development tools.
3. Anytime, the operator may online edit and customize the program.
4. Operators now have access to cloud resources that are available at any point and are web-based.
5. On-demand self-services are provided via cloud computing, and a link to a cloud-based specialized co-op is not required.
6. Typically, cloud computing operates efficiently. Since it won't be used to its full potential, it is very cost-effective.
7. The load balancing capability in cloud computing makes it more dependable.
8. Load balancing, a component of cloud computing, is a sign of its dependability.

2.1.4 Need for Cloud Computing

1. It provides assistance every day of the week.
2. Cloud registration with pay as you go.
3. The total cost of owner transportation is cheaper.
4. Using cloud computing offers dependability, adaptability, and maintainability.
5. Secure Storage Management Expenditure is provided.
6. Freeing up internal resources is appropriate.
7. These frameworks are regarded as being highly automated.
8. Utility has been a need for these systems.
9. It provides quick and easy deft deployments.
10. These frameworks are independent of location and device.

Figure 2.5 shows the need of cloud computing.

2.1.5 Agriculture Precision

Precision Agriculture (PA) is a form of farm management that takes into consideration the diversity in crops that occurs both across and within individual fields. Beginning in the latter half of the 1980s, advancements were first made in both the theory and practice of PA. Researchers working in the area of precision agriculture have been hard at work developing a decision support system for the entire management of farms. Their

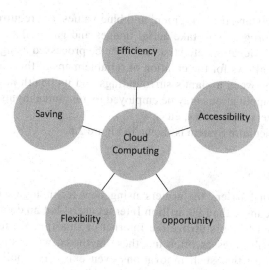

FIGURE 2.5 Need for cloud computing.

overarching objective is to design a system that would maximize revenues from inputs while minimizing waste. One of these approaches is called phytogeomorphology, and it compares the physical characteristics of the landscape to the growth of crops in terms of both their shape and their consistency over the course of several growing seasons. The hydrology of agricultural land is often controlled by the geomorphology component; thus, the phytogeomorphological approach is crucial.

Farmers have been able to more effectively use precision farming thanks to advancements in satellite navigation systems such as global positioning system (GPS) and global navigation satellite system. After a specific location has been identified by a researcher or a farmer, generating maps of the spatial variability allows for the measurement of a variety of characteristics, including crop yields, pH, topographical features, nitrogen levels, ML, and more. The harvesters that are equipped with GPS have sensor arrays so that they may collect the same data. Real-time sensors and multispectral photography, which are both components of the arrays, may be used to monitor the amounts of water and chlorophyll that are present in plants and crops in real time. This is possible thanks to the arrays. Using satellite photos and data collected from seeders, sprayers, etc., variable rate technology makes efficient use of the available resources. However, thanks to technological advancements, real-time sensors that can communicate wirelessly without human intervention may now be installed in the soil.

Precision farming is now a practical option because of the increasing use of unmanned aerial vehicles. Multiple big images acquired by multispectral or red, green, and blue sensors mounted on agricultural drones may be stitched together using different photogrammetric methods to create an orthophoto. In order to properly analyze and evaluate vegetative indices like normalized difference vegetation index maps, multispectral images, which include several values per pixel in addition to the standard red, green, and blue values, are required. Drones are used by certain programs to take aerial images and gather data on topography (including height, elevation, etc.) that may then be processed using algebraic map operations. This allows for the creation of reliable maps of their locations. After the connection between a plant's surroundings and its health has been mapped out, a variety of applications may be employed to guarantee inputs like water and chemical fertilizers, herbicides, etc.

Precision Agriculture System is shown in Figure 2.6.

2.1.6 Challenges

One of the main barriers to farmers using new technologies is connectivity issues. There are more than 90 million Internet users, but no detailed data exists on whether or not farmers utilize the Internet or smartphones for farming purposes. Large agribusinesses often use these methods, while small farmers are frequently unaware that such innovations even exist. The poll included 1600 American farmers, just 68% of whom had any familiarity with the IoT. It will be more difficult for smaller farms growing a wider variety of crops to use these technologies. However, putting in the work to create and maintain such advances would be quite expensive. The high cost of autonomous system and hardware is a direct result of wear and tear from exposure to the elements, such as heat, cold,

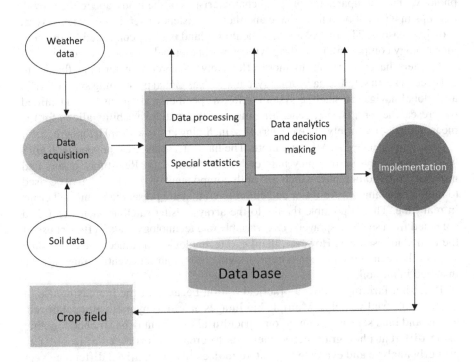

FIGURE 2.6 Precision agriculture system.

Impact of Technology on Smart Healthcare and Agricultural Solutions 33

wind, sand, and physical impact. Consequently, deciding on the most suitable company approach would be a daunting task. Being well-prepared is crucial to the success of the IoT.

Realizing any agricultural goal requires a strong Internet of Things business plan. The second issue is the security of our digital data. It is crucial to monitor and protect all linked devices due to the risk of theft and abuse as well as the potential for altering product prices and spending. The last barrier is a dearth of employees with the necessary skills. It would be difficult to evaluate and comprehend the data that these gadgets collected and produced. IoT devices might be used on even the tiniest farms, however, provided farm workers are given the right training.

The deployment of IoT devices and data analysis techniques by agricultural equipment manufacturers has led to a rise in output. There is still a heavy reliance on antiquated technology in the agricultural business, which seldom provides adequate data backup or protection. For instance, certain drones used for field monitoring might be connected to agricultural equipment. When these devices are linked to public networks or the web, they lack even the most fundamental types of security, such as the capacity to monitor user activity or the need for a second authentication factor for remote access.

2.2 LITERATURE REVIEW

Numerous studies have been conducted up to this point in an attempt to secure cloud services. Some of these are covered in this chapter. After reviewing previous studies on many different subjects, including IoT, communications and knowledge management, Access Points (AP), and wireless sensors network (WSN), the focus of this investigation was determined. It was in this part that we discussed the following academic works:

In [1], researchers spoke about CC and how to use fog computing to provide cloud-based services to the network's periphery. The scaling of computing from CC to Fog Computing (FC) has been explored in [2], and according to them, it is becoming more and more significant and beneficial in the modern world. Fog computing is a novel technique that has been introduced for the first time as a means of overcoming issues related to cloud computing. Guan and Zhou in [3, 4] have examined the advances in cloud computing technology and determined that this technology causes significant changes in the IT sector. A variety of cloud-based technologies has been developed as a result of the cloud computing industry's rapid expansion. The authors in [5] were the first to introduce the fog computing technique. They took into account the fact that this technology will assist CC in the years to come. According to them, even if CC has increased access to computing resources through the Internet on a cost-per-use basis, fog computing will advance the simplicity and adaptability of computing. In [6], the researchers speak on the rise of cloud computing across all industries, not just in the field of information technology. They said that the cloud can provide us with highly flexible computer resources at a cheap access cost. The capacity to create a new link between corporate IT departments and business divisions is another benefit of cloud computing.

The authors in [7] have previously shown that in the context of equipment sharing, this point technology has a crucial objective of high resource utilization. Therefore, in reality, utilization stays customarily low due to difficulties in appropriately anticipating combined application performance. We provide an experimental approach for testing in conjunction with an n-tier use demonstration at its highest application. Because of reshapable dimensions, problems of this kind have already come to fruition. The authors in [8] state that protecting sensitive data in wireless sensor networks was a primary area of study. In order to facilitate dynamic key management using the second-level cluster-to-node authentication matrix, this research combines an asymmetric public key system with a threshold key scheme. Cluster-based WSN were suggested by the authors in [9]. It is advised that the hierarchical cluster structure of a sensor network be taken into account while designing a key management and distribution strategy that employs pair-wise and group-based threshold key cryptography. In a recent article [10], the authors with an elliptic curve cryptographic-based data transmission route and an Author-required key transfer mechanism mentioned that WSNs are becoming more common. Key distribution security was ensured by isolating it from the transmission channel. Abdallah and his colleagues [11] were concerned that the vulnerability of WSN might be a problem in a wide variety of contexts. Compared to past options, this system was more successful and less challenging since it needed fewer communications and easier processing to accomplish key exchange. WSNs with a limited amount of resources were first developed by Patel and Gheewala [12]. As per them, target tracking, battlefield surveillance, and even environmental monitoring are all possible applications. This means they may potentially maximize embedded subscriber identity module (ESIM) while minimizing power consumption by picking CH dynamically for each round and sending messages through Channel (CH). This has the potential to improve the ESIM and make it more power-efficient. They were advised by Hemapriya and Gomathy [13] to focus emphasis on clustering techniques. The author created a clustering method that takes into account the decisions made by each node. For this study, we zeroed in on a comprehensive kind of control message reduction applicable to MANETs' cluster-based point to point file searching. In the beginning, Raut et al. expressed concern over India's growing population. The study's population peaked at 1.2 billion within 25–30 years and was steadily increasing. A serious food scarcity may occur, prompting a hastening of agricultural progress [14]. Soil moisture probe, Soil Temperature (ST) measuring equipment, ambient temperature sensor, CO_2 sensor, and sunshine intensity device were all provided by Keswani[15]. The WSN architecture is considered in this investigation.

Feng et al. [16] considered several different agriculturally relevant situations in their tests. This study aimed to provide workable and widely accepted wireless communication systems for PA. Thakur et al. [17] examined applications in engineering, research, farming, and possibly military spying. The purpose of this research was to identify the various types of WSN technologies in use in PA and to assess their impact on modern agricultural practices. The authors in [18] noted that WSNs have uses outside engineering and research, such as in agricultural and military monitoring. They make important contributions to the area by giving a cluster-based approach to management. Gómez-Chabla et al. [19] looked at how the IoT has helped farmers catch up to the rest of the digital world. Increased water efficiency and input/

treatment effectiveness will allow for future productivity advances in eco-friendly food production. Using moth-flame optimization, Mittal [20] devised a reliable clustered routing approach that conserves energy. Khanna and Kaur [21] claim that IoT has altered the practice of precision farming. Secure hybrid hierarchical group key agreements for massive wireless ad hoc networks were presented by Naresh et al. [22]. A recent study by the authors in [23] looked at the value of decisions made during IoT data collection for precision agriculture. Internet-of-things-based sensors were studied by the authors in [24] for their potential use in precision farming. In [25], the major emphasis was on cloud-based context-aware middleware solutions for incorporating precision farming equipment into IoT for agriculture. Key management, authentication, and trust management procedures in wireless sensor networks were extensively studied by Gautam and Kumar [26]. Senthil et al. [27] suggested a method for energy-efficient cluster-based routing for WSN using hybrid particle swarm optimization. The authors in [28] presented a method in which fog nodes would share the burden equally. One major advantage of this method was that it allowed delay-sensitive queries to be handled even when all available fog nodes were already in use. Further, the authors in [29] created a more effective routing approach by encrypting both the picture and the data connected with it. By creating a way to compare different facial recognition and identification algorithms, Ara et al. [30] made online transactions more secure and reliable.

2.3 PROPOSED WORK

2.3.1 Problem Statement

The administration of IoT for precision agriculture and healthcare is the subject of several studies. However, using cloud computing is essential. An IoT-powered web interface will provide consumers access to the practical implementation of IoT in fields as diverse as agriculture and medicine [31–35]. Data transmission has to be safe from outside threats. In the process of agricultural precision, the data obtained from sensors are sorted according to criteria such as the presence of cattle, soil nutrition, and soil moisture. In the event that any unfavorable conditions exist, an alarm signal will be produced. On the other hand, if the conditions are satisfactory, the surgery will not be carried out [36–42].

2.3.2 Proposed Research Methodology

In precision agriculture, sensor data is categorized based on the presence of livestock, soil nutrition, and soil moisture. An alarm signal is issued in the event of any unfavorable circumstance. In contrast, no procedure is done when the conditions are favorable. Methods from Cloud Computing, IoT, and wearable robotics have been used in a variety of recently conducted Smart Healthcare and Agriculture Solution-related research articles [43–49]. Research has often focused on the problem-solving aspects of the field. In spite of this, the Implication of Cloud Computing, IoT, and wearable robotics for Smart Healthcare and Agriculture Solutions presents a number

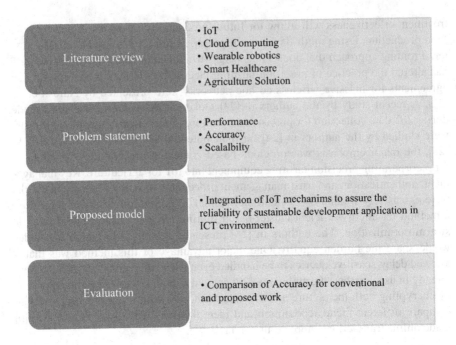

FIGURE 2.7 Research methodology.

of obstacles, one of which is the need to include an accuracy mechanism in order to ensure the integrity of Smart Healthcare and Agriculture Solutions while operating in a Cloud Computing, IoT and wearable robotics environment. In addition to this, traditional techniques of research need to have a greater capacity for accuracy [50–57].

The Research Methodology is shown in Figure 2.7.

2.4 RESULTS AND DISCUSSION

During agriculture precision, data captured from the sensor is classified on the basis of the presence of cattle, soil nutrition, and soil moisture. In the case of any unfavorable conditions, an alert signal is generated. On the other hand in the case of favorable conditions, no operation is performed. MATLAB, a programming environment with many helpful capabilities, was used to replicate the research procedure. Table 2.1 presents the matrix of alert and normal data transmissions in the case of unfiltered IoT.

Results:
TP: 1764
Overall Accuracy: 88.2%

Considering Table 2.1, the accuracy parameter of the unfiltered IoT model has been considered in Table 2.2.

Table 2.3 presents the matrix of alert and normal data transmission in cases of filtered IoT.

TABLE 2.1
Confusion matrix for unfiltered IoT mechanism.

	Alert	Normal
Alert	891	109
Normal	873	127

TABLE 2.2
Accuracy Parameter of Unfiltered IoT Mechanism

Class	n (truth)	n (classified)	Accuracy	Precision	Recall	F1 Score
1	1018	1000	88.2%	0.89	0.88	0.88
2	982	1000	88.2%	0.87	0.89	0.88

TABLE 2.3
Confusion Matrix for Filtered IoT Mechanism

	Alert	Normal
Alert	957	43
Normal	61	939

TABLE 2.4
Accuracy Parameter of Filtered IoT Mechanism

Class	n (truth)	n (classified)	Accuracy	Precision	Recall	F1 Score
1	1018	1000	94.8%	0.96	0.94	0.95
2	982	1000	94.8%	0.94	0.96	0.95

Results
TP: 1896
Overall Accuracy: 94.8%
Considering Table 2.3, the accuracy parameter of the filtered IoT model has been considered in Table 2.4.

2.4.1 Comparison Analysis

1. **Accuracy** (comparison and graphical representation are described in Table 2.5 and Figure 2.8 respectively).

TABLE 2.5
Comparison of Accuracy of Filtered and Unfiltered IoT Mechanism

Class	Unfiltered IoT Mechanism	Filtered IoT Mechanism
1	88.2%	94.8%
2	88.2%	94.8%

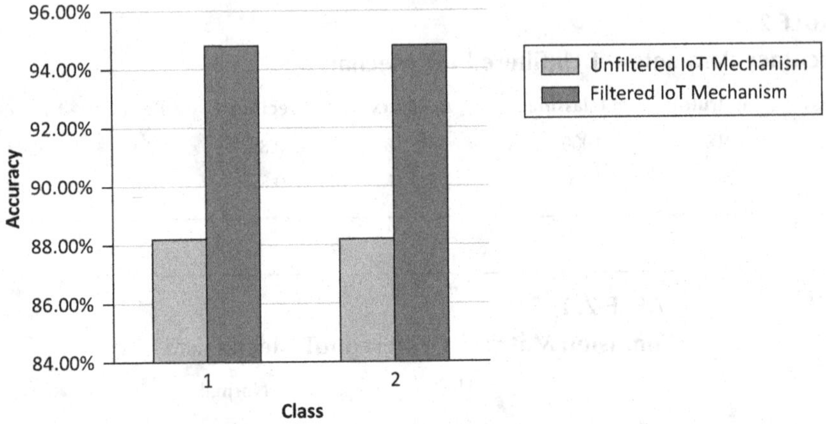

FIGURE 2.8 Graphical comparison of Aacuracy of Filtered and unfiltered IoT mechanism.

TABLE 2.6
Comparison of Precision of Filtered and Unfiltered IoT Mechanism

Class	Unfiltered IoT Mechanism	Filtered IoT Mechanism
1	0.89	0.96
2	0.87	0.94

2. **Precision** (comparison and graphical representation are described in Table 2.6 and Figure 2.9 respectively).
3. **Recall Value** (comparison and graphical representation are described in Table 2.7 and Figure 2.10 respectively).
4. **F1-Score** (comparison and graphical representation are described in Table 2.8 and Figure 2.11 respectively).

2.5 CONCLUSION

Sensor data in precision agriculture is sorted according to three factors: the presence or absence of animals, the availability of soil nutrients, and the level of soil moisture.

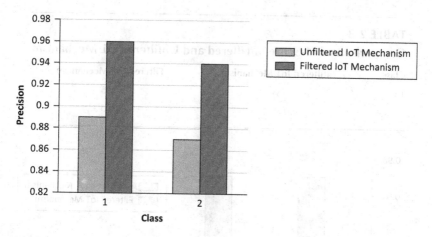

FIGURE 2.9 Graphical comparison of precision of filtered and unfiltered IoT mechanism.

TABLE 2.7
Comparison of Recall Value of Filtered and Unfiltered IoT Mechanism

Class	Unfiltered IoT Mechanism	Filtered IoT Mechanism
1	0.88	0.94
2	0.89	0.96

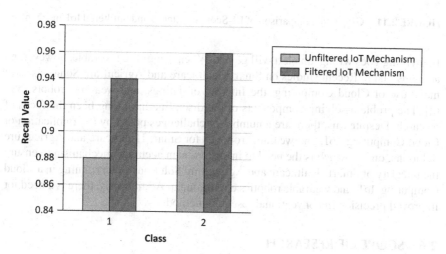

FIGURE 2.10 Graphical comparison of recall value of filtered and unfiltered IoT mechanism.

TABLE 2.8
Comparison of F1-Score of Filtered and Unfiltered DL Mechanism

Class	Unfiltered IoT Mechanism	Filtered IoT Mechanism
1	0.88	0.95
2	0.88	0.95

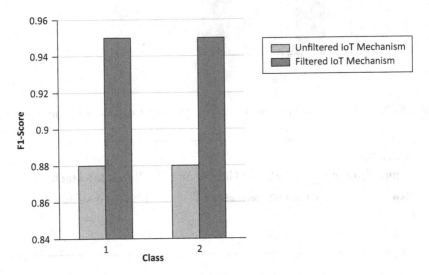

FIGURE 2.11 Graphical comparison of F1-Score of filtered and unfiltered IoT mechanism.

If anything goes wrong, an alert will go out. When things are favorable, however, no action is taken. Recent studies on Smart Healthcare and Agriculture Solutions have made use of Cloud Computing, the Internet of Things, and wearable robots [58–61]. The problem-solving components of the discipline have long been the focus of research. Despite this, there are a number of challenges posed by the Implication of Cloud Computing, IoT, and wearable robotics for Smart Healthcare and Agriculture Solutions, one of which is the need to incorporate an accuracy mechanism to ensure the integrity of Smart Healthcare and Agriculture Solutions when running in a Cloud Computing, IoT, and wearable robotics environment. Additionally, there is a need for improved precision in conventional research methods.

2.6 SCOPE OF RESEARCH

The use of pesticides and fertilizers, as well as tillage and irrigation water, may be more efficiently used thanks to precision agriculture. It's possible that more efficient use of inputs might boost agricultural productivity and/or quality while simultaneously reducing their overall impact on the environment. Such a study is necessary if agricultural precision is to be elevated to a position of importance in the decision-making process. In addition, the categorization of threats using a method known as DL helps to ensure the safe transfer of data, and the careful selection of cluster heads helps to ensure improved performance [62]. The user-friendly web interface is built on.NET may gather photographs from a variety of Internet of Things sensor nodes that have been put in an agricultural setting.

REFERENCES

1. Abbasi, B. Z., and Shah, M. A. (2017). Fog Computing: Security Issues, Solutions and Robust Practices, Proceedings of 23rd International Conference on Automation and Computing, University of Huddersfield, Hudders Field, UK, 7–8 September 2017.
2. Abubaker, N., Dervishi, L., and Ayday, E. (2017). Privacy-Preserving Fog Computing Paradigm, The 3rd IEEE Workshop on Security and PrivaAcy in Cloud (SPC 2017).
3. Guan, Y., Shao, J., and Wei, G. (2018). Data Security and Privacy in Fog Computing, 0890-8044/18/$25.00 © 2018 IEEE.
4. Zhou, J., Wang, T., Bhuiyan, M. Z. A., and Liu, A. (2017). A Hierarchic Secure Cloud Storage Scheme Based on Fog Computing, 2017 IEEE 15th International Conference on Dependable, Autonomic and Secure Computing, 15th International Conference on Pervasive Intelligence and Computing, 3rd International Conference on Big Data Intelligence and Computing and Cyber Science and Technology Congress.
5. Firdhous, M., Ghazali, O., and Hassan, S. (2014). Fog Computing: Will It Be Future of Cloud Computing? ISBN: 978-1-941968-00-0 ©2014 SDIWC.
6. Georgescu, M., and Matei, M. (2013). The Value of Cloud Computing in Business Environment. *The USV Annals of Economics and Public Administration*, *13*(1), 222–228.
7. Malkowski, S., Kanemasa, Y., Chen, H., Yamamoto, M., Wang, Q., Jayasinghe, D., Pu, C., and Kawaba, M. (2012). Challenges and Opportunities in Consolidation at High Resource Utilization: Non-monotonic Response Time Variations in n-Tier Applications. In *Fifth IEEE International Conference on Cloud Computing*, Honolulu, pp. 162–169.
8. Zhao, Q., and Liu, X. (2014b, May 1). Cluster Key Management Scheme for Wireless Sensor Networks. www.atlantis-press.com; Atlantis Press. https://doi.org/10.2991/ictcs-14.2014.23.
9. Diop, A., Qi, Y., and Wang, Q. (2014). Efficient Group Key Management Using Symmetric Key and Threshold Cryptography for Cluster Based Wireless Sensor Networks. *International Journal of Computer Network and Information Security*, 6(8), 9–18. https://doi.org/10.5815/ijcnis.2014.08.02.
10. Bao, X., Liu, J., She, L., and Zhang, S. (2014, June 1). A Key Management Scheme Based on Grouping within Cluster. *IEEE Xplore*. https://doi.org/10.1109/WCICA.2014.7053290.
11. Abdallah, W., Boudriga, N., Kim, D., and An, S. (2015, July 1). An Efficient and Scalable Key Management Mechanism for Wireless Sensor Networks. *IEEE Xplore*. https://doi.org/10.1109/ICACT.2015.7224913.
12. Patel, V., and Gheewala, J. (2015, June 1). An Efficient Session Key Management Scheme for Cluster Based Wireless Sensor Networks. *IEEE Xplore*. https://doi.org/10.1109/IADCC.2015.7154847.
13. Hemapriya, K., and Gomathy, K. (2007). IJARCCE A Survey Paper of Cluster Based Key Management Techniques for Secured Data Transmission in Manet. *International Journal of Advanced Research in Computer and Communication Engineering ISO*, *3297*. https://doi.org/10.17148/IJARCCE.2016.510102.

14. Raut, R., Varma, H., Mulla, C., and Pawar, V. R. (2017). Soil Monitoring, Fertigation, and Irrigation System Using IoT for Agricultural Application. In *Intelligent Communication and Computational Technologies*, pp. 67–73. https://doi.org/10.1007/978-981-10-5523-2_7.
15. Keswani, B., Mohapatra, A. G., Mohanty, A., Khanna, A., Rodrigues, J. J. P. C., Gupta, D., and de Albuquerque, V. H. C. (2018). Adapting Weather Conditions Based IoT Enabled Smart Irrigation Technique in Precision Agriculture Mechanisms. *Neural Computing and Applications*, *31*(S1), 277–292. https://doi.org/10.1007/s00521-018-3737-1.
16. Feng, X., Yan, F., and Liu, X. (2019). Study of Wireless Communication Technologies on Internet of Things for Precision Agriculture. *Wireless Personal Communications*, *108*(3), 1785–1802. https://doi.org/10.1007/s11277-019-06496-7.
17. Thakur, D., Kumar, Y., Kumar, A., and Singh, P. K. (2019). Applicability of Wireless Sensor Networks in Precision Agriculture: A Review. *Wireless Personal Communications*, *107*(1), 471–512. https://doi.org/10.1007/s11277-019-06285-2.
18. Lalitha, T., and Umarani, R. (2011b). Energy Efficient Cluster Based Key Management Technique for Wireless Sensor Networks. *Oriental Journal of Computer Science and Technology*, *4*(2), 293–304. http://www.computerscijournal.org/vol4no2/energy-efficient-cluster-based-key-management-technique-for-wireless-sensor-networks/.
19. Gómez-Chabla, R., Real-Avilés, K., Morán, C., Grijalva, P., and Recalde, T. (2019). IoT Applications in Agriculture: A Systematic Literature Review. In R. Valencia-García, G. Alcaraz-Mármol, J. del Cioppo-Morstadt, N. Vera-Lucio, and M. Bucaram-Leverone (Eds.), *Springer Link*. Springer International Publishing. https://doi.org/10.1007/978-3-030-10728-4_8.
20. Mittal, N. (2018). Moth Flame Optimization Based Energy Efficient Stable Clustered Routing Approach for Wireless Sensor Networks. *Wireless Personal Communications*, *104*(2), 677–694. https://doi.org/10.1007/s11277-018-6043-4.
21. Khanna, A., and Kaur, S. (2019). Evolution of Internet of Things (IoT) and Its Significant Impact in the Field of Precision Agriculture. *Computers and Electronics in Agriculture*, *157*, 218–231. https://doi.org/10.1016/j.compag.2018.12.039.
22. Naresh, V. S., Reddi, S., and Murthy, N. V. E. S. (2019). A Provably Secure Cluster-Based Hybrid Hierarchical Group Key Agreement for Large Wireless Ad hoc Networks. *Human-Centric Computing and Information Sciences*, *9*(1). https://doi.org/10.1186/s13673-019-0186-5.
23. Dewi, C., and Chen, R.-C. (2019). Decision Making Based on IoT Data Collection for Precision Agriculture. In *Intelligent Information and Database Systems: Recent Developments*, pp. 31–42. https://doi.org/10.1007/978-3-030-14132-5_3.
24. Gsangaya, K. R., Hajjaj, S. S. H., Sultan, M. T. H., and Hua, L. S. (2020). Portable, Wireless, and Effective Internet of Things-Based Sensors for Precision Agriculture. *International Journal of Environmental Science and Technology*. https://doi.org/10.1007/s13762-020-02737-6.
25. Symeonaki, A., and Piromalis, D. (2020). A Context-Aware Middleware Cloud Approach for Integrating Precision Farming Facilities into the IoT toward Agriculture 4.0. *Applied Sciences*, *10*(3), 813. https://doi.org/10.3390/app10030813.
26. Gautam, A. K., and Kumar, R. (2021). A Comprehensive Study on Key Management, Authentication and Trust Management Techniques in Wireless Sensor Networks. *SN Applied Sciences*, *3*(1). https://doi.org/10.1007/s42452-020-04089-9.
27. Senthil, G. A., Raaza, A., and Kumar, N. (2021). Internet of Things Energy Efficient Cluster-Based Routing Using Hybrid Particle Swarm Optimization for Wireless Sensor Network. *Wireless Personal Communications*. https://doi.org/10.1007/s11277-021-09015-9.

28. Ajay, P., Sharma, A., Gowda V, D., Sharma, A., Kumaraswamy, S., and Arun, M. R. (2022, May 1). Priority Queueing Model-Based IoT Middleware for Load Balancing. *IEEE Xplore*. https://doi.org/10.1109/ICICCS53718.2022.9788218.
29. Dankan Gowda, V., Sharma, A., Nagabushanam, M., Govardhana Reddy, H. G., and Raghavendra, K. (2022). Vector Space Modelling-Based Intelligent Binary Image Encryption for Secure Communication. *Journal of Discrete Mathematical Sciences and Cryptography*, 1–15. https://doi.org/10.1080/09720529.2022.2075090.
30. Ara, A., Sharma, A., and Yadav, D. (2022). An Efficient Privacy-Preserving User Authentication Scheme Using Image Processing and Blockchain Technologies. *Journal of Discrete Mathematical Sciences and Cryptography*, 1–19. https://doi.org/10.1080/09720529.2022.2075089.
31. Anand, S., and Sharma, A. (2021, December 1). Hybrid Security Mechanism to Enhance the Security and Performance of IoT System. *IEEE Xplore*. https://doi.org/10.1109/TRIBES52498.2021.9751455.
32. Anand, S., and Sharma, A. (2022). An Advanced and Efficient Cluster Key Management Scheme for Agriculture Precision IoT Based Systems. *International Journal of Electrical and Electronics Research*, *10*(2), 264–269. https://doi.org/10.37391/ijeer.100235.
33. Anand, S., and Sharma, A. (2020). WITHDRAWN: Assessment of Security Threats on IoT Based Applications. *Materials Today: Proceedings*. https://doi.org/10.1016/j.matpr.2020.09.350.
34. Sindhwani, N., Anand, R., Niranjanamurthy, M., Verma, D. C., and Valentina, E. B. (Eds.) (2022). *IoT Based Smart Applications*. Springer Nature.
35. Bommareddy, S., Khan, J. A., and Anand, R. (2022). A Review on Healthcare Data Privacy and Security. In *Networking Technologies in Smart Healthcare*. CRC, Publishers, pp. 165–187.
36. Attkan, A., and Ranga, V. (2022). Cyber-Physical Security for IoT Networks: A Comprehensive Review on Traditional, Blockchain and Artificial Intelligence Based Key-Security. *Complex and Intelligent Systems*. https://doi.org/10.1007/s40747-022-00667-z.
37. Ghani, A., Mansoor, K., Mehmood, S., Chaudhry, S. A., Rahman, A. U., and NajmusSaqib, M. (2019). Security and Key Management in IoT-Based Wireless Sensor Networks: *An Authentication Protocol Using Symmetric Key*. *International Journal of Communication Systems*, *32*(16), e4139. https://doi.org/10.1002/dac.4139.
38. Gupta, D. N., and Kumar, R. (2021). Multi-layer and Clustering-Based Security Implementation for an IoT Environment. *International Journal of System Dynamics Applications (IJSDA)*, *11*(2), 1–21. https://doi.org/10.4018/IJSDA.20220701.oa3.
39. Wang, Q., and Li, H. (2022). Application of IoT Authentication Key Management Algorithm to Personnel Information Management. *Computational Intelligence and Neuroscience*, *2022*, e4584072. https://doi.org/10.1155/2022/4584072.
40. Khan, A., Ahmad, A., Ahmed, M., Sessa, J., and Anisetti, M. (2022). Authorization Schemes for Internet of Things: Requirements, Weaknesses, Future Challenges and Trends. *Complex and Intelligent Systems*. https://doi.org/10.1007/s40747-022-00765-y.
41. Hwang, C.-E., Lee, S.-H., and Jeong, J.-W. (2019). VisKit: Web-Based Interactive IoT Management with Deep Visual Object Detection. *Journal of Sensor and Actuator Networks*, *8*(1), 12. https://doi.org/10.3390/jsan8010012.
42. Gupta, R. S., Nassa, V. K., Bansal, R., Sharma, P., Koti, K., and Koti, K. (2021). Investigating Application and Challenges of Big Data Analytics with Clustering. In *2021 International Conference on Advancements in Electrical, Electronics, Communication, Computing and Automation (ICAECA)*, pp. 1–6. https://doi.org/10.1109/ICAECA52838.2021.9675483.

43. Veeraiah, V., Khan, H., Kumar, A., Ahamad, S., Mahajan, A., and Gupta, A. (2022). Integration of PSO and Deep Learning for Trend Analysis of Meta-verse. In *2022 2nd International Conference on Advance Computing and Innovative Technologies in Engineering (ICACITE)*, pp. 713–718. https://doi.org/10.1109/ICACITE53722.2022.9823883.
44. Pandey, D., Pandey, B. K., Sindhwani, N., Anand, R., Nassa, V. K., and Dadheech, P. (2022). An Interdisciplinary Approach in the Post-COVID-19 Pandemic Era. In *An Interdisciplinary Approach in the Post-COVID-19 Pandemic Era*. Nova publishers, pp. 1–290.
45. Veeraiah, V., Kumar, K. R., LalithaKumari, P., Ahamad, S., Bansal, R., and Gupta, A. (2022). Application of Biometric System to Enhance the Security in Virtual World. In *2022 2nd International Conference on Advance Computing and Innovative Technologies in Engineering (ICACITE)*, pp. 719–723. https://doi.org/10.1109/ICACITE53722.2022.9823850.
46. Bansal, R., Gupta, A., Singh, R., and Nassa, V. K. (2021). Role and Impact of Digital Technologies in E-learning amidst COVID-19 Pandemic. In *2021 Fourth International Conference on Computational Intelligence and Communication Technologies (CCICT)*, pp. 194–202. https://doi.org/10.1109/CCICT53244.2021.00046.
47. Shukla, A., Ahamad, S., Rao, G. N., Al-Asadi, A. J., Gupta, A., and Kumbhkar, M. (2021). Artificial Intelligence Assisted IoT Data Intrusion Detection. In *2021 4th International Conference on Computing and Communications Technologies (ICCCT)*, pp. 330–335. https://doi.org/10.1109/ICCCT53315.2021.9711795.
48. Kaur, J., Sindhwani, N., Anand, R., and Pandey, D. (2022). Implementation of IoT in Various Domains. In *IoT Based Smart Applications*. Cham: Springer International Publishing, pp. 165–178.
49. Dushyant, K., Muskan, G., Annu, Gupta, A., Pramanik, S. (2022). Utilizing Machine Learning and Deep Learning in Cybesecurity: An Innovative Approach. In *Cyber Security and Digital Forensics: Challenges and Future Trends*. Wiley, 2022, pp. 271–293, doi: 10.1002/9781119795667.ch12.
50. Tripathi, A., Sindhwani, N., Anand, R., and Dahiya, A. (2022). Role of IoT in Smart Homes and Smart Cities: Challenges, Benefits, and Applications. In *IoT Based Smart Applications*. Cham: Springer International Publishing, pp. 199–217.
51. Bansal, B., Jenipher, N., Jain, R., Dilip, R., Kumbhkar, M., Pramanik, S., Sandip, R., and Gupta, A. (2022). Big Data Architecture for Network Security. In *Cyber Security and Network Security*. Wiley, pp. 233–267. https://doi.org/10.1002/9781119812555.ch11.
52. Meelu, R., and Anand, R. (2011). Performance Evaluation of Cluster-Based Routing Protocols Used in Heterogeneous Wireless Sensor Networks. *International Journal of Information Technology and Knowledge Management*, 4(1), 227–231.
53. Sindhwani, N., Maurya, V. P., Patel, A., Yadav, R. K., Krishna, S., and Anand, R. (2022). Implementation of Intelligent Plantation System Using Virtual IoT. In *Internet of Things and its Applications*, pp. 305–322.
54. Veeraiah, V., Rajaboina, N. B., Rao, G. N., Ahamad, S., Gupta, A., and Suri, C. S. (2022). Securing Online Web Application for IoT Management. In *2022 2nd International Conference on Advance Computing and Innovative Technologies in Engineering (ICACITE)*, pp. 1499–1504. https://doi.org/10.1109/ICACITE53722.2022.9823733.
55. Veeraiah, V., Gangavathi, P., Ahamad, S., Talukdar, S. B., Gupta, A., and Talukdar, V. (2022). Enhancement of Meta Verse Capabilities by IoT Integration. In *2022 2nd International Conference on Advance Computing and Innovative Technologies in Engineering (ICACITE)*, pp. 1493–1498. https://doi.org/10.1109/ICACITE53722.2022.9823766.

56. Gupta, N., Janani, S., Dilip, R., Hosur, R., Chaturvedi, A., and Gupta, A. (2022). Wearable Sensors for Evaluation Over Smart Home Using Sequential Minimization Optimization-Based Random Forest. *International Journal of Communication Networks and Information Security (IJCNIS)*, *14*(2), 179–188. https://doi.org/10.17762/ijcnis.v14i2.5499.
57. Keserwani, H., Rastogi, H., Kurniullah, A. Z., Janardan, S. K., Raman, R., Rathod, V. M., and Gupta, A. (2022). Security Enhancement by Identifying Attacks Using Machine Learning for 5G Network. *International Journal of Communication Networks and Information Security (IJCNIS)*, *14*(2), 124–141. https://doi.org/10.17762/ijcnis.v14i2.5494.
58. Pandey, D., Pandey, B. K., Noibi, T. O., Babu, S., Patra, P. M., Kassaw, C., ... Canete, J. J. O. (2020). Covid-19: Unlock 1.0 Risk, Test, Transmission, Incubation and Infectious Periods and Reproduction of Novel Covid-19 Pandemic. *Asian Journal of Advances in Medical Science*, 23–28.
59. Pandey, D., George, S., Aremu, B., Wariya, S., and Pandey, B. K. (2021). Critical Review on Integration of Encryption, Steganography, IOT and Artificial Intelligence for the Secure Transmission of Stego Images.
60. Juneja, S., Juneja, A., and Anand, R. (2019, April). Reliability Modeling for Embedded System Environment Compared to Available Software Reliability Growth Models. In *2019 International Conference on Automation, Computational and Technology Management (ICACTM)*, pp. 379–382. IEEE.
61. Shukla, R., Dubey, G., Malik, P., Sindhwani, N., Anand, R., Dahiya, A., and Yadav, V. (2021). Detecting Crop Health Using Machine Learning Techniques in Smart Agriculture System. *Journal of Scientific and Industrial Research*, *80*(08), 699–706.
62. Kumar Pandey, B., Pandey, D., Nassa, V. K., Ahmad, T., Singh, C., George, A. S., and Wakchaure, M. A. (2022). Encryption and Steganography-Based Text Extraction in IoT Using the EWCTS Optimizer. *The Imaging Science Journal*, 1–19.

3 Role of Human-Robot Interaction in Building Automated Mixed Palletization Based on Industrial Applications

K. Hamela, Vivek Veeraiah, Parimita,
Dharmesh Dhabliya, V. Vidya Chellam,
S. Praveenkumar, and Nidhi Sindhwani

3.1 INTRODUCTION

Every situation in which humans and robots are co-located is covered under Human-Robot Interaction (HRI). In order to teach the robot what it needs to know and recognize, such data will need to be annotated by humans. The term "industrial robot" refers to the class of robots used mostly in factories for making and producing goods.

3.1.1 Human-Robot Interaction

HRI is the study of robots and robot systems designed to work with or alongside humans. AI, CS, robotics, and HCI are all a part of it, as are branches of psychology, cognitive science, the social sciences, and a host of others. This all-encompassing term encompasses any scenario in which human beings and robots share the same physical space. The various kinds of HRI are shown in Figure 3.1. HRI's principal goal is to train robots in social interaction with humans. Humans should not be pushed to acquire new ways of communicating with robots as technology becomes more prevalent [1–5]. When people interact with robots, the robots will learn from their human counterparts, adapting to their needs.

There are two broad kinds of communication and interaction:

- **Remote interaction** – As far as space and time are concerned, there is no direct physical connection between the two individuals involved.
- **Proximate interaction** – People and robots share the same space.

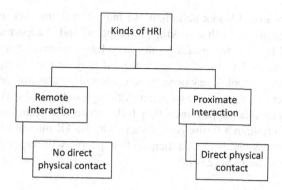

FIGURE 3.1 Kinds of HRI.

In order to supply robots with all of the abilities they need to successfully interact with people, they need a lot of training data. In order to teach the robot what it needs to know and recognize, such data will need to be annotated by humans. In order to train such a robot, a lot of text annotation would be necessary to train the natural language processing (NLP) skills so the robot learns which words have the greatest significance. A robot's ability to recall and recognize a person's facial characteristics is dependent on the use of landmark annotations in facial recognition technology [6–11].

Since the early 1990s, robots have undergone drastic changes in both behavior and appearance, and this trend shows no signs of stopping. Robot aesthetics span a wide range, from mechanoid to zoomorphic to humanoid machines and, at the other end of the spectrum, android. There is a vast amount of room for creativity in how robots are supposed to seem, act, and think. The hardware and software of one robot may be incompatible with those of another robot or even an earlier version of the same robot. Keep in mind that robots are not like other living things since they have not evolved to suit their settings through time. To fully grasp the anatomy, biology, behavior, and other aspects of a species, one must be familiar with its evolutionary history due to the complex relationships that exist between different generations of that species. A robot is a machine built and operated by humans. Learning robots, too, have been programmed to acquire knowledge in a certain order and at a particular age [12–18].

What we mean by "robot" now will be quite different from what we mean a century from now. What we mean when we say "robot" is evolving all the time. Consequently, there is a massive amount of work involved in studying interactions with robots and gaining general insights on HRI applicable across a variety of platforms. The 'H' in HRI ("user studies") focuses on humans, while the 'R', the robot's technical and robotic attributes, is often overlooked; in-depth studies of both humans and interactive robots are necessary if we are to find a way for them to cohabit in the same space. HRI investigations are hampered by the inability to compare results obtained with different robots. It would be ideal to conduct each HRI experiment using a variety of robots and their accompanying behaviors [19–22].

Using the Theatrical Robot paradigm, we may start doing user research within the system design phase. With a working prototype in hand, it's less probable that the TR method will provide any useful results since future research may be conducted on the "real" system. On the other hand, the TR may be used independently as a useful tool to study people's reactions to others based on appearance, or to a robot that looks and acts very much like a human. Although android robots can come close to approximating human appearance, they lack human-like gestures, behavior, and cognition, which remain a future goal. As a result, the TR may shorten the lengthy development process and help us anticipate how people will respond to robots that are very similar to humans [23–29].

3.1.2 Various Sensors Used in Robots

Below are some examples of sensors (shown in Figure 3.2) that might be utilized in robotics:

a) **Light sensors** – The amount of light that hits a light sensor causes a voltage difference, and this difference is proportional to the amount of light that hits the sensor. Two of the most common types of light sensors used on robots are solar panels that use a photo-resistor as a light detector and photodiodes, photoresistors, phototransistors, and photovoltaic light sensors [30].

b) **Sound sensors** – These sensors are microphones which sense sound and return a voltage differential that corresponds to the volume of sound. An example of a sound sensor is instructing a robot via clap.

c) **Temperature sensor** – As the ambient temperature changes, the sensors that track it respond appropriately. In response to a change in temperature, a voltage difference is generated. LM34, TMP35, TMP36, and LM37 are all examples of integrated circuits used as thermal sensors [31–36].

d) **Proximity sensor** – A proximity sensor may detect nearby objects without coming into direct contact with them. The following types of proximity sensors are often found in robots: ultrasonic sensors, photoresistors, and infrared (IR) transceivers [37–42].

e) **Acceleration sensor** – An accelerometer is a device that can measure acceleration and maybe tilt to accommodate the data it collects.

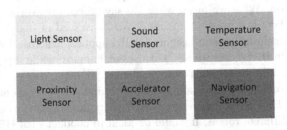

FIGURE 3.2 Various sensors used in robots.

f) **Navigation sensor** – These sensors may be used to pinpoint the precise position of a robot. Global Positioning System, DC, and Localization System are all examples of navigation sensors.

3.2 INDUSTRIAL ROBOTS AND THEIR TYPES

Manufacturing and production are the primary functions of industrial robots. The following are brief descriptions of several common types of industrial robots, which are used in a wide variety of settings depending on their purpose [43–48].

1) **Cartesian:** the Cartesian coordinate system is used by the Cartesian robot (X, Y, and Z). Three linear joints characterize this class of robots. They may also have a wrist capable of rotating.
2) **Polar:** it is a sort of robot that may be made up of a rotating base and a pivoting height. Only one arm of the polar robot is capable of doing numerous tasks.
3) **SCARA:** the abbreviation "Selective Compliance Assembly Robot Arm" means that in addition to a vertical movement, the Sacra robot is capable of three linear motions. The Z axis is fixed, whereas the X and Y axes are movable [49, 50].
4) **Delta:** spider-like robots with parallel arms that link to universal joints make up the form of these robots.
5) **Cylindrical:** the cylindrical robot has two types of joints: a rotatory joint for doing rotational transactions and a prismatic joint for making linear ones.
6) **Articulated:** articulated robots may have anywhere from two to ten or more rotatory joints, depending on their complexity.

3.3 HRI AND SAFETY

In order for robots to work alongside humans in a variety of settings, including the home, business, and other public places, they must ensure their users' safety. As a result, common solutions to robot safety in the industry (such as warning noises and flashing lights) may not be acceptable or viable. To put it simply, human life should be protected from harm caused by robots or parts of robots in any interaction between humans and machines. Safety may be improved through new advancements in the robot's technological characteristics (such as the robot's control and sensors), as well as the robot's materials (soft, lightweight, etc.). It's important to evaluate research that emphasizes several methods for human-robot safety, from design and sensors to software and planning, since different situations call for different approaches and measurements. The analysis and design of safety features, the design of safety for robots through the development of specific mechanical and actuator systems or by exploiting new materials, the design of low and medium-level controllers for safe compliance via direct force compliance, and the development of high-level cognition, control, and decision-making aspects are all explored in this field of study [51–56].

However, even non-aggressive encounters may not be seen as pleasant. As a result, both factual and subjective measures of physical safety can be considered. Over time, as a user becomes more familiar with the robot and learns more about its capabilities and limits, their expectations of the robot's behavior are likely to improve [57, 58]. Very little research has been done on the topic of using social signals to make human-robot interactions safer. It is rare for safety studies in human-robot interaction to address human behavioral and social dynamics, preferring instead to concentrate on technological safety criteria. Even in potentially dangerous situations (e.g., when walking down a corridor), humans are able to communicate effectively with robots through a variety of vocal and nonverbal coordination techniques.

When it comes to using social cues to make robots more secure, there are two key considerations:

1) It is possible for a robot to convey social cues and behavior that alert the user to the possibility of a potentially harmful action being taken by the robot. In this instance, the individual would have to make the effort to change their own behavior to guarantee safe engagement with the robot.
2) If the user's behavior and activities are unpredictable, it can utilize knowledge from previous encounters with the user to forecast what the robot will do next. The robot can then adjust its own behaviors to prevent potentially harmful interactions. Instead, the robot takes charge and strives to keep things under control so that the human may converse freely with it.

Robot control and sensing systems are more technically difficult, but both techniques have the potential to make robots safer to use in human-centered environments. Human-robot cooperation in anticipating and responding to each other's present and projected activities would be a more "natural" option, as would the use of both "safety-aware" humans and robots in concert to avoid dangerous circumstances. Complex activities requiring advanced perception and prediction on the part of the robot would need to be used in dynamic and realistic surroundings.

3.3.1 Palletizing

Robotic palletizing is the process of transporting products on a pallet by utilizing an industrial robot to put and stack them. One of the most frequent ways to move large, heavy things is to employ palletizing robots that are specifically designed for this purpose. Palletizing is stacking items, such as cardboard boxes, in a certain arrangement on a pallet or similar device. Unloading the loaded item in the reverse manner is called depalletizing.

A palletizing robot solution of some form is used in many companies and facilities today. By automating the palletizing process, companies may enhance production and profitability while also enabling them to keep their items on the shelf for longer.

Using a robot control system that has a palletizing feature built-in, an object may be loaded and unloaded quickly and efficiently. Robotic workstations are a great addition to any project. The recent advancement of end-of-arm-tooling has led to the widespread use of robot-operated palletizing work cells on numerous factory floors (EOAT). There are several factors to think about while choosing a robot, including payload, reach, duty, cycle time, and available floor space. Your end-of-arm product and dress out should be taken into consideration while deciding on a payload range.

Some robotic technologies are revolutionizing the order fulfillment sector of the material handling business. An organization's productivity, adaptability, and reliability may all see significant boosts through the use of layer forming and inline palletizing, layer depalletizing, pallets, and mixed case palletizing.

3.3.2 Inline Palletizing

Automation of end-of-line palletizing for like-products has matured into a reliable practice during the last decade. Possible throughput, availability of the system, required floor space, and complete production line utilization rate are major factors in determining the economic logic of a palletizing system. Robots set new norms in Inline Palletizing, especially for high throughput requirements like those seen in the beverage industry. Articulated robots allow for the implementation of novel, inventive robot-based solutions for inline palletizing at high speeds. When packages are received, they are forcibly pushed into their designated slots on a mesh belt conveyor using gripper mechanisms custom-designed to suit each individual box. This method of layer building is reliable since robot motion is synced with conveyor speed. Because the objects are actively seized, the process is independent of friction and packaging characteristics, which may be a concern with shrink-wrapped goods.

A very small gap is established between the packages during the layer-building procedure to prevent any accidental collisions. When a layer is finished, a mechanical end stop is lowered and the layer is sent to the spot where the layer pickup robot will collect it. The layer's limit is increased as soon as it reaches it, allowing the construction of the next layer to begin.

The robot's built-in puller allows it to load a whole layer onto the gripper. By the time the layer reaches the gripper's core, it's compact enough to be placed into the intended pallet without error. The layer gripper may also use a slip sheet if necessary. A robot-based inline palletizing system may be tuned to the required system performance dependent on the pace of the manufacturing line. An additional layer-forming robot, for instance, could potentially triple output.

3.3.3 Layer Depalletizing and Palletizing

Optimal utilization of established equipment is the only way to accomplish cost-effective automation. In layer-picking systems, this means that you can regulate a process so that the technology implemented can handle the largest proportion of your whole product package variety.

Automated depalletizing and order picking at a distribution center are challenging due to the wide diversity of product and packaging types. A roll-up principle has been included in a universal grasping mechanism since the majority of package types have flat bottoms. With this technology, you may handle many goods at the same time.

The pallet layer is pushed against by two servo-driven rubber rollers on either side. As a result, the crates are raised and rolled onto two symmetrical plates of transport. Depending on the package weight and dimensions, software that regulates the gripper may alter the rollers' contact pressure and height. Thus, total selection efficiency may be greatly improved by utilizing the roll-up technique. Layer count is restricted if either the maximum payload mass or the maximum payload height is exceeded.

3.3.4 Mixed Case Palletizing

True mixed-type package palletizing has matured into a well-established practice during the last decade. Possible throughput (cases per hour), system availability, required floor space, and complete production line utilization rate are major factors in determining the economic logic of a palletizing system. This has gotten increasingly time-consuming as the number of SKUs in circulation has grown.

The pressure on automated order picking systems in warehouses is growing daily. Palletizing orders, products, and a wide variety of containers (on pallets or pushcarts) must be done exactly on time and with minimal mistakes.

Whether it's open or closed trays or drinks wrapped in PET film, the variety of products and the size and materials of the packaging units are almost infinite. As a result of these needs, we built a custom application module called Pallet-MIX to do automatic palletizing. The primary objective was to develop an application module using innovative software and gripper technologies on commercially available, proven jointed-arm robots.

We opted for a six-degrees-of-freedom, jointed-arm robot specifically to avoid having to rotate the packing unit from the outside. The advantages over gantry systems are significant. The primary goals were performance and dependability in palletizing the goods with the final goal being to achieve high throughput rates.

3.3.5 HRI for Industrial Robots

There are a variety of ways in which humans and robots might interact in the workplace, whether fixed or mobile. Articles on multimodal interaction describe a method for combining several modalities. These days, touch displays are the standard human-robot interface in industrial settings, and they often come packaged with the robot. Apps and programming blocks help make these more user-friendly. Users can't use gestures or speech recognition to interface with devices currently. Researchers have proposed a variety of interfaces, including gestures, speech recognition, and multimodal techniques. As part of the experiment, four actions are tested: wave left/right/rise/put down. In order to recognize hand motions, the method combines depth data with a regular Camshft tracking technique, using a Kinect and HMM for dynamic gesture identification. Robot guiding can be used to refine the approximate poses specified by the gestures. The worker is also supported with a mobile AR interface that shows the postures and trajectories. Body and hand motions were included in the lexicon they devised. They use an RGB-d camera to detect movement of the whole body, and a jump motion to detect movement of the hands.

The robot arm may be moved in a variety of directions using any of these movements.

When using gesture recognition for industrial robots, the most common usage is to move robots into particular positions or change their trajectory. It is possible that voice commands will become the primary method for operating robots. Voice commands are often paired with additional modalities, like as gestures or eye gazing, to improve recognition accuracy. A semantic approach may improve the consistency and fluency of communication between humans and industrial robots in everyday settings. The method relies on being able to recognize spoken commands and hand movements that indicate a need for processing. The efficacy and efficiency of deburring wax components and disassembling tasks from the casting process are examined. Researchers have observed that people's extroversion and attitude towards robots affect the temporal dynamics of social signals (i.e., glance towards the robot's face and voice) during a human-robot interaction task in which the two parties must physically work together to construct an item.

3.4 LITERATURE REVIEW

Before initiating this research work, many past studies and research articles were studied by us. Especially those pertaining to HRI, palletizing, and industrialization, a brief description of some of them is given below.

Charalambous, G., Fletcher, S., and Webb, P. (2016) Having been created to boost efficiency and product quality in manufacturing, the notion of industrial HRC is becoming more prevalent. However, until recently, insufficient study has been devoted to understanding the critical HF that must be addressed in order for industrial HRC to be successfully implemented. HF at both the organizational and individual levels has recently been identified by the authors' recent study. In this study, the authors summarize their findings and provide a path for the effective deployment of industrial HRC based on their research. With this roadmap, automation experts and MSE will be able to comprehend the main HF that must be taken into account for cooperation between people and industrial robots to maximize efficiency and productivity, which will have substantial repercussions [1].

Hayes, B., and Scassellati, B. (2016) reviewed. In order for humans and robots to collaborate successfully, problems including multi-agent planning, SE, and goal inference must be addressed. The lack of precise task models makes it difficult to find solutions to many of these difficulties. In this research, they introduce a novel kind of HTN called a CC-HTN and provide a method for automatically producing them. Since no unique symbolic knowledge of motor primitives or environmental representation is required, this technique may be used for a broad range of human-robot interaction situations. To demonstrate the value of their approach, they provide both a multi-resolution goal inference challenge and a transfer learning application [2].

Stumm, S., Braumann, J., Von Hilchen, M., and Brell-Cokcan, S. (2017) said that industrial robots' programming tends to be static and optimized for a specific environment. On the other hand, service robots concentrate on operating in dynamic and unpredictable contexts. Robotics in the construction industry must be able to adapt to dynamic and unstructured situations in order to be effective. However, a great deal of data has been gathered throughout the planning stage. On the building site, this information is frequently not utilized effectively or immediately. In order to achieve acceptable quality in assembly operations, they plan to combine force torque sensors and human-robot interaction in the form of haptic programming to enable intuitive on-site robot programming [3].

Salam, H., Celiktutan, O., Hupont, I., Gunes, H., and Chetouani, M. (2017) Presented designing IS that can adapt to characteristics of their users necessitates active user participation. Human-robot interaction is the topic of this article, which examines how to automatically classify interactions between people and robots using their personality characteristics. In this study, two volunteers interact with a humanoid robot to see whether their interest levels can be predicted based on the robot's and their own personalities. The completely autonomous system is developed by first predicting the Big Five traits of each participant by extracting individual and interpersonal features from nonverbal behavioral indicators. The engagement classification method takes into account personality predictions as one of its inputs. Third, they analyze the impact of shared and divergent traits on the classification of group engagement, which we define as the extent to which all participants interact with the robot. The proposed automatic personality prediction module is effective because (i) the F-measure for using automatically predicted personality labels for engagement classification was equivalent to that of using the manually annotated personality labels and (ii) the use of individual and interpersonal features without utilizing personality information is insufficient for engagement classification [4].

Park, S. Il, and Lee, S. J. (2017) looked at a system that employs Bluetooth beacons based on Industry 4.0 technology to track a worker's whereabouts and improve security. Manufacturing and logistics operations are becoming more diverse as a result of the transition to a Smart Factory. Adapting to these shifts will need new and inventive approaches, particularly in the areas of safety and security. Bluetooth-enabled devices, such as Smart Pads, smartphones, and other mobile devices, may currently communicate with each other through an app that measures the distance between them and the beacon. The server will receive the distance information and use it to operate other manufacturing devices. If you're concerned about security or other factors that would make it difficult to use Smart Pads in a production environment, this strategy may not be the best option for you. As a solution, a beacon was affixed to each worker's safety helmet, synchronized with their unique ID, and then used to verify both signals to determine whether a worker was allowed entry into the workplace or whether security should intervene. There are several advantages to using a system like this to keep employees safe and prevent them from accessing dangerous areas, as well as to keep production running smoothly by monitoring workers' movements and estimating their locations in the case of an accident. There will be no need for extra application installs, equipment, or an external network or GPS in order to construct an autonomous network at factories that are far from cities [5].

Salmi, T., Ahola, J. M., Heikkilä, T., Kilpeläinen, P., and Malm, T. (2018) comprised studies that discussed sensor-based robotics in construction in conjunction with human-robot collaboration technologies. Safety and control technologies for human-robot collaboration, as well as sensor-assisted industrial robot control and a dynamic safety system for industrial robots, are discussed at length. The feasibility of sensor-based robotics in the construction industry, as well as the potential of robots generally, is also investigated [6].

Richert, A., Müller, S., Schröder, S., and Jeschke, S. (2018) introduced the fourth industrial revolution as being characterized by new types of AI on one hand, and pervasive networking of "everything with everything" on the other. There is a shift in our knowledge of human-machine interaction, as well as a shift in the way we think about manufacturing, as a consequence of this. "So canalizing with robots" is an empirical research that seeks to learn more about the circumstances and procedures of hybrid human-machine teams. In this project, the virtual environment was used to carefully monitor the behaviors and interactions between humans and robots. They were all unique in form and personality (reliable or faulty). In order to accomplish a goal, participants were given the task of working together as a team. Using the results of an experiment, the authors combine ideas from the fields of social robotics and psychology to discuss the phenomenon of anthropomorphization. Anthropomorphic phonation and mechanization are examined in the context of social robotics, which is an inter- and transdisciplinary discipline [7].

Kaiser, L., Schlotzhauer, A., and Brandstötter, M. (2018) presented that sensitive robots have been utilized in industry, and in certain circumstances, they collaborate directly with people on shared workstations. This form of human-machine contact seems to be fraught with danger at first glance. It is possible to make such cooperative settings safe for humans, however, with the inclusion of additional devices, enhanced functionality, and risk mitigation measures. Cooperative methods of operation, workspace organization, end effectors, human-machine interfaces, and ergonomics are all critical components of any system. Human-robot cooperation has many elements, and this study sheds light on some of them. When a robot's possible hidden hazards are appropriately reduced and communicated, a human may feel confident using the robot [8].

Kretschmer, V., Plewan, T., Rinkenauer, G., and Maettig, B. (2018) focused on the most important logistical processes. Palletization relies heavily on the efficiency and knowledge of its workers. In order to gauge the potential benefits of using an AR device for palletization, it was compared against both a paper-based pick list and a tablet computer. Workload assessments suggest that the AR device provides the least difficult way to help participants in palletization, whereas the tablet computer surpassed the AR device and the choise list in terms of usability. AR gadgets are suitable for helping logistics employees with palletizing, however their usability has to be significantly enhanced [9].

Silva, R., Rocha, L. F., Relvas, P., Costa, P., and Silva, M. F. (2018) on the increased usage of robotic palletizing devices in the so-called "Fast Moving Consumer Products" market. Adaptability in packaging for different products is very low due to the great performance and efficiency of the solutions that have been created thus far. Since many companies are switching to high-variety, low-volume production, this is a critical problem. In this setting, robots need to be set up more often, which has increased the need for offline programming tools that shorten the amount of downtime each robot requires. In this work, we provide an offline, automated technique for building robot algorithms that avoid collisions when used in palletizing tasks [10].

Moura, F. M., and Silva, M. F. (2018) claimed that providing current market needs necessitates a wide range of manufacturing and logistical flexibility, speed, and repeatability. An increasingly sought-after solution for dynamic enterprises and engineering processes will be robotics, which is commonly connected with the notions of a fourth industrial revolution that is heavily reliant on smart machines, storage systems, and manufacturing facilities cooperating. Because of this, the global operating stock of industrial robots has grown steadily over the last decades and is likely to continue growing in this manner. In the European market, handling operations are the most common use of robots in industrial applications. Due to the increasing importance of palletizing in today's supply chains, it has become a crucial handling activity. This project intends to contextualize and create a palletizing robot application in this setting. The palletizing operations of a robot may be automatically programmed using this application and an offline programming software. The program was created in Rapid Application Development (RAPID) and features a simple XML-based user interface [11].

Jabbar, S., Khan, M., Silva, B. N., and Han, K. (2018) to facilitate communication amongst warehouse goods, the rest framework is introduced as a web-oriented design. In the recommended procedure, the smart warehouse comprises a data collection module and an administrative module. First group consists of radio frequency identification sensors, wireless sensors used for acquiring operational data, and actuators. Its primary function is to record the warehouse's environmental conditions, such as temperature, humidity, air quality, and pressure, as well as product and inventory information. The smart warehouse's brain analyzes and organizes information, creates events, and performs autonomous action. A user at a warehouse may utilize the Internet to operate a variety of sensors, gadgets, appliances, and a high-fidelity sound system. An event decision system is built on top of the web infrastructure to control the instantaneous processing of sensors, objects, and so on. The suggested architecture for the smart gateway consists of two tiers: TM and service module. By fusing TM and DSM, a device controller may be made for embedded devices using non-standard protocols. The suggested system's response time, transmission failure rate, and time to discovery are analyzed under varying conditions [12].

3.5 PROPOSED WORK

3.5.1 Problem Statement

Human-robot interaction research has been done extensively to automate mixed palletization of commodities for industrial use. Smart places may benefit greatly from these services. But the problem with the currently available study is that it has a sluggish performance and is inaccurate. Therefore, an optimal technique has to be proposed in order to gain better accuracy while maintaining high performance.

3.5.2 Proposed Methodology

The functioning of the current Human-Robot Interaction for automated Mixed Palletization based on Industrial Applications has been taken into consideration in the suggested study work. Additionally, the elements that are impacting the performance and accuracy have been taken into account. For the goal of categorization, many different automation systems are being examined; nevertheless, it is necessary to include an optimization mechanism. In this line of research, an optimizer is being considered as a means of filtering the dataset before training time. It is anticipated that an optimized technique coupled with Human-Robot Interaction for automated Mixed Palletization would increase the accuracy of event categorization for services that are relevant in smart cities. In conclusion, an evaluation of the suggested work in comparison with the usual method would be carried out, focusing on performance and accuracy. The research methodology is shown in Figure 3.3.

3.6 RESULTS AND DISCUSSION

It is necessary to stake into consideration the factors that impact the model's performance and accuracy in order to gain an understanding of how Human-Robot Interaction for automated Mixed Palletization based on Industrial Applications works for a smart city. This will allow one to comprehend how the system works. Several different approaches to automation are being investigated for the purpose of classification, but it is essential that an optimization mechanism be included in the overall design as well. An improvement in accuracy is anticipated to result from the use of human-robot interaction in the process of automated Mixed Palletization to event classification for smart city services. In conclusion, the purpose of the study is to improve accuracy parameters.

The accuracy comparison is shown in Table 3.1 and Figure 3.4 while the error rate comparison is shown in Table 3.2 and Figure 3.5. It is obvious that the proposed method is far superior to the existing work in terms of accuracy as well as error rate.

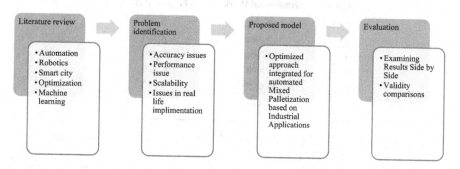

FIGURE 3.3 Research methodology.

TABLE 3.1
Comparison of Accuracy

Class	Traditional Work	Proposed Work
1	87.48%	93.11%
2	87.51%	93.24%
3	87.99%	93.51%
4	88.12%	93.83%
5	88.32%	94.01%
6	88.51%	94.28%

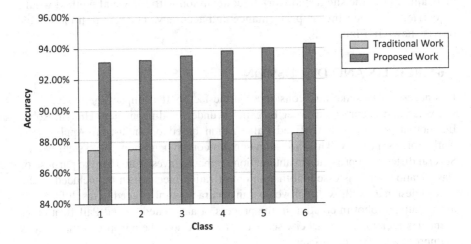

FIGURE 3.4 Comparison of accuracy.

TABLE 3.2
Comparison of Error Rate

Class	Traditional Work	Proposed Work
1	12.52%	6.89%
2	12.49%	6.76%
3	12.01%	6.49%
4	11.88%	6.17%
5	11.68%	5.99%
6	11.49%	5.72%

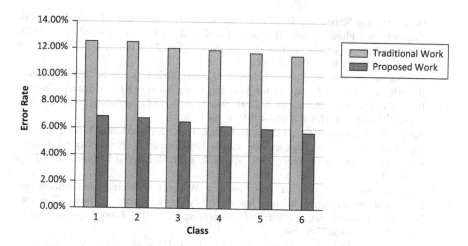

FIGURE 3.5   Comparison of error rate.

3.7 CONCLUSION

The results of the simulation lead one to the conclusion that optimization has a key influence in the selection of datasets that are appropriate. When compared to more traditional techniques, the optimized dataset provides much higher levels of accuracy. Additionally, the suggested approach has decreased the amount of time spent performing training operations by decreasing the size of the dataset.

3.8 FUTURE SCOPE

The process of industrial automation requires the use of a variety of devices, including PLCs, SCADAs, HMIs, VFDs, and DCSs. Their mission is to decrease the amount of human involvement and the mistakes that result from it. Additionally, there is a considerable decrease in the amount of time required for manufacturing, in addition to an improvement in both dependability and productivity. When considering the future of robotics, the development of artificial intelligence will lead to an increase in the number of ways in which people and robots may work together. As sensors, artificial intelligence (AI), and technologies that recognize and analyze voices continue to advance, we will be able to connect with machines in a way that is more natural and intuitive for humans [59–61].

REFERENCES

1. Charalambous, G., Fletcher, S., and Webb, P. (2016). Development of a Human Factors Roadmap for the Successful Implementation of Industrial Human-Robot Collaboration. *Advances in Intelligent Systems and Computing*, 490, 195–206. https://doi.org/10.1007/978-3-319-41697-7_18

2. Hayes, B., and Scassellati, B. (2016). Autonomously Constructing Hierarchical Task Networks for Planning and Human-Robot Collaboration. In *Proceedings - IEEE International Conference on Robotics and Automation*, 2016-June, 5469–5476. https://doi.org/10.1109/ICRA.2016.7487760
3. Stumm, S., Braumann, J., Von Hilchen, M., and Brell-Cokcan, S. (2017). On-Site Robotic Construction Assistance for Assembly Using A-Priori Knowledge and Human-Robot Collaboration. *Advances in Intelligent Systems and Computing*, 540, 583–592. https://doi.org/10.1007/978-3-319-49058-8_64
4. Salam, H., Celiktutan, O., Hupont, I., Gunes, H., and Chetouani, M. (2017). Fully Automatic Analysis of Engagement and Its Relationship to Personality in Human-Robot Interactions. *IEEE Access*, 5(8), 705–721. https://doi.org/10.1109/ACCESS.2016.2614525
5. Park, S., and Lee, S. J. (2017). A Study on Worker's Positional Management and Security Reinforcement Scheme in Smart Factory Using Industry 4.0-Based Bluetooth Beacons. *Lecture Notes in Electrical Engineering*, 421, 1059–1066. https://doi.org/10.1007/978-981-10-3023-9_164
6. Pandey, B. K., Pandey, D., Wairya, S., Agarwal, G., Dadeech, P., Dogiwal, S. R., and Pramanik, S. (2022). Application of Integrated Steganography and Image Compressing Techniques for Confidential Information Transmission. *Cyber Security and Network Security*, 169–191.
7. Pandey, B. K., Pandey, D., and Agarwal, A. (2022). Encrypted Information Transmission by Enhanced Steganography and Image Transformation. *International Journal of Distributed Artificial Intelligence (IJDAI)*, 14(1), 1–14.
8. Kaiser, L., Schlotzhauer, A., and Brandstötter, M. (2018). Safety-Related Risks and Opportunities of Key Design-Aspects for Industrial Human-Robot Collaboration. *Lecture Notes in Computer Science (Including Subseries Lecture Notes in Artificial Intelligence and Lecture Notes in Bioinformatics)*, 11097 LNAI, 95–104. https://doi.org/10.1007/978-3-319-99582-3_11
9. Kretschmer, V., Plewan, T., Rinkenauer, G., and Maettig, B. (2018). Smart Palletisation: Cognitive Ergonomics in Augmented Reality Based Palletising. *Advances in Intelligent Systems and Computing*, 722, 355–360. https://doi.org/10.1007/978-3-319-73888-8_55
10. Silva, R., Rocha, L. F., Relvas, P., Costa, P., and Silva, M. F. (2018). Offline Programming of Collision Free Trajectories for Palletizing Robots. *Advances in Intelligent Systems and Computing*, 694, 680–691. https://doi.org/10.1007/978-3-319-70836-2_56
11. Pandey, B. K., Pandey, D., Gupta, A., Nassa, V. K., Dadheech, P., and George, A. S. (2023). Secret Data Transmission Using Advanced Morphological Component Analysis and Steganography. In *Role of Data-Intensive Distributed Computing Systems in Designing Data Solutions* (pp. 21–44). Springer International Publishing.
12. Jabbar, S., Khan, M., Silva, B. N., and Han, K. (2018). A REST-Based Industrial Web of Things' Framework for Smart Warehousing. *Journal of Supercomputing*, 74(9), 4419–4433. https://doi.org/10.1007/s11227-016-1937-y
13. Tellaeche, A., Maurtua, I., and Ibarguren, A. (2015). Human Robot Interaction in Industrial Robotics. *IEEE International Conference on Emerging Technologies and Factory Automation, ETFA*, 2015-October. https://doi.org/10.1109/ETFA.2015.7301650
14. Beneficiary, L., and Update, L. (2020). D7. 1 – Human Factors Modelling in Human-Robot Interaction, 1–46.
15. Farouk, M. (2022). Studying Human Robot Interaction and Its Characteristics. *International Journal of Computations, Information and Manufacturing (IJCIM)*, 2(1), 38–49. https://doi.org/10.54489/ijcim.v2i1.73
16. Pandey, B. K., Pandey, D., Wariya, S., Aggarwal, G., and Rastogi, R. (2021). Deep Learning and Particle Swarm Optimisation-Based Techniques for Visually Impaired Humans' Text Recognition and Identification. *Augmented Human Research*, 6, 1–14.

17. Pandey, D., George, S., Aremu, B., Wariya, S., and Pandey, B. K. (2021). Critical Review on Integration of Encryption, Steganography, IOT and Artificial Intelligence for the Secure Transmission of Stego Images., Multidisciplinary Approach to Modern Digital Steganography
18. Kumar, M. S., Sankar, S., Nassa, V. K., Pandey, D., Pandey, B. K., and Enbeyle, W. (2021). Innovation and Creativity for Data Mining Using Computational Statistics. In *Methodologies and Applications of Computational Statistics for Machine Intelligence* (pp. 223–240). IGI Global.
19. Joe, W. Y., and Song, S. Y. (2019). Applying Human-Robot Interaction Technology in Retail Industries. *International Journal of Mechanical Engineering and Robotics Research*, 8(6), 839–844. https://doi.org/10.18178/IJMERR.8.6.839-844
20. Szafr, D., and Szafr, D. A. (2021). Connecting Human-Robot Interaction and Data Visualization. *Academic Medicine/IEEE International Conference on Human-Robot Interaction*, 281–292. https://doi.org/10.1145/3434073.3444683
21. Johannsmeier, L., and Haddadin, S. (2017). A Hierarchical Human-Robot Interaction-Planning Framework for Task Allocation in Collaborative Industrial Assembly Processes. *IEEE Robotics and Automation Letters*, 2(1), 41–48. https://doi.org/10.1109/LRA.2016.2535907
22. BoyrazBaykas, P., Bayraktar, E., and Bora Yigit, C. (2020). Safe Human-Robot Interaction Using Variable Stiffness, Hyper-redundancy, and Smart Robotic Skins. *Service Robotics*. https://doi.org/10.5772/intechopen.92693
23. Ghidoni, S., Terreran, M., Evangelista, D., Menegatti, E., Eitzinger, C., Villagrossi, E., Pedrocchi, N., Castaman, N., Malecha, M., Mghames, S., Castri, L., Hanheide, M., and Bellotto, N. (2022). From Human Perception and Action Recognition to Causal Understanding of Human-Robot Interaction in Industrial Environments. Ital-IA 2022, 1–4. https://eprints.lincoln.ac.uk/id/eprint/48515/
24. Ghosh, A., Soto, D. A. P., Veres, S. M., and Rossiter, A. (2020). Human Robot Interaction for Future Remote Manipulations in Industry 4.0. *IFAC-PapersOnLine*, 53(2), 10223–10228. https://doi.org/10.1016/j.ifacol.2020.12.2752
25. Costa, G. de M., Petry, M. R., and Moreira, A. P. (2022). Augmented Reality for Human–Robot Collaboration and Cooperation in Industrial Applications: A Systematic Literature Review. *Sensors*, 22(7). https://doi.org/10.3390/s22072725
26. Hjorth, S., and Chrysostomou, D. (2022). Human–Robot Collaboration in Industrial Environments: A Literature Review on Non-destructive Disassembly. *Robotics and Computer-Integrated Manufacturing*, 73(December 2020), 102208. https://doi.org/10.1016/j.rcim.2021.102208
27. deCarvalho, P. R. V., and Elhedhli, S. (2021). A Data-Driven Approach for Mixed-Case Palletization with Support. *Optimization and Engineering*, 89, 01234567. https://doi.org/10.1007/s11081-021-09673-5
28. Liu, Y., Habibnezhad, M., and Jebelli, H. (2021). Brainwave-Driven Human-Robot Collaboration in Construction. *Automation in Construction*, 124(November 2020), 103556. https://doi.org/10.1016/j.autcon.2021.103556
29. Saenz, J., Behrens, R., Schulenburg, E., Petersen, H., Gibaru, O., Neto, P., and Elkmann, N. (2020). Methods for Considering Safety in Design of Robotics Applications Featuring Human-Robot Collaboration. *International Journal of Advanced Manufacturing Technology*, 107(5–6), 2313–2331. https://doi.org/10.1007/s00170-020-05076-5
30. Brosque, C., Galbally, E., Khatib, O., and Fischer, M. (2020). Human-Robot Collaboration in Construction: Opportunities and Challenges. HORA 2020 - 2nd International Congress on Human-Computer Interaction, Optimization and Robotic Applications, Proceedings. https://doi.org/10.1109/HORA49412.2020.9152888
31. Kováč, J., Jenčík, R., Andrejko, P., Hajduk, M., Pilat, Z., Tomči, P., Varga, J., and Bezák, M. (2020). Integrated Palletizing Workstation with an Industrial Robot and a

Cobot. *Advances in Intelligent Systems and Computing*, 980, 202–209. https://doi.org/10.1007/978-3-030-19648-6_24

32. Kopp, T., Baumgartner, M., and Kinkel, S. (2020). Success Factors for Introducing Industrial Human-Robot Interaction in Practice: An Empirically Driven Framework. *International Journal of Advanced Manufacturing Technology*. https://doi.org/10.1007/s00170-020-06398-0

33. Berg, J., and Lu, S. (2020). Review of Interfaces for Industrial Human-Robot Interaction. *Current Robotics Reports*, 1(2), 27–34. https://doi.org/10.1007/s43154-020-00005-6

34. Petruck, H., Faber, M., Giese, H., Geibel, M., Mostert, S., Usai, M., Mertens, A., and Brandl, C. (2019). Human-Robot Collaboration in Manual Assembly – A Collaborative Workplace. In *Advances in Intelligent Systems and Computing* (Vol. 825). Springer International Publishing. https://doi.org/10.1007/978-3-319-96068-5_3

35. Neto, P., Simão, M., Mendes, N., and Safeea, M. (2019). Gesture-Based Human-Robot Interaction for Human Assistance in Manufacturing. *International Journal of Advanced Manufacturing Technology*, 101(1–4), 119–135. https://doi.org/10.1007/s00170-018-2788-x

36. Gopinath, V., and Johansen, K. (2019). Understanding Situational and Mode Awareness for Safe Human-Robot Collaboration: Case Studies on Assembly Applications. *Production Engineering*, 13(1), 1–9. https://doi.org/10.1007/s11740-018-0868-2

37. Anand, R., Singh, B., and Sindhwani, N. (2009). Speech Perception and Analysis of Fluent Digits' Strings Using Level-by-Level Time Alignment. *International Journal of Information Technology and Knowledge Management*, 2(1), 65–68.

38. Bonini, M., and Echelmeyer, W. (2018). A Method for the Design of Lean Human-Robot Interaction. *Proceedings - 2018 11th International Conference on Human System Interaction*, HSI 2018, 457–464. https://doi.org/10.1109/HSI.2018.8430879

39. Meelu, R., and Anand, R. (2010, November). Energy Efficiency of Cluster-Based Routing Protocols Used in Wireless Sensor Networks. In *AIP Conference Proceedings* (Vol. 1324, No. 1, pp. 109–113). American Institute of Physics.

40. Tripathi, A., Sindhwani, N., Anand, R., and Dahiya, A. (2022). Role of IoT in Smart Homes and Smart Cities: Challenges, Benefits, and Applications. In *IoT Based Smart Applications* (pp. 199–217). Springer International Publishing.

41. Gupta, R. S., Nassa, V. K., Bansal, R., Sharma, P., Koti, K., and Koti, K. (2021). Investigating Application and Challenges of Big Data Analytics with Clustering. In *2021 International Conference on Advancements in Electrical, Electronics, Communication, Computing and Automation (ICAECA)* (pp. 1–6). https://doi.org/10.1109/ICAECA52838.2021.9675483

42. Veeraiah, V., Khan, H., Kumar, A., Ahamad, S., Mahajan, A., and Gupta, A. (2022). Integration of PSO and Deep Learning for Trend Analysis of Meta-Verse. In *2022 2nd International Conference on Advance Computing and Innovative Technologies in Engineering (ICACITE)* (pp. 713–718). https://doi.org/10.1109/ICACITE53722.2022.9823883

43. Pandey, B. K., Pandey, D., Nassa, V. K., George, S., Aremu, B., Dadeech, P., and Gupta, A. (2023). Effective and Secure Transmission of Health Information Using Advanced Morphological Component Analysis and Image Hiding. In Gupta, M., Ghatak, S., Gupta, A., Mukherjee, A .L. (eds) *Artificial Intelligence on Medical Data. Lecture Notes in Computational Vision and Biomechanics* (Vol. 37). Springer. https://doi.org/10.1007/978-981-19-0151-5_19

44. Veeraiah, V., Kumar, K. R., LalithaKumari, P., Ahamad, S., Bansal, R., and Gupta, A. (2022). Application of Biometric System to Enhance the Security in Virtual World. In *2022 2nd International Conference on Advance Computing and Innovative Technologies in Engineering (ICACITE)* (pp. 719–723). https://doi.org/10.1109/ICACITE53722.2022.9823850

45. Kaur, J., Sindhwani, N., Anand, R., and Pandey, D. (2022). Implementation of IoT in Various Domains. In *IoT Based Smart Applications* (pp. 165–178). Springer International Publishing.
46. Shukla, A., Ahamad, S., Rao, G. N., Al-Asadi, A. J., Gupta, A., and Kumbhkar, M. (2021). Artificial Intelligence Assisted IoT Data Intrusion Detection. In *2021 4th International Conference on Computing and Communications Technologies (ICCCT)* (pp. 330–335). https://doi.org/10.1109/ICCCT53315.2021.9711795
47. Pathania, V., Babu, S. Z. D., Ahamad, S., Thilakavathy, P., Gupta, A., Alazzam, M. B., and Pandey, D. (2023). A Database Application of Monitoring COVID-19 in India. In Gupta, M., Ghatak, S., Gupta, A., Mukherjee, A. L. (eds) *Artificial Intelligence on Medical Data: Lecture Notes in Computational Vision and Biomechanics*, 37. Springer. https://doi.org/10.1007/978-981-19-0151-5_23
48. Dushyant, K., Muskan, G., Annu, Gupta, A., and Pramanik, S. (2022). Utilizing Machine Learning and Deep Learning in Cybesecurity: An Innovative Approach. In *Cyber Security and Digital Forensics: Challenges and Future Trends* (pp. 271–293). Wiley. https:doi.org/10.1002/9781119795667.ch12
49. Babu, S. Z. D., Pandey, D., Naidu, G. T., Sumathi, S., Gupta, A., Bader Alazzam, M., and Pandey, B. K. (2023). Analysation of Big Data in Smart Healthcare. In Gupta, M., Ghatak, S., Gupta, A., Mukherjee, A. L. (eds) *Artificial Intelligence on Medical Data. Lecture Notes in Computational Vision and Biomechanics*, 37. Springer. https://doi.org/10.1007/978-981-19-0151-5_21
50. Bansal, B., Jenipher, N., Jain, R., Dilip, R., Kumbhkar, M., Pramanik, S., Sandip, R., and Gupta, A. (2022). Big Data Architecture for Network Security. In *Cyber Security and Network Security* (pp. 233–267). Wiley. https://doi.org/10.1002/9781119812555.ch11
51. Gupta, A., Kaushik, D., Garg, M., and Verma, A. (2020). Machine Learning Model for Breast Cancer Prediction. In *2020 Fourth International Conference on I-SMAC (IoT in Social, Mobile, Analytics and Cloud) (I-SMAC)* (pp. 472–477). https://doi.org/10.1109/I-SMAC49090.2020.9243323
52. Sreekanth, N. et al. (2022). Evaluation of Estimation in Software Development Using Deep Learning-Modified Neural Network. *Applied Nanoscience*. https://doi.org/10.1007/s13204-021-02204-9
53. Veeraiah, V., Rajaboina, N. B., Rao, G. N., Ahamad, S., Gupta, A., and Suri, C. S. (2022). Securing Online Web Application for IoT Management. In *2022 2nd International Conference on Advance Computing and Innovative Technologies in Engineering (ICACITE)* (pp. 1499–1504). https://doi.org/10.1109/ICACITE53722.2022.9823733
54. Veeraiah, V., Gangavathi, P., Ahamad, S., Talukdar, S. B., Gupta, A., and Talukdar, V. (2022). Enhancement of Meta Verse Capabilities by IoT Integration. In *2022 2nd International Conference on Advance Computing and Innovative Technologies in Engineering (ICACITE)* (pp. 1493–1498). https://doi.org/10.1109/ICACITE53722.2022.9823766
55. Gupta, N., Janani, S., Dilip, R., Hosur, R., Chaturvedi, A., and Gupta, A. (2022). Wearable Sensors for Evaluation Over Smart Home Using Sequential Minimization Optimization-Based Random Forest. *International Journal of Communication Networks and Information Security (IJCNIS)*, 14(2), 179–188. https://doi.org/10.17762/ijcnis.v14i2.5499
56. Keserwani, H., Rastogi, H., Kurniullah, A. Z., Janardan, S. K., Raman, R., Rathod, V. M., and Gupta, A. (2022). Security Enhancement by Identifying Attacks Using Machine Learning for 5G Network. *International Journal of Communication Networks and Information Security (IJCNIS)*, 14(2), 124–141. https://doi.org/10.17762/ijcnis.v14i2.5494
57. Anand, R., Shrivastava, G., Gupta, S., Peng, S. L., and Sindhwani, N. (2018). Audio Watermarking with Reduced Number of Random Samples. In *Handbook of Research on Network Forensics and Analysis Techniques* (pp. 372–394). IGI Global.

58. Srivastava, A., Gupta, A., and Anand, R. (2021). Optimized Smart System for Transportation Using RFID Technology. *Mathematics in Engineering, Science and Aerospace (MESA)*, 12(4), 953–965.
59. Chaudhary, A., Bodala, D., Sindhwani, N., and Kumar, A. (2022, March). Analysis of Customer Loyalty Using Artificial Neural Networks. In *2022 International Mobile and Embedded Technology Conference (MECON)* (pp. 181–183). IEEE.
60. Anand, R., Sindhwani, N., and Juneja, S. (2022). Cognitive Internet of Things, Its Applications, and Its Challenges: A Survey. In *Harnessing the Internet of Things (IoT) for a Hyper-connected Smart World* (pp. 91–113). Apple Academic Press.
61. Jain, S., Sindhwani, N., Anand, R., and Kannan, R. (2022, March). COVID Detection Using Chest X-Ray and Transfer Learning. In *Intelligent Systems Design and Applications: 21st International Conference on Intelligent Systems Design and Applications (ISDA 2021) Held During December 13–15, 2021* (pp. 933–943). Springer International Publishing.

4 Emerging Soft Computing and Machine Learning Techniques Applicable in Industrial and Automated Applications

*Jay Kumar Pandey, Vivek Veeraiah,
M. Ranjith Kumar, P. Lalitha Kumari,
Jayasri Kotti, M.K. Dharani, and Ankur Gupta*

4.1 INTRODUCTION

4.1.1 Soft Computing

Unlike hard (traditional) computing, soft computing is the opposite of hard. Artificial intelligence (AI) and natural selection are the foundations of this collection of computer approaches. It gives cost-effective solutions to difficult real-world situations for which there is no hard-computational answer. Soft computing, a branch of mathematics and computer science, has been around since the early 1990s. The human mind's ability to approximate real-world answers to issues was the source of inspiration for this project. Approximate computations are used in soft computing to give viable answers to complicated computational issues. With this strategy, it is possible to address issues that are either intractable or just too time-consuming to handle with existing technology. Computational intelligence is also referred to as "soft computing" [1–6].

Proble-solving can be done without computers if you use soft computing. Soft computing, like the human mind, accepts partial facts, ambiguity, imprecision, and approximation. This is in contrast to standard computer models, which don't.

When there isn't enough knowledge to answer an issue, the strategy known as "possibility" is utilized instead of "soft computing." When the problem is too vague to be solved using normal arithmetic and computer procedures, soft computing is

DOI: 10.1201/9781003317456-4

applied. Real-world applications for soft computing may be found across a wide range of industries and settings, including the home, the workplace, and government [7–12].

4.1.2 Features of Soft Computing

The characteristics of soft computing are elaborated below.

1. In real-world situations, soft computing delivers an approximation yet exact answer.
2. Because the algorithms used in soft computing are adaptable, any changes to the surrounding environment have no effect on the activity now underway.
3. Soft computing is founded on the idea of learning from the results of experiments. In other words, no mathematical model is required to solve a problem in soft computing.
4. To tackle real-world issues, soft computing provides approximate outcomes that traditional and analytical models cannot solve.
5. Fuzzy logic, genetic algorithms, machine learning (ML), and artificial neural networks Artificial Neural Network (ANNs) are used in its design.

4.1.3 Need for Soft Computing

Some real-world issues cannot be solved by standard computers or analytical methods. To get an approximation, we'll need to use a technique like soft computing.

- Solving mathematical problems that necessitate an exact solution necessitates the use of hard computing. It doesn't address some real-world issues. As a result, soft computing is useful in solving real-world issues that lack a clear answer.
- Soft computing can be used when standard mathematical and analytical models fail, for example, to map even the human mind [13, 14].
- Mathematical issues can be solved using analytical models, which are valid in the ideal situation. Real-world issues, on the other hand, do not exist in an ideal setting.
- Soft computing isn't only a theoretical construct; it may also provide light on real-world issues.
- Soft computing, like all of the previous reasons, helps to map the human mind.

4.1.4 Soft Computing Elements

A developing discipline of conceptual intelligence has soft computing as one of its foundational components. As a complement to hard computing, soft computing makes use of techniques such as fuzzy logic and machine learning. Moreover, neural networks and probabilistic reasoning are considered for evolutionary calculation (as

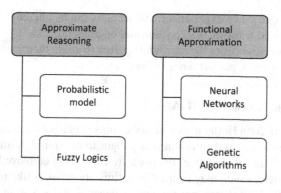

FIGURE 4.1 Elements of soft computing.

shown in Figure 4.1); when it comes to solving complicated problems, soft computing uses these methods [15–21].

This set of tools can be used to fix just about anything. Soft computing employs a variety of approaches, including the following:

o Fuzzy based Logic
o ANN
o Genetic Algorithms

4.1.5 Fuzzy Logic

Fuzzy logic is mathematical reasoning that utilizes an open and imperfect range of facts to solve issues. A wide range of exact conclusions may be drawn quickly and easily. The goal of fuzzy logic is to find the best feasible solution to a difficult issue using all of the knowledge and input data that is accessible to the system. The best solution finders are said to be fuzzy logics [22].

4.1.6 Artificial Neural Network (ANN)

In the 1950s, the development of neural networks paved the way for the application of soft computing to real-world issues that computers cannot address on their own. We are all aware that a computer is unable to accurately express the conditions that exist in the actual world like a human brain can.

In an ANN, the neurons that make up a human brain are replicated in a computer simulation. Thus, a computer or a machine may be trained to make judgments in the same way that a human brain does. Regular computer programming is used to develop ANNs that are mutually related to brain cells. It functions in a similar way to the brain [23–25].

Features of ANN models:

- Parallel Distributed processing of information.
- A high degree of interconnectivity between the most fundamental parts.

- Based on one's own experiences, connections can be altered.
- Unsupervised learning is a constant and ongoing process.
- Uses information from the surrounding area to guide their learning.
- With fewer units, performance suffers.

4.1.7 Genetic Algorithms (GA)

In 1965, Professor John Holland presented the concept of a genetic algorithm. Natural selection concepts are used in evolutionary algorithms to find optimal solutions to problems. Two forms of objective functions are commonly employed: an ant colony and swarm particle, which may be used in optimization issues like maximizing and minimization. Biological mechanisms such as genetics and evolution play a role in it. Almost all of the genetic algorithms' inspiration comes from nature. Natural selection and the idea of genetics are not the basis for a genetic algorithm that uses search-based algorithms. It is also important to note that genetic algorithms are a part of a broader field of computation [26–32].

4.1.8 Soft Computing vs. Hard Computing

Existing mathematical techniques are used in hard computing to tackle specific challenges. An accurate and precise solution is provided. Hard computing may be seen in any numerical problem [33].

Soft computing, on the other hand, uses a different method. Solving existing complicated issues is the goal of soft computing. Also, the outcomes of soft computing are not always exact. They have a hazy and inaccurate quality to them [34–37].

A computer paradigm is shown in Figure 4.2.

Table 4.1 presents the difference between soft and hard computing.

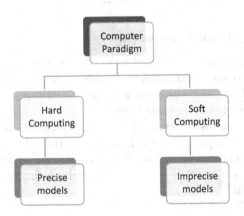

FIGURE 4.2 Elaborating computing paradigm.

TABLE 4.1
Soft and Hard Computing Comparison

Parameters	Soft Computing	Hard Computing
Computation time	Less	More
Dependency	Approximation and dispositional	Numerical systems and binary logic
Computation type	Parallel	Sequential
Results	Approximate	Exact and precise
Example	Neural Networks	Computers may be used to solve any numerical or conventional problem.

4.1.9 Applications of Soft Computing

Soft computing may be seen in a variety of settings. The following is a list of some of them:

1) **Home Appliances:** since we currently use part of this, this is a highly intriguing application to look at more. Artificial intelligence, machine learning, and fuzzy logic are making ordinary products like refrigerators, microwaves, and washing machines smarter. They can not only communicate with the users about their utilization, but they can also alter their settings in response to the current workload. Microwaves and rice cookers, for example. Refrigerators, air conditioners, and washing machines are among the most commonly used home appliances.

2) **Robotics:** fuzzy logic and expert systems, two aspects of soft computing, are being used in this discipline, which is only getting started. It aids in the effective operation of industries, including production and inventory management. Robots with soft computing incorporated are already being used by some of the largest e-commerce organizations to assist in handling the significant load of items that pass through the warehouse on a regular basis. Besides these other uses, it is also employed in the field of robotics [38–42].

3) **Communication:** communication necessitates a constantly shifting environment since the need for it might arise at any time, with little or no warning. An artificial neural network and fuzzy logic are used in soft computing to detect abrupt increases in demand and distribute resources accordingly. Reduced bandwidth utilization not only saves money, but also frees up resources that may be used in other areas with greater demand [43].

4) **Transportation:** soft computing is also being used extensively in the transportation industry, thanks to the proliferation of connected vehicles. Fuzzy logic and evolutionary computing [44] are widely employed, from the manufacturing of automobiles in the factory to on-the-road navigation, traffic

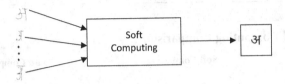

FIGURE 4.3 Soft computing for handwriting recognition.

prediction, troubleshooting, and diagnostics of the vehicle. Lifts employ similar strategies for managing several lifts with a single system.
5) Poker and Checker are two of the most popular games to use it.
6) Soft computing may also be used for image processing and data compression.
7) It is utilized for handwriting recognition (as shown in Figure 4.3).

4.1.10 Advantages of Soft Computing

Soft computing has a number of advantages. Here are just a few examples:

- Methodologies are tolerant of ambiguity and inconsistency.
- It has been able to deal with the inherent ambiguity of issues.
- In addition, it is capable of creating and recognizing "linguistic variables."
- An approximation of a solution can be derived.
- It is capable of dealing with problems using data that is not statistical in nature.
- Instead of equations with clearly defined bounds, it might use a range of overlapping values.

4.1.11 Disadvantages of Soft Computing

The disadvantages of soft computing are:

- If a little error happens, the entire system will cease operating, and correcting the entire system from the beginning is a time-consuming procedure.
- It offers an estimated output value.

4.1.12 Machine Learning

Automated machines that can learn and improve themselves without any human involvement or programming are known as machine learning (ML) [45, 46]. Data is provided to machines, which examine it for patterns and use them as examples in their learning. They learn to make decisions and changes to their activities on their own.

A key benefit of machine learning is that it allows businesses to see patterns in consumer behavior and operational patterns, and it also helps them design new products.

Emerging Soft Computing and Machine Learning Techniques in Industry 71

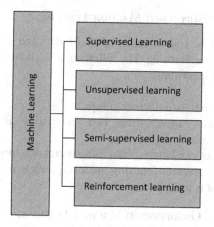

FIGURE 4.4 Types of ML.

If Facebook, Google, and Uber are any indication, machine learning is becoming an increasingly important part of the world's largest organizations. Machine learning has been shown to be a significant competitive advantage for many businesses.

4.1.13 DIFFERENT TYPES OF MACHINE LEARNING

One typical approach to categorizing classical machine learning is by how an algorithm increases its predictive accuracy. The four most prevalent approaches are supervised learning, unsupervised learning, semi-supervised learning, and reinforcement learning (as shown in Figure 4.4). For data scientists, the type of algorithm they utilize is dependent on the sort of data they are attempting to forecast.

- **Supervised learning:** machine learning algorithms are trained using labeled training data and the variables that are to be evaluated for correlations.
- **Unsupervised learning:** unlabeled data is used to train algorithms in this sort of machine learning. The program looks for correlations between various datasets. Algorithms' training data and output predictions and suggestions are fixed.
- **Semi-supervised learning:** this method of machine learning is a hybrid of the first two. Even if a data scientist feeds an algorithm primarily labeled training data, the model is still free to explore data set and create its own knowledge of dataset itself.
- **Reinforcement learning:** reinforcement learning is a common technique employed by data scientists to train a computer to carry out a multi-step process with well-stated rules. In order for an algorithm to learn how to complete a job, data scientists instruct it to look for certain patterns in either positive or negative feedback. The algorithm, on the other hand, makes the majority of these decisions for itself.

4.1.14 Working of Supervised Machine Learning

An algorithm can only be trained using labeled inputs and desired outcomes when using supervised machine learning. The following tasks benefit greatly from the use of supervised learning algorithms:

- Classifying data into two distinct groups.
- Classification into more than two categories is referred to as multi-classification.
- Continuous values can be predicted using regression modeling.
- When numerous machine learning models' predictions are assembled, an accurate forecast is produced.

4.1.15 Working of Unsupervised Machine Learning

No labeled data is required for unsupervised machine learning methods. Their job is to seek patterns in unlabeled data in order to divide it into manageable chunks for further analysis. Neural networks, for example, are an unsupervised method. To do the following tasks, unsupervised learning algorithms are ideal:

- Data is divided into groups depending on their degree of resemblance (clustering).
- Anomaly detection: Identifying data anomalies.
- For instance, you may use association mining to identify groupings of objects in a data collection that are commonly found together.
- When a dataset has too many variables, dimensionality reduction (also known as data curation) might help.

4.1.16 Working of Semi-Supervised Learning

Researchers in the field of data science use semi-supervised learning to train an algorithm using only a minimal quantity of labeled training data. Dataset dimensions are learned from this, which the algorithm may subsequently apply to additional unlabeled data. It is generally accepted that the efficiency and accuracy of algorithms may be enhanced by training on labeled datasets. Data labeling, on the other hand, can be time- and money-consuming. A happy medium exists between the effectiveness of supervised learning and the efficacy of unsupervised learning: semi-supervised learning. Semi-supervised learning is utilized in a variety of contexts, including:

- Learning algorithms to translate languages with less than a full lexicon of words is known as "machine translation".
- Spotting fraudulent activity when there aren't many good instances to go on.
- Data sets can teach algorithms how to classify bigger datasets automatically, so that they can do so on their own.

4.1.17 WORKING OF REINFORCEMENT LEARNING

Programming algorithm having an explicit objective and a set of rules for reaching a particular objective is the primary method of reinforcing learning. Algorithms are also programmed by data scientists to seek good incentives and avoid negative consequences. The following are just a few examples of where reinforcement learning is frequently employed:

- Robotics: robots may be taught to do tasks in the real world using this method.
- Many video games have been taught to play using reinforcement learning in robots.
- Reinforcement learning may be used to assist organizations in determining how to distribute resources when they have limited resources and a clear objective.

4.2 APPLICATION OF ML

Machine learning, in a sense, is like statistics boosted to the next level. A wide range of industries are now using machine learning. Facebook's news feed recommendation engine is one of the best-known examples of ML in action.

Facebook's algorithms employ ML to tailor the news feeds of its users. Recommended posts from certain groups appear earlier in the feed when a user often visits that group's posts.

The engine is working behind the scenes to promote the member's online habits. A member's habits may shift in the next several weeks, and the news stream will be re-evaluated to reflect this.

There are a number of other uses for machine learning outside recommendation engines, including:

- Taking care of the customer's internet service. It is possible to use ML algorithms in CRM to identify significant emails and prioritize them for sales agents. Systems that are much more sophisticated are able to make recommendations for effective solutions as well.
- The company's details. Analytics businesses employ machine learning to look for patterns and anomalies in data that may contain important information.
- Human resources management information systems. Human resource information systems (HRIS) may use machine learning models to sift through applications and choose the most qualified applicants for open positions.
- Vehicles that are not driven by humans. Automated vehicles can recognize and alert drivers to objects that are only partially visible, using machine learning techniques.
- Intuitive virtual helpers. Smart assistants frequently utilize a combination of supervised and unsupervised ML models. These are used to analyze spoken speech and give context.

4.2.1 Automation

Technology applications where human intervention is minimal are referred to as "automation." This encompasses BPA, IT automation, consumer applications like home automation.

4.2.2 Types of Automation

All the types of automation are shown in Figure 4.5.

i. Basic automation: automating simple, elementary tasks is the goal of basic automation. It's about digitizing work by leveraging technologies to expedite and consolidate everyday processes, such as using a common message system instead of having data in unconnected silos. There are two sorts of basic automation, these are process management for business and robotic process automation.
ii. Process automation: process automation is a tool for streamlining and standardizing corporate operations. Dedicated software and business apps are normally in charge of this. Increasing productivity and efficiency in your company may be achieved via the implementation of process automation. Furthermore, it has the ability to provide fresh perspectives on current business issues and make recommendations for potential solutions. This is a sort of process automation that includes process mining and workflow automation.
iii. Automated integration: machines may be programmed to replicate human acts and repeat them once the rules are defined by humans. The "digital worker" is one example. Recently, the term "digital worker" has been used interchangeably with the term "software robot." They can be "hired" to work in groups and bring a certain set of abilities to the table.
iv. Automated by artificial intelligence: AI automation is the most challenging form of automation. For robots to "learn" and make judgments, artificial intelligence (AI) must be implemented. Customer service is an example

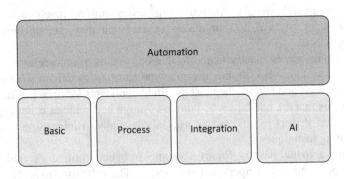

FIGURE 4.5 Types of automation.

where enabled virtual assistants may decrease costs while empowering consumers and human agents, resulting in the best possible customer service experience.

4.2.3 Industrial Automation

There are a wide range of businesses and sectors that employ industrial automation to produce more efficient operations via the use of a number of technological advancements. An increasing number of tedious chores are being replaced by industrial applications. This makes it possible for employees to do repetitive and precise jobs with fewer mistakes, freeing them up to engage in higher-level activities.

4.2.4 Industries Using Industrial Automation

Processes for industrial automation can assist a wide range of sectors. Manufacturing, oil and gas, paper mills, and steel mills are among the major ones. Listed below are a few instances of how these firms are putting it to use.

- There are many different ways that manufacturing organizations use industrial automation. Assembling or creating a product, monitoring maintenance chores, and managing inventory levels are all examples of how they may employ technology.
- In the oil and gas business, industrial automation is particularly beneficial because of the distant locations and offshore stations involved. There are fewer unsafe and unpleasant excursions for personnel because of sensors and other monitoring devices.
- Paper mills: It may be used in paper mills to manage batch production, as well as control instrumentation, plant devices, and equipment in the paper mill. This gives operators a clear view of the whole manufacturing process.
- Steel mills: Hierarchical industrial automation is also being used in steel mills. The steel mill may be managed and controlled as a whole with the help of this new technology.
- Autopilot controls on commercial planes have long been a part of industrial automation in the aviation sector. Self-driving vehicles, both for commercial and personal usage, will utilize industrial automation technologies in the near future.
- The distribution business takes over when the manufacturing process is complete and the items are ready to be shipped. The need for speedier delivery is growing in all sectors. As a result, it will continue to deploy distribution-related solutions to expedite product shipment and delivery.

4.2.5 Benefits of Automation

Improve safety, save time, increase quality output, and minimize production costs with the aid of industrial automation technologies. Profitability and productivity go hand in hand because of all of these advantages. The advantages of automation are shown in Figure 4.6.

4.2.5.1 Enhance Safety

For the safety of your employees, suppliers, and customers, workplace safety is critical. Retrofitting outdated equipment can assist in improving safety procedures. This eliminates the need for as much human involvement in the work that the machines do.

4.2.5.2 Save Time

Production lines and other activities in facilities may often benefit from increased efficiency and reliability thanks to industrial automation technologies. That implies that your organization will be able to meet or surpass the expectations of your customers more swiftly.

4.2.5.3 Boost Quality Production

No matter how rapidly you can produce a product, if you bring faults or quality concerns into the process, your efforts will be in vain. It not only speeds up the production process, but also decreases the risk of future problems. Your items will be of a high quality and there will be fewer mistakes made during production as a consequence.

4.2.5.4 Reduce Monitoring

Monitoring assets and equipment is considerably easier and more cost-effective with industrial automation. Without the need for human intervention, technology can often conduct round-the-clock surveillance and capture data at predetermined times.

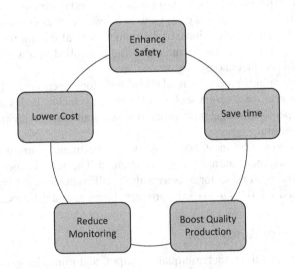

FIGURE 4.6 Benefits of industrial automation.

That implies that maintenance workers may focus on more complex jobs, while the quality of their work increases.

4.2.5.5 Lower Costs

Your company's expenditures will be reduced if industrial automation minimizes the number of emergency repair requests and downtime. Aside from that, it frequently increases output and efficiency, allowing you to generate more with fewer resources.

4.3 LITERATURE REVIEW

After reading a large number of research papers on machine language, soft computing, and industrial automation, it was decided to perform a study on "Emerging Soft Computing and Machine Learning Techniques Applicable in Industrial and Automated Applications." As a starting point, here is a brief review of the articles that will help us achieve our study objectives.

H. Garg. (2017) focused on issues for system analysts to forecast and improve a system's performance to the desired level of accuracy due to imprecise information. As a result, the primary goal is to reduce decision-makers' uncertainty so that they can make more informed decisions in a reasonable amount of time. This research addressed these challenges by quantifying issues in data in the form of fuzzy numbers and addressed various dependability parameters of the industrial system, which showed the behavior of the system. By developing and solving a nonlinear optimization model, the appropriate membership functions of a system's parameters are calculated. When the generated results were compared to existing and traditional methods and outcomes, it was discovered that they had a smaller range of uncertainty during the study. Finally, through a case study of a calf feed facility, a repairable industrial system, a solution was demonstrated [1].

S. K. S. Fan (2018) observed that traffic data can be easily acquired from numerous sources. As the capacity for automation and processing grows, sensors and surveillance cameras might be among these sources. To get insights from the mountains of data at our disposal, we needed a way to filter and identify patterns in those databases. In this study, researchers estimate travel times on Taiwanese highways using data collected through electronic toll booths. This information is then fed into a machine learning model contained inside a big data analytics platform, which is itself based on the random forests technique and Apache Hadoop. Thereafter, many models were built to provide drivers with estimated and updated route time information based on historical and real-time data from highway traffic [2].

L. Gauerhof, P. Munk, and S. Burton (2018) observed that the certification of highly automated driving vehicles was a significant difficulty for the automobile industry. This is so because, even if the system has no bugs on the inside, its actual behavior may be different from what was expected. These deviations from intended operation may be traced back to both the unpredictability of external events and the inherent uncertainties of the Machine Learning (ML) methods used to make sense of this enormous input field. In the research, they developed a safety assurance case for a pedestrian detection function, which is a crucial piece of foundational functionality for any autonomous vehicle. The graphical structuring notation (GSN) was used to express their safety assurance case, which incorporates our arguments against underspecification the semantic gap, and the deductive gap [3].

D. H. Kim, T. J. Y. Kim, X. Wang, M. Kim, Y. J. Quan, J. W. Oh, S. H. Min, H. Kim, B. Bhandari, I. Yang, and S. H. Ahn (2018) suggested artificial intelligence, quantum computing, nanotechnology, biotechnology, robotics, 3D printing, autonomous cars, and the Internet of Things were all part of the Fourth Industrial Revolution. It's like the internet for the real world. The advent of machine learning was a turning point in the history of artificial intelligence since it made it possible for machines to learn, develop, and carry out a job without being explicitly programmed. Using machine learning, manufacturers may improve output while also keeping tabs on system health and fine-tuning design and process parameters. The term "smart machining" was coined to describe a new approach to manufacturing in which all machine tools were integrated into a single cyber-physical system. Their research not only provided an overview of the machining industry but also evaluated and explained machining processes that make use of machine learning algorithms [4].

A. Fernandes, V. Gomes, and P. Melo-Pinto (2018) presented the difficult measurement of enological parameters. These parameters are sugar content, pH, and anthocyanin content. Samples containing a small number of grape berries with the goal of assessing grape ripeness will be the focus of this review. Multiclass issues of identifying plant varieties and clones have been considered. During this presentation, we will share the findings from many research initiatives that have been conducted in these areas. This article provides a concise introduction to spectroscopic principles, followed by a systematic literature evaluation that accounts for the number of berries included in each sample and the total number of samples used in each study. The model will account for a wide range of varieties, years, and harvesting regions. Particular care was taken to compare and contrast the various validation techniques used. There were several recommendations given to facilitate the future comparability of reported findings [5].

C. Expósito-Izquierdo (2018) looked at maritime logistics, which had been a very appealing topic of research for applying the broad frameworks of soft computing due to the constant expansion in international seaborne trade over the last few decades. There was a severe shortage of efficient ways in this environment for achieving accurate solutions to a large variety of optimization issues. These issues arise in this subject and are classed as hard from the standpoint of complexity theory. These optimization challenges necessitate the development of novel computing algorithms. It is capable of reporting inexact solutions. This is performed by leveraging uncertainty, tolerance for imprecision, and partial truth, among other things, in order to achieve tractability. In this chapter, they presented a review of the most prominent soft computing approaches used in marine logistics and related domains, as well as some potential to expand on knowledge chevalier [6].

B. S. Maya (2018) introduced stroke, a cerebrovascular disease that affects the blood supply to the brain and affects people over the age of 65. Considering 2D cerebrum CT images, the paper suggests an automatic technique. These techniques are used for perceiving and orchestrating the types of strokes. There were four steps to the approach. Preprocessing the image in the first stage might remove unnecessary noise by using median filtering. Distinct texture-based features are extracted and classified using wavelet packet transform (WPT) as a second step. Linear Discriminant Analysis (LDA) is used to reduce the dimensionality of characteristics in the next stage. Finally, supervised learning algorithms for the categorization of normal and diseased regions are linked to the reduced group of features. This research aims to develop a framework for reliably extracting the stroke location from CT scans. It would aid clinicians in making diagnosis decisions. In comparison to other neural system-based classifiers, the suggested technique significantly improved stroke classification precision [7].

M. Rath and B. K. Pattanayak (2018) presented the current reality of congested activity in smart cities as being inextricably unclear, making the development of a brilliant movement control framework a difficult task. Because of the growing number of vehicles at movement hubs in densely populated areas, there were unexpected delays in travel, the risk of accidents rose, excessive fuel consumption was a problem, and unsanitary conditions due to pollution further degraded the health of ordinary city dwellers. To overcome such problems, several forward-thinking metropolitan communities were currently implementing enhanced activity control frameworks based on movement robotization to avoid the problems mentioned above. The fundamental test entails applying continual analysis of online mobility data to a specific activity stream. An improved activity management framework named Soft Computing based Intelligent Communication System (SCICS) was provided, which leverages swarm knowledge as a delicate registering technique with astute correspondence between smart cars and tower focal movement. In nanorobots, it employed an improved course correction mechanism with executed logic. Communication between wise sensors is done jointly by nanorobots in a vehicular ad hoc network (VANET). The findings of the reenactment using the Ns2 test system show encouraging outcomes. Results are presented in terms of better execution to control problems [8].

I. Candanedo (2018) stated that the data created by sensor networks in an Industry 4.0 environment necessitates machine learning and data analysis approaches. As a result, organizations confront new opportunities as well as obstacles, one of which is predictive analysis, which uses computer tools to find patterns in studied data using the same criteria that can be used to construct predictions. Heating, ventilation, and air conditioning (HVAC) systems regulate indoor climate, air temperature, humidity, and pressure. These are resulting in an optimal production environment in a variety of sectors [9].

J. S. Ortiz, J. S. Sánchez, P. M. Velasco, W. X. Quevedo, and C. P. Carvajal (2018) described the creation of virtual environments for managing pneumatic controls in industrial processes. The objective of the research is to improve training and teaching-learning processes. The developed program allows for multi-user immersion along with interaction. The objective is to complete predetermined tasks in lab environments and virtualized scenes for industrial processes [10].

S. Seyedzadeh (2018) presented that the largest contribution to greenhouse gas emissions was the ever-growing population and increasing municipal business demands for new building construction. Therefore, cutting down on carbon emissions and fossil fuel use requires boosting construction's energy efficiency. The incorporation of energy efficiency features from the outset of the design process proved to be one of the most effective means of lowering CO_2 emissions and energy consumption in newly constructed structures. However, older stock may have its energy performance enhanced by astute refurbishments and careful energy management. All of these choices need a reliable energy forecast in order to make informed choices. Artificial intelligence (AI) in general, and machine learning (ML) approaches in particular, have been developed in recent years for forecasting energy use and performance in buildings. Artificial neural networks, support vector machines, Gaussian-based regressions, and clustering are the four main ML approaches employed in this research to anticipate and improve building energy efficiency [11].

M. Hildebrandt, S. S. Sunder, S. Mogoreanu, I. Thon, V. Tresp, and T. Runkler (2019) cited that when developing complex engineering solutions, they investigated numerous techniques to provide recommendations. The team's objective was to make use of data-driven statistical patterns with high predictive power that are also much more malleable than rigid, deterministic rules. They proposed a generalized technique for recommending complex industrial solutions by integrating users' past actions with semantic data in a single database. Knowledge graphs, and multi-relational data descriptions, were produced as a result of this process. Predicting an edge in this graph corresponds to speculating on a consumer's choice of a product. Their recommender system appeared to be powerful, despite its ease of data preparation and maintenance, as demonstrated in comprehensive experiments with real-world data, where this model surpasses numerous state-of-the-art systems. Furthermore, once their model was trained, they could quickly recommend new things. This ensured that their method could assist users in configuring new solutions in real time [12].

4.4 PROPOSED WORK

4.4.1 Problem Statement

Research efforts have been made in the field of soft computing services on several occasions. Machine Language and Automation are two fields that often make use of these services. But the problem with the currently available study is that it has a sluggish performance and is inaccurate. Previous studies overlooked the need for optimization before training and testing procedures. Therefore, an optimal technique has to be proposed in order to gain better accuracy while maintaining high performance.

4.4.2 Proposed Methodology

The suggested study effort takes into consideration the elements impacting the performance and accuracy of current research in the fields of machine learning, soft computing, and industrial and automated applications. While several machine learning algorithms may be used for categorization, an optimization strategy [47–49] should be included. In order to refine the dataset before it's time to train, researchers are thinking about using a soft computing technique. The goal of combining a soft computing method with machine learning is to increase classification accuracy in automated and industrial settings. Finally, the suggested work will be compared to the standard method based on performance and precision.

Research Methodology is shown in Figure 4.7.

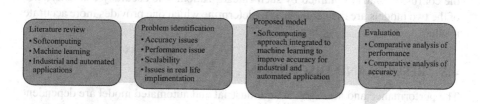

FIGURE 4.7 Research methodology.

4.4.3 PROCESS FLOW OF NOVEL APPROACH

Figure 4.8 shows the process flow of the unique technique, including training and testing on both the optimized and non-optimized datasets. The optimization method of choice for scientific studies is PSO. To filter the data, PSO uses a particle swarm optimization process to locate the optimal answer. To train and test the dataset after filtering, it is used for training. Accuracy parameters may be obtained by processing

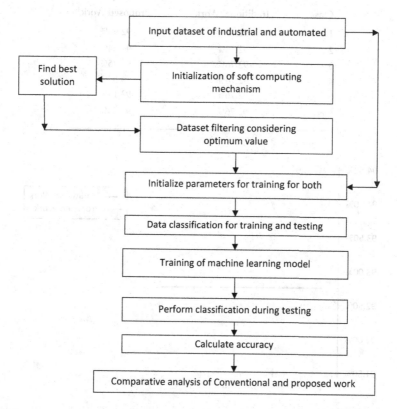

FIGURE 4.8 Flow chart of novel approach.

the confusion matrix obtained by such categorization. The accuracy characteristics of the two models are then compared to determine which one provides more accurate results.

4.5 RESULTS AND DISCUSSION

The performance and accuracy of an industrial and automated model are dependent on a number of factors that must be taken into account while trying to grasp how such a model operates. For classification, researchers are exploring several machine learning approaches; however, they must also include some kind of optimization process. Automatic and industrial applications of event classification using machine learning are anticipated to improve accuracy.

The accuracy comparison is shown in Table 4.2 and Figure 4.9, the error rate comparison is shown in Table 4.3 and Figure 4.10 and the time consumption is shown in

TABLE 4.2
Comparison of Accuracy

Class	Traditional Work	Proposed Work
1	92.76%	92.96%
2	92.76%	94.64%
3	92.76%	94.75%
4	92.76%	94.53%
5	92.76%	93.14%
6	92.76%	93.23%

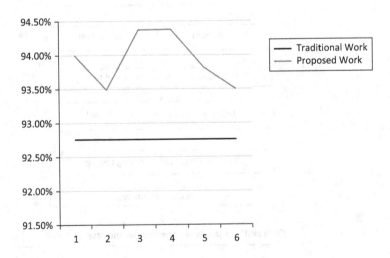

FIGURE 4.9 Comparison of accuracy.

TABLE 4.3
Comparison of Error Rate

Class	Traditional Work	Proposed Work
1	7.24%	7.04%
2	7.24%	5.36%
3	7.24%	5.25%
4	7.24%	5.47%
5	7.24%	6.86%
6	7.24%	6.77%

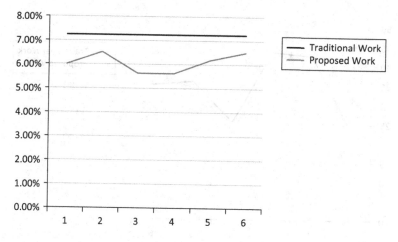

FIGURE 4.10 Comparison of error rate.

Table 4.4 and Figure 4.11. It is obvious that the proposed method is far superior to the existing work in terms of accuracy, error rate, and time consumption.

4.6 CONCLUSION

The simulation results show that soft computing is crucial in choosing the right dataset. When compared to traditional procedures, the precision provided by the filtered dataset is much higher. Moreover, the presented work has very less time consumption during training operations by reducing the amount of data in the dataset.

4.7 FUTURE SCOPE

To put it simply, soft computing is a set of computational techniques that draws inspiration from both AI and natural selection. It provides affordable and realistic solutions to complex real-world situations that have no easy computational solution. Soft computing

TABLE 4.4
Comparison of Time Consumption

Class	Traditional Work	Proposed Work
1	23.234	22.81546
2	42.34	41.84121
3	43.33	41.9764
4	25.23	22.43955
5	35.65	35.04173
6	34.78	34.40877

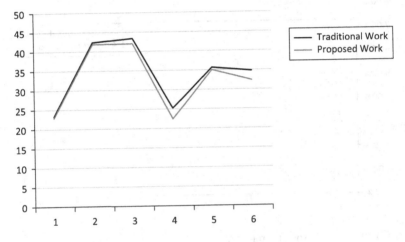

FIGURE 4.11 Comparison of time consumption.

is a subfield of both mathematics and computer science that has been around since the early 1990s. The ability of the human brain to approximate reality and come up with workable answers was the impetus for this venture. The discipline of soft computing relies on approximations to solve challenging computational issues. The term "machine learning" is used to describe the method by which automated systems may learn and improve themselves with little to no human input or specialized programming (ML) [47–49]. Machines are fed data and trained to look for patterns; these patterns are then used as templates for further learning. They get the independence to choose their own actions and make necessary changes to their routines.

REFERENCES

1. H. Garg, "Performance analysis of an industrial system using soft computing based hybridized technique," *J. Braz. Soc. Mech. Sci. Eng.*, vol. 39, no. 4, pp. 1441–1451, 2017, doi: 10.1007/s40430-016-0552-4.

2. S. K. S. Fan, C. J. Su, H. T. Nien, P. F. Tsai, and C. Y. Cheng, "Using machine learning and big data approaches to predict travel time based on historical and real-time data from Taiwan electronic toll collection," *Soft Comput.*, vol. 22, no. 17, pp. 5707–5718, 2018, doi: 10.1007/s00500-017-2610-y.
3. L. Gauerhof, P. Munk, and S. Burton, *Structuring Validation Targets of a Machine Learning Function Applied to Automated Driving* (Vol. 11093). LNCS. Springer International Publishing, 2018. doi: 10.1007/978-3-319-99130-6_4.
4. D. H. Kim et al., "Smart machining process using machine learning: A review and perspective on machining industry," *Int. J. Precis. Eng. Manuf. Green Technol.*, vol. 5, no. 4, pp. 555–568, 2018, doi: 10.1007/s40684-018-0057-y.
5. A. Fernandes, V. Gomes, and P. Melo-Pinto, "A review of the application to emergent subfields in viticulture of local reflectance and interactance spectroscopy combined with soft computing and multivariate analysis," *Stud. Fuzziness Soft Comput.*, vol. 358, pp. 87–115, 2018, doi: 10.1007/978-3-319-62359-7_5.
6. C. Expósito-Izquierdo, B. Melián-Batista, and J. M. Moreno-Vega, "A review of soft computing techniques in maritime logistics and its related fields," *Stud. Fuzziness Soft Comput.*, vol. 360, pp. 1–23, 2018, doi: 10.1007/978-3-319-64286-4_1.
7. B. S. Maya and T. Asha, "Automatic detection of brain strokes in CT images using soft computing techniques," *Lect. Notes Comput. Vis. Biomech.*, vol. 25, pp. 85–109, 2018, doi: 10.1007/978-3-319-61316-1_5.
8. M. Rath and B. K. Pattanayak, "SCICS: A soft computing based intelligent communication system in VANET," *Commun. Comput. Inf. Sci.*, vol. 808, pp. 255–261, 2018, doi: 10.1007/978-981-10-7635-0_19.
9. I. Candanedo, E. Hernández, S. Rodríguez, T. Santos, and A. González, "Machine learning predictive model," *Springer*, vol. 1, pp. 501–510, 2018, doi: 10.1007/978-3-319-95204-8.
10. J. S. Ortiz et al., "Virtual training for industrial automation processes through pneumatic controls," *Lect. Notes Comput. Sci. (including Subser. Lect. Notes Artif. Intell. Lect. Notes Bioinformatics)*, vol. 10851 LNCS, pp. 516–532, 2018, doi: 10.1007/978-3-319-95282-6_37.
11. S. Seyedzadeh, F. P. Rahimian, I. Glesk, and M. Roper, "Machine learning for estimation of building energy consumption and performance: A review," *Vis. Eng.*, vol. 6, no. 1, 2018, doi: 10.1186/s40327-018-0064-7.
12. M. Hildebrandt, S. S. Sunder, S. Mogoreanu, I. Thon, V. Tresp, and T. Runkler, *Configuration of Industrial Automation Solutions Using Multi-Relational Recommender Systems* (Vol. 11053). LNAI. Springer International Publishing, 2019. doi: 10.1007/978-3-030-10997-4_17.
13. P. Chawla and R. Anand, "Micro-switch design and its optimization using pattern search algorithm for applications in reconfigurable antenna," *Mod. Antenna Syst.*, 10, pp. 189–210, 2017.
14. R. Anand and P. Chawla, "Optimization of inscribed hexagonal fractal slotted microstrip antenna using modified lightning attachment procedure optimization," *Int. J. Microw. Wirel. Technol.*, 12(6), pp. 519–530, 2020.
15. A. Kumar and N. Sachdeva, "Cyberbullying detection on social multimedia using soft computing techniques: A meta-analysis," *Multimed Tool. Appl.*, vol. 78, no. 17, pp. 23973–24010, 2019, doi: 10.1007/s11042-019-7234-z.
16. H. Rahmanifard and T. Plaksina, "Application of artificial intelligence techniques in the petroleum industry: A review," *Artif. Intell. Rev.*, vol. 52, no. 4, pp. 2295–2318, 2019, doi: 10.1007/s10462-018-9612-8.
17. W. P. Rogers et al., "Automation in the mining industry: Review of technology, systems, human factors, and political risk," *Min. Metall. Explor.*, vol. 36, no. 4, pp. 607–631, 2019, doi: 10.1007/s42461-019-0094-2.

18. S. Meivel, N. Sindhwani, R. Anand, D. Pandey, A. A. Alnuaim, A. S. Altheneyan, ... M. E. Lelisho, "Mask detection and social distance identification using internet of things and faster R-CNN algorithm," *Comp. Intell. Neurosci.*, 2022, 2022.
19. M. Abdel-Basset, G. Manogaran, A. Gamal, and V. Chang, "A novel intelligent medical decision support model based on soft computing and IoT," *IEEE Internet Things J.*, vol. 7, no. 5, pp. 4160–4170, 2020, doi: 10.1109/JIOT.2019.2931647.
20. R. Gupta, G. Shrivastava, R. Anand, and T. Tomažič, "IoT-based privacy control system through android," in *Handbook of E-business Security*. Auerbach Publications, pp. 341–363, 2018.
21. S. Zhang, W. Huang, and H. Wang, "Crop disease monitoring and recognizing system by soft computing and image processing models," *Multimed Tool. Appl.*, vol. 79, no. 41–42, pp. 30905–30916, 2020, doi: 10.1007/s11042-020-09577-z.
22. D. P. Penumuru, S. Muthuswamy, and P. Karumbu, "Identification and classification of materials using machine vision and machine learning in the context of industry 4.0," *J. Intell. Manuf.*, vol. 31, no. 5, pp. 1229–1241, 2020, doi: 10.1007/s10845-019-01508-6.
23. S. T. Park, G. Li, and J. C. Hong, "A study on smart factory-based ambient intelligence context-aware intrusion detection system using machine learning," *J. Ambient Intell. Humaniz Comput.*, vol. 11, no. 4, pp. 1405–1412, 2020, doi: 10.1007/s12652-018-0998-6.
24. Y. Li, S. Carabelli, E. Fadda, D. Manerba, R. Tadei, and O. Terzo, "Machine learning and optimization for production rescheduling in Industry 4.0," *Int. J. Adv. Manuf. Technol.*, vol. 110, no. 9–10, pp. 2445–2463, 2020, doi: 10.1007/s00170-020-05850-5.
25. P. Singh, O. Kaiwartya, N. Sindhwani, V. Jain, and R. Anand (Eds.). *Networking Technologies in Smart Healthcare: Innovations and Analytical Approaches.* CRC Press, 2022.
26. I. H. Sarker, "Machine learning: Algorithms, real-world applications and research directions," *SN Comput. Sci.*, vol. 2, no. 3, pp. 1–21, 2021, doi: 10.1007/s42979-021-00592-x.
27. R. Singh Gupta, V. K. Nassa, R. Bansal, P. Sharma, K. Koti, K. Koti, "Investigating application and challenges of big data analytics with clustering," *2021 International Conference on Advancements in Electrical, Electronics, Communication, Computing and Automation (ICAECA)*, pp. 1–6, 2021, doi: 10.1109/ICAECA52838.2021.9675483.
28. V. Veeraiah, H. Khan, A. Kumar, S. Ahamad, A. Mahajan, and A. Gupta, "Integration of PSO and deep learning for trend analysis of meta-verse," *2022 2nd International Conference on Advance Computing and Innovative Technologies in Engineering (ICACITE)*, pp. 713–718, 2022, doi: 10.1109/ICACITE53722.2022.9823883.
29. B. K. Pandey, D. Pandey, V. K. Nassa, S. George, B. Aremu, P. Dadeech, and A. Gupta, "Effective and secure transmission of health information using advanced morphological component analysis and image hiding," in Gupta, M., Ghatak, S., Gupta, A., Mukherjee, A.L. (eds) *Artificial Intelligence on Medical Data: Lecture Notes in Computational Vision and Biomechanics*, 37. Springer, 2023, doi: 10.1007/978-981-19-0151-5_19.
30. V. Veeraiah, K. R. Kumar, P. LalithaKumari, S. Ahamad, R. Bansal, and A. Gupta, "Application of biometric system to enhance the security in virtual world," *2022 2nd International Conference on Advance Computing and Innovative Technologies in Engineering (ICACITE)*, pp. 719–723, 2022, doi: 10.1109/ICACITE53722.2022.9823850.
31. R. Bansal, A. Gupta, R. Singh, and V. K. Nassa, "Role and impact of digital technologies in e-learning amidst COVID-19 pandemic," *2021 Fourth International Conference on Computational Intelligence and Communication Technologies (CCICT)*, pp. 194–202, 2021, doi: 10.1109/CCICT53244.2021.00046.
32. A. Shukla, S. Ahamad, G. N. Rao, A. J. Al-Asadi, A. Gupta, and M. Kumbhkar, "Artificial intelligence assisted IoT data intrusion detection," *2021 4th International Conference on Computing and Communications Technologies (ICCCT)*, pp. 330–335, 2021, doi: 10.1109/ICCCT53315.2021.9711795.

33. V. Pathania, S. Z. D. Babu, S. Ahamad, P. Thilakavathy, A. Gupta, M. B. Alazzam, D. Pandey, "A database application of monitoring COVID-19 in India," in Gupta, M., Ghatak, S., Gupta, A., Mukherjee, A.L. (eds) *Artificial Intelligence on Medical Data. Lecture Notes in Computational Vision and Biomechanics*, 37. Springer, 2023, doi: 10.1007/978-981-19-0151-5_23.
34. K. Dushyant, G. Muskan, Annu, A. Gupta, & S. Pramanik, "Utilizing machine learning and deep learning in cybesecurity: An innovative approach," in *Cyber Security and Digital Forensics: Challenges and Future Trends*. Wiley, pp. 271–293, 2022, doi: 10.1002/9781119795667.ch12.
35. S. Z. D. Babu, D. Pandey, G. T. Naidu, S. Sumathi, A. Gupta, M. Bader Alazzam, B. K. Pandey, "Analysation of big data in smart healthcare," in Gupta, M., Ghatak, S., Gupta, A., Mukherjee, A.L. (eds) *Artificial Intelligence on Medical Data. Lecture Notes in Computational Vision and Biomechanics*, 37. Springer, 2023, doi: 10.1007/978-981-19-0151-5_21.
36. V. BijenderBansal, N. Jenipher, R. Jain, R. Dilip, M. Kumbhkar, S. Pramanik, S. Roy, A. Gupta, "Big Data Architecture for Network Security," in *Cyber Security and Network Security*. Wiley, pp. 233–267, 2022, doi: 10.1002/9781119812555.ch11.
37. A. Gupta, D. Kaushik, M. Garg, and A. Verma, "Machine learning model for breast cancer prediction," *2020 Fourth International Conference on I-SMAC (IoT in Social, Mobile, Analytics and Cloud) (I-SMAC)*, pp. 472–477, 2020, doi: 10.1109/I-SMAC49090.2020.9243323.
38. N. Sreekanth et al, "Evaluation of estimation in software development using deep learning-modified neural network," *Appl. Nanosci.*, 2022, doi: 10.1007/s13204-021-02204-9.
39. V. Veeraiah, N. B. Rajaboina, G. N. Rao, S. Ahamad, A. Gupta, and C. S. Suri, "Securing online web application for IoT management," *2022 2nd International Conference on Advance Computing and Innovative Technologies in Engineering (ICACITE)*, pp. 1499–1504, 2022, doi: 10.1109/ICACITE53722.2022.9823733.
40. V. Veeraiah, G. P., S. Ahamad, S. B. Talukdar, A. Gupta, and V. Talukdar, "Enhancement of meta verse capabilities by IoT integration," *2022 2nd International Conference on Advance Computing and Innovative Technologies in Engineering (ICACITE)*, pp. 1493–1498, 2022, doi: 10.1109/ICACITE53722.2022.9823766.
41. A. Tripathi, N. Sindhwani, R. Anand, and A. Dahiya, "Role of IoT in smart homes and smart cities: Challenges, benefits, and applications," in *IoT Based Smart Applications*. Springer International Publishing, pp. 199–217, 2022.
42. H. Keserwani, H. Rastogi, A. Z. Kurniullah, S. K. Janardan, R. Raman, V. M. Rathod, and A. Gupta, "Security enhancement by identifying attacks using machine learning for 5G network," *Int. J. Commun. Netw. Inf. Sec. (IJCNIS)*, vol. 14, no. 2, pp. 124–141, 2022, doi: 10.17762/ijcnis.v14i2.5494.
43. R. Anand and P. Chawla, "Bandwidth optimization of a novel slotted fractal antenna using modified lightning attachment procedure optimization," in *Smart Antennas: Latest Trends in Design and Application*. Springer International Publishing, pp. 379–392, 2022.
44. S. K. Chauhan, P. Khanna, N. Sindhwani, K. Saxena, and R. Anand, "Pareto optimal solution for fully fuzzy bi-criteria multi-index bulk transportation problem," in *Mobile Radio Communications and 5G Networks. Proceedings of Third MRCN 2022*. Springer Nature Singapore, pp. 457–470, 2023.
45. N. Sindhwani, A. Rana, and A. Chaudhary, "Breast cancer detection using machine learning algorithms," in *2021 9th International Conference on Reliability, Infocom Technologies and Optimization (Trends and Future Directions)(ICRITO)*, September. IEEE, pp. 1–5, 2021.

46. A. Chaudhary, D. Bodala, N. Sindhwani, and A. Kumar, "Analysis of customer loyalty using artificial neural networks," in *2022 International Mobile and Embedded Technology Conference (MECON)*, March. IEEE, pp. 181–183, 2022.
47. D. Pandey, S. George, B. Aremu, S. Wariya, and B. K. Pandey, *Critical Review on Integration of Encryption, Steganography, IOT and Artificial Intelligence for the Secure Transmission of Stego Images*, 2021.
48. M. S. Kumar, S. Sankar, V. K. Nassa, D. Pandey, B. K. Pandey, and W. Enbeyle, "Innovation and creativity for data mining using computational statistics," in *Methodologies and Applications of Computational Statistics for Machine Intelligence*. IGI Global, pp. 223–240, 2021.
49. B. K. Pandey, D. Pandey, and A. Agarwal, "Encrypted information transmission by enhanced steganography and image transformation," *Int. J. Distrib. Artif. Intell. (IJDAI)*, vol. 14, no. 1, pp. 1–14, 2022.

5 PSO-Based Nature-Inspired Mechanisms for Robots during Smart Decision-Making for Industry 4.0

Harinder Singh, Veera Talukdar, Huma Khan, Dharmesh Dhabliya, Rohit Anand, Abhra Pratip Ray, and Sanjiv Kumar Jain

5.1 INTRODUCTION

Industry 4.0 uses smart machinery and smart factories to create items more effectively and profitably across the value chain. Particle swarm optimization (PSO) tries to determine the global maximum or minimum. Particle Swarm Optimization uses the mobility and intelligence of swarms. Robots replicate human motions and perform complicated jobs based on predefined instructions. PSO's particle swarms are like robotic swarms. So, PSO is a natural choice for controlling swarm robotic systems.

5.2 INDUSTRY 4.0

It's the fourth industrial revolution of the 21st century. Connected businesses that use data to drive smart physical activities may be created when top-notch manufacturing and operations methods are combined with digital technology. More efficient and profitable production all the way down the value chain is the goal of Industry 4.0, which employs smart technology and smart factories. As a result, manufacturers will have more leeway to tailor products to individual customers' preferences via mass customization, eventually aiming for efficiency with a lot size of one. Gathering more data from the factory floor and linking it with corporate operational data might lead to greater transparency and improved judgment in smart factories. Robotics, artificial intelligence, quantum computing, additive manufacturing, and IoT are just a few examples of smart and networked technologies that are being incorporated into organizations, assets, and even persons through wearable gadgets (IoT) [1–7].

DOI: 10.1201/9781003317456-5

5.2.1 Characteristics of a Smart Factory under Industry 4.0

The term "Smart Factory" is used to describe a modern manufacturing facility that is heavily networked and digitalized, allowing for the collection and analysis of relevant data to optimize production, cut costs, and shorten timelines. These automated plants can run themselves with little or no oversight from humans. Because of the vast amounts of data they process, they can easily adjust production to meet fluctuating demands. That is to say, they are able to detect and respond to new information in real time [8–11]. The characteristics of Smart Factory are depicted in Figure 5.1.

a) **Data analysis for optimal decision-making**: in the manufacturing industry, big data is generated in large quantities via embedded sensors and networked machines. Manufacturers may use data analytics to look into past trends, spot patterns, and make more informed decisions. In order to get even more information, smart factories can include data from other sections of the company and its wider ecosystem of suppliers and distributors. Manufacturers may make informed judgments about production and staffing levels based on information gleaned from human resources, sales, and warehousing. It is possible to construct a digital "twin" that is an exact replica of how things normally work [12–16].

b) **IT-OT integration**: connectivity is essential to the network architecture of a smart factory. On the factory floor, sensors, devices, and machines provide data in real time that can be ingested and used instantly by other industrial assets and shared throughout the enterprise software stack, which includes enterprise resource planning (ERP) and other business management applications [17–21].

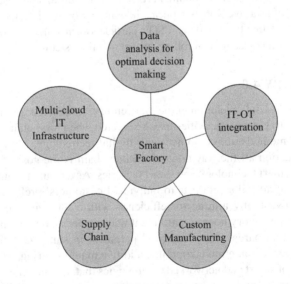

FIGURE 5.1 Characteristics of a smart factory.

c) **Custom manufacturing**: smart factories can lower the production costs of making products specifically for each consumer. "Lot size of one" manufacturing is sought after by manufacturers in numerous industries. Using advanced simulation software, innovative materials, and technologies like 3-D printing, manufacturers can rapidly make low-quantity, bespoke items for particular customers. A key difference between the first industrial revolution and Industry 4.0 is that the latter is focused on mass customization rather than mass manufacturing [22, 23].

d) **Supply chain**: Industry 4.0 emphasizes the need to integrate industrial processes with supply chain management for maximum efficiency. Manufacturers' procurement of raw materials and distribution of completed goods will be drastically altered as a result of this. Sharing certain production data with suppliers might help manufacturers better schedule delivery times. For instance, delivery might be rerouted or postponed when a manufacturing line is down to save resources. Predictive shipping allows businesses to ship out products when demand is highest by evaluating data from sources like the weather, shipping partners, and merchants. One of the most important uses for blockchain technology is increasing visibility in the supply chain.

e) **Multi-cloud IT infrastructure**: businesses need to undertake a digital transformation, which involves the development of a hybrid multi-cloud IT infrastructure, to take advantage of the potential given by Industry 4.0. One definition of "hybrid multi-cloud" is an environment in which a business makes use of both public and private clouds to meet its diverse computing requirements. Because certain cloud environments are better suited to or more cost-effective for particular workloads, this allows businesses to optimize their workloads across all of their clouds. When looking to accomplish digital transformation and an open, secure environment, manufacturers should think about transferring their present workloads to the best cloud environment currently accessible [24–28].

5.2.2 Nature-Inspired Computing (NIC)

Computing that draws inspiration from the natural world is known as nature-inspired computing (NIC). It's incredible how nature works. There are often several unseen factors at work behind seemingly random occurrences. For ages, philosophers and scientists have studied natural occurrences like these in an effort to better comprehend, explain, adapt, and reproduce manmade systems. The complexity of the interplay between the living and nonliving worlds, with all its myriad agents and forces, is beyond the scope of human understanding. These forces shape the natural world and keep the balance, beauty, and vitality of life in check by working together and sometimes at odds with one another. This concept, known as dialectics of nature, is central to the notion of natural evolution. There is a particular pattern to the increase in complexity seen in nature. Even in the absence of a single overseer, natural systems are capable of carrying out information processing in a decentralized, self-organized, and optimum fashion. All of these categories – mechanical, physical, chemical, biological, and social – are ordered from least complicated to most complex. This order is a structural and historical expression of their interdependence and connection. Due to the altered conditions, the actions take on a new form. Emerging in the fields of science, technology, and computing are efforts to understand the principles and mechanisms of natural, physical, chemical, and biological organisms that accomplish complex tasks in an appropriate manner with limited resources and capability, as well as problem-solving techniques inspired by nature. Scientists have had an ongoing dialogue with the natural world, which has led to the emergence of novel concepts, methods, and academic subfields. Ingenuity has been used ever since to learn more about the natural world. The field of nature-inspired computing involves combining the knowledge of computing science with that of other scientific disciplines, such as physics, chemistry, and engineering, to create new computational tools, such as algorithms, hardware, or wetware, for problem-solving, synthesis of patterns, behaviors, and organisms [29–33].

5.2.3 Concepts Relating To NIC

The paradigm of computers that takes its cues from nature is a sizable one. Despite the fact that science and technology have progressed over the course of hundreds of years and provide numerous sophisticated instruments and procedures for their solution, there are still many problems that need to be solved, phenomena that need to be synthesized, and questions that need to be answered. In most cases, natural computing methods should be looked at when:

- A vast number of variables and/or solutions are involved, and/or the problem has more than one goal to achieve.
- Complex pattern recognition and classification tasks are not adequate models for the situation at hand.
- Optimal solutions are either impossible to get or cannot be guaranteed using conventional methods, although a quality metric does exist to facilitate comparison of different approaches.
- Many other approaches can be taken successfully, or should be used, to solve the issue.

Meta-heuristic algorithms that are based on, or at least largely inspired by, the sorts of natural occurrences documented in the natural sciences are collectively referred to as NIC. All meta-heuristic algorithms that take their cues from the natural world employ a combination of rule-based and random processes to mimic the behavior of the real world. Several nature-inspired computer models have emerged in recent years [34–39]. PSO is one example.

5.2.4 Particle Swarm Optimization

The term "particle swarm optimization" was first proposed by scientists Kennedy and Eberhart in 1995. According to sociobiologists, flocking animals and birds "may profit

PSO-Based Nature-Inspired Mechanisms for Robots

from the experience of all other individuals." Flocks of birds, for example, may coordinate their foraging by sharing the locations of successful food finds. Since each bird in a flock contributes to the search for the best solution in a high-dimensional solution space, we may assume that the best solution identified by all the birds in the flock is the best answer. Due to a lack of contrary evidence, we are forced to rely on a heuristic. However, in reality, we see that PSO's solution frequently approaches the global optimum [40–44].

Profit maximization and loss avoidance are the focus of this section. This motivates our efforts to optimize by exploring different values for a function. The range of a function might be bounded by many local maxima and minima. No matter what, there is only one absolute maximum and absolute minimum on Earth. Finding the global maximum of a complicated function can be a significant challenge. PSO searches for a global optimum or optima. It may not be perfect, but it comes close to capturing the highest and lowest points on the planet. In this sense, PSO is a heuristic model. Sometimes a higher-dimensional vector space or more variables are required when trying to express a mathematical function that is grounded in the actual world. Bacterial growth in a jar may be affected by factors such as temperature, humidity, the container, the solvent, etc. However, determining the global maximum and minimum of such a function is more difficult [45].

5.2.5 Movement toward a More Hospitable Zone in the Search for the Global Optimum:

A. In a group of particles, each individual's flight speed is dynamically affected by the speeds of other particles in the group as well as the individual's own flight history.
B. The particles are constantly making efforts to:
   1. For that individual, it's the greatest possible result; we call this a "pbest."
   2. A particle's global best (gbest) value is its best possible value in the whole universe.
C. Every particle changes location in response to:
   1. the present state
   2. present speed
   3. a measure of how far away its existing location is from where it would be optimally placed
   4. how far away from gbest it is right now

PSO is a technique used in computational science that entails a computer making repeated, small adjustments to a candidate solution in an effort to increase its quality according to some metric of interest. Using a simple formula based on the particles' current locations and velocities, particle swarm optimization PSO produces a large number of particles (possible solutions) and moves them around in the search space. An individual's best-known position at any given time is used as a guiding principle, and this position is updated when other particles discover more advantageous positions in the search space. This is being constructed with the intention that the hive mind will select the optimal design [46–49].

Originally, it was meant to be a stylized representation of the social behavior of a group of animals, such as a flock of birds or a school of fish. Simpler approaches to optimization have been found to be effective. Throughout their research, Kennedy and Eberhart consider several philosophical questions about PSOs and swarm intelligence. Poli conducts in-depth analyses of U.S. public safety operations software.

PSO is a metaheuristic that searches through a large number of solutions rapidly since it makes no assumptions about the situation at hand. This indicates that PSO may be applied to optimization problems that are not differentiable, in contrast to gradient descent and quasi-newton approaches. Even the finest metaheuristics, such as PSO, can't guarantee you'll always find the optimal solution.

The stochastic optimization technique known as PSO is inspired by the cooperative intelligence and flexibility of insect swarms. Social contact may be harnessed via PSO to aid in issue-solving. Particles (agents) are used in a swarm to get the optimal answer. Every individual particle in the swarm is searching for the location in the solution space that corresponds to the optimal answer it has found so far. A pbest is an individual's best performance and represents their greatest potential. The PSO also keeps tabs on another top-tier ranking: gbest. That's the highest recorded figure for particles that are close together [50–52].

5.2.6 PSO WITH INDUSTRY 4.0

Among the many areas where PSO has proven useful in the context of industry 4.0 are:

- As it relates to the design and planning of material flows and infrastructure in the subject of mechanical engineering.
- With the tools at hand, the best route for the fluid to take may be determined in a thermodynamic application.
- In the early phase of plant commissioning, when several raw material sources are in play, facility planners must optimize the installation design.
- PSO's optimization capabilities have applications in a wide variety of fields, including telecommunications, control, design, combinatorial optimization etc.
- PSO algorithms may be developed to address a wide variety of situations, including those with constraints, those with multiple objectives, and those with dynamically shifting environments.

5.2.7 ADVANTAGES AND DISADVANTAGES OF PARTICLE SWARM OPTIMIZATION

5.2.7.1 Advantages

These points highlight the merits of PSO.

- When it comes to PSO, it relies on brainpower. It has potential in both academic study and practical engineering. There is no overlap or mutation calculation in PSO. Using the particle's velocity, a search may be conducted.

PSO-Based Nature-Inspired Mechanisms for Robots

- The speed of research is quite high, and the Optimist particle may transfer knowledge to the other particles.
- In PSO, the computation is trivial. It requires less time and effort to perform than other optimization computations used in the development process.

5.2.7.2 Disadvantages

These points highlight the disadvantages of PSO:

- It's easy for the approach to fall prey to overconfidence, which in turn makes it less precise in controlling speed and course.
- The strategy is ineffective in resolving scattering and optimization issues.

5.2.7.3 Robotics

The study and development of robots and other intelligent machines that can function independently is known as robotics, a branch of artificial intelligence. Robotics crosses traditional academic boundaries since robots are hybrid mechanical/electrical/programmable devices. Robotics is often seen as a subset of AI despite the fact that the two sciences seek different aims and have different applications (AI). Robots equipped with AI have the potential to appear and behave just like humans. Robots are robots that can perform complicated activities and replicate human motions in response to predefined instructions while appearing to be human. Some of the many applications of robots in the industry include medication compounding, automobile manufacturing, order picking, industrial cleaning, and the sage automation gantry.

5.2.7.4 Applications of Robotics

Robots have applications in a wide variety of industries (as shown in Figure 5.2). The following are some of the most prominent applications of robots today:

- **Robotics in defense sectors:** without a doubt, a country's armed forces are an integral element of its overall infrastructure. An efficient military is a goal shared by all nations. Robots are helpful during war because they may penetrate dangerous and inaccessible regions. DRDO's Daksh robot is capable of eliminating threats without endangering anyone. They help

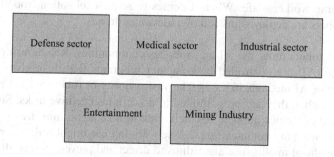

FIGURE 5.2 Application areas of robotics.

military personnel stay alive and, on the job, even when operating in hostile conditions. Robots serve numerous purposes in the military, including combat support, anti-submarine operations, fire support, and even laying machines [53–58].

- **Robotics in medical sectors:** the employment of robots has improved the quality of care in several areas of medicine, including laparoscopy, neurosurgery, orthopedic surgery, hospital cleaning, and medication administration.
- **Robotics in the industrial sector:** cutting, welding, assembly, disassembly, picking and placing for printed circuit boards, packing and labeling, palletizing, product inspection and testing, color coating, drilling, polishing, and handling are just some of the many industrial processes that benefit from the employment of robots [59–61]. In addition to lowering the risk of injury to people, robotics also boosts productivity and revenue. The industrial robot's many benefits include the following:
  - Accuracy
  - Flexibility
  - Cost of labor reduction
  - Silent performance
  - Reduced losses in production
  - Rapid progress in productivity
- **Robotics in entertainment:** over the past decade, robots' roles in the entertainment industry have grown increasingly prominent. There are several examples of robots serving as comic relief in popular media. It's convenient to have robots around when there are several repetitions of the same work. Without becoming tired or frustrated, a robotic camera operator may reshoot a scene as many times as needed. Disney, a Hollywood powerhouse, has unleashed a legion of robots against the film industry.
- **Robots in the mining industry:** robots have various uses in the mining industry, including but not limited to dozing, excavation, hauling, mapping, surveying, drilling, handling explosives, etc. [62]. The cameras and other sensors on a mining robot give it the ability to look beyond the surface of the water and track out valuable minerals. To give a more in-depth example, robots help in excavation by sniffing out dangerous gases and chemicals, keeping workers safe. When it comes to space exploration, robotic rock climbers are deployed instead of underwater drones.

AI-enabled robots unite the AI and robotics communities. Robots using artificial intelligence (AI) are controlled by computer programs that themselves employ a wide range of AI methods [63, 64]. The vast majority of robots are not artificially intelligent; rather, they are programmed to do certain repetitive tasks. Such robots have limited capacities, though. Artificial intelligence algorithms are used to train the robot to do more complex tasks. A self-driving car might utilize a number of different artificial intelligence algorithms to detect and prevent potentially harmful scenarios.

5.2.8 PSO with Robotics

The field of research known as "swarm robotics" investigates the potential of a large number of tiny robots to carry out complex tasks that would be too difficult for a single robot to do alone. The strength, adaptability, and scalability of a swarm robotic system are its distinguishing features. Swarm robotic systems have broad use, with potential uses including spatial organization, collective motion, and distributed decision-making. In PSO, particle swarms function similarly to robot swarms in the real world. As a result, it's not unexpected that PSO has been studied as a potential mechanism of command for robotic swarms. Therefore, many different iterations of the PSO algorithm have been created. The main goal of these algorithms is to enhance robot mobility by including obstacle avoidance. Examples of swarm intelligence algorithms that have been proposed for use as swarm controllers in target localization tasks include the GSO algorithm, ABCO algorithm, BFO algorithm, the Firefly algorithm, and the Bees algorithm.

Generally speaking, the following issues arise when PSO is used as a robot controller in swarm robotic systems:

1. Particles in PSO are considered to have unrestricted velocity and acceleration, which is not the case for real-world robots.
2. Robot control and state updates happen in parallel. The robot's control system determines how quickly the velocity is updated.

Consequently, the properties of the physical system will change if the loop delay of a controller is altered, even if the PSO parameters utilized are kept constant. Therefore, the principles for parameter tuning that work well when PSO is used for numerical parameter optimization cannot be used for swarm robotics. Also, when applied to source localization activities, PSO has the following issues:

3. PSO operates on the assumption that the surrounding conditions are static. It's because PSO doesn't think about whether or not the cost (fitness) of previous places could shift over time. When applied to robotics in the real world, where the source and environment are both subject to change, it becomes immediately apparent that this is not compatible.
4. Collision avoidance is incompatible with swarm behavior because the swarm cannot determine the position of a source until a particle has passed right over it. As a result of these limitations, PSO has not seen widespread adoption in swarm robots.

Incorporating several features of mobile robotics systems into the PSO algorithm is necessary for our reduced microscopic model of multi-robot search to succeed.

1) **Discrete versus Continuous Movement and Robot Collisions:** collisions and obstacle avoidance between robots and between robots and the target or barriers are some of the key issues of multi-robot search that will impact the

performance. If we use the default PSO particle displacement, here we will be unable to identify any collisions that may occur. This means that we need to mimic the robots' continuous movement by segmenting the displacement into many stages and performing collision checks at each. Particles are said to have "collided" with their targets when they get within a certain distance from them, which is determined by the sizes of the targets and the particles, as well as the detection range of the particles' proximity sensors. A particle's velocity remains the same but is redirected when it collides with an object.

The collision reaction will change based on the type of object involved:
- **Environment Boundary** – the particle's motion is "reflected" from the barrier by canceling out the proportion of its velocity that is perpendicular to it.
- **Search Target** – the speed at which the particle is moving away from the target will be altered. Another Particle – this will cause the speeds of both particles to be changed such that they move perpendicular to one another. As a rule, multi-robot systems need some amount of communication time so that the robots can coordinate their course corrections in order to avoid obstacles. To get a rough idea, we disregard this delay and assume that particles may instantly reverse course.

2) **Low Dimensionality:** in practice, PSO is employed to optimize functions with a large number of variables (anywhere from around ten to several thousand). As a result, the dimensions of the PSO virtual search space tend to be rather large. Since our multi-robot search scenario takes place on a flat surface, we can only consider two dimensions. The ideal settings for the system's many parameters may change as a result of this.

3) **Real-World Noise:** the randomness of their parts and the world around them is a constant challenge for robotic systems in the real world. Even the most precisely calibrated sensors and actuators will experience some level of noise in the real world.

5.3 LITERATURE REVIEW

The authors in [Azizi, A 1] introduced the design of an artificial neural network model to optimize the dynamic behavior of robotic arms, which is one example of how AI is being used to improve the long-term viability of the Fourth Industrial Revolution. In order to determine the best settings for the gimbal joints, this study used the Genetic Algorithm (GA), a popular evolutionary method. It has been suggested that the GA's behavior may be modeled using an ANN architecture to cut down on the lengthy process of adopting the GA. This demonstrates that the suggested ANN model may be utilized to identify the best gimbal joint settings, as opposed to the more complicated and time-consuming GA. In [2], the authors focused on robot gear needing to be sturdy and versatile enough to operate on land, sea, or in the air. In autonomous robots, localization to the origin is a crucial decision-making issue. Using the provided coordinate system, the localization process identifies its position and orientation in Cartesian space. Locating the robot at all times is crucial for efficient autonomous navigation. As a population-based, worldwide search technique, PSO was very effective. The suggested approach for SLM is a refinement of micro-PSO. By comparing the outcomes to those of classic PSO and PSO, the efficacy of this technique can be determined. Further, the authors in [3] discussed robotic arms in digital manufacturing processes: some examples from the Fourth Industrial Revolution. The use of robots in formerly manual processes including pick-and-place, welding, subtractive, and additive manufacturing has increased in recent years as the concept of Industry 4.0 has gained popularity. This article describes the several uses and combinations put to a KUKA KR16 robotic arm at PUC-Rio. The objective was to demonstrate the potential for a robot arm to be used in conjunction with Industry 4.0, 3D printing, milling, and other technologies to verify their impact on engineering and industry. The opportunities and risks in artificial intelligence, cyber security, and the Fourth Industrial Revolution have been presented in [4]. From the manufacturer's perspective, they also provide an evaluation of their applicability to OT. In addition to introducing the reader to the core concepts and motivating factors behind Industry 4.0, the essay delves into the emergence of AI in production systems and the specific set of issues this trend presents. Several applications were covered, and a suite of AI-based approaches for keeping an eye on and fine-tuning output was offered. Following this study of how this change may impact current security measures, we discuss the technical, operational, and security challenges that arise as a result of it. The study concludes with an outlook on potential future challenges in terms of cybersecurity for the Industry 4.0 sector. The article begins with a discussion of the need for keeping tabs on how people and machines interact, before moving on to discuss how to build adversarial/robust AI and orchestrate dispersed detection approaches.

In [5], the authors considered synergistic pneumatic artificial muscle controlled by a neural network and a proportional-integral-deriv. PSO Algorithm was used. To regulate Pulse Amplitude Modulation (PAM's) location, the authors of this research suggest a novel hybrid Deep Neural Networks (DNN)/PID controller. The controller's effectiveness is improved by the study's application of an optimization pattern to the issue. PSO has been utilized to fine-tune the PID-DNN parameters. The proposed controller was put into use in a real-world production setting and then compared to a conventional sliding mode controller. Simulation and experimental results demonstrate that the proposed controller is able to track the reference signals over the whole PAM. In [6], the authors have reviewed the literature on the challenges, tactics, and solutions related to CPS implementation in manufacturing within the context of Industry 4.0. The studies were then evaluated on a finer scale, with the titles, abstracts, and complete texts being examined for inclusion in the prospective research list. We reviewed 78 publications on the future research on CPS for manufacturing in Industry 4.0 from a pool of 626 primary retrieved from the relevant literature. We began by extracting the articles' contexts to determine Industry 4.0's broad and fine-grained components. They next reviewed many studies that used synthesized matrices to define the challenges, tactics, and procedures that are the foundation for the CPS in the manufacturing sector of Industry 4.0. HC, CC, and PC as well as their HC-CC, CC-PC, and HC-PC interfaces must all be standardized for Industry 4.0 to be a success. It has been argued in [7] that the inverse kinematics of a serial 7-DOF robotic manipulator may be solved using the quantum-behaved particle swarm approach as a metaheuristic. Estimating the position of the robotic manipulator's end effector with respect to the work area was performed with the help of swarm algorithms like Quantum PSO. Accordingly, we decided that the Euclidean fitness function best reflected our goals. The goal of this process was to quantify the error in where the manipulator's end effector was thought to be located. These algorithms are tested in two different settings in this study. In the first situation, the values are only accurate to the hundredth place, whereas in the second case, they are only accurate to the one-hundredth place. In reality, the second case validates the first by confirming the high quality of the quantum particle swarm optimization algorithm (QPSO) in the inverse kinematic solution. The results demonstrate the superiority of quantum-behaved PSO over traditional PSO, ABC, and FA. Increased particle count, lower iteration count, and shorter computation times were all advantages of the new method. The authors in [8] have introduced ROS2's integrated drive system, RS, horizontal integration of a robotic network, human-robot friendly and NI, and DL robots are all part of the conceptual framework for collaborative and intelligent robotic systems: the transition from Industry 4.0 to Robotics 4.0. Robotics 4.0 aims to combine motion, computation, vision, and cognition seamlessly to address a wide range of industry and social requirements. To illustrate the present effort to realize the concept of Robotics 4.0, a road map and case studies will be provided.

A Hybrid GA-PSO Algorithm has been used in [9] to find the Best Route for a Salesperson on the Road. When trying to identify the best answer to the traveling salesman issue, heuristic search methods are a useful tool (TSP). TSP has several practical uses, including in the semiconductor industry, the logistics industry, and the transportation industry. This chapter's goal is to provide a heuristic approach to TSP by fusing the widely used optimization techniques of GA and PSO. This hybrid GA-PSO method uses crossover and mutation operators of GA and PSO mechanisms to generate new individuals. Ten conventional TSPs were used to compare the performance of the hybrid GA-PSO method to that of GA and PSO in terms of time to locate the best solution and computing effort. Computational findings show that the hybrid GA-PSO approach is much better than GA and PSO for TSPs. The outline of particle swarm optimization's potential applications in sustainable water management has been discussed in [10]. The performance of the PSO variants was compared to that of a number of EAs and mathematical optimization approaches. The performance of PSO versions is compared and contrasted with that of conventional EAs and mathematical approaches in terms of correct convergence to optimum Pareto fronts, quicker convergence rate, and variety of computed solutions. In [11], the potential of the robots has been discussed to improve the rollout of Industry 4.0. In this study, we look at the promising future of robotics in the industrial sector and related fields. There are 18 main uses of robotics in Industry 4.0 that are covered in this study. Robots, which can go far closer to a component than any of the other equipment in a plant, are perfect for gathering this kind of cryptic data. Automation, high temperatures, continuous operation, and extended shifts in hazardous conditions are all made possible by this technology, making it ideal for use in assembly lines. Artificial intelligence is used by many robots in smart industries to carry out complex tasks. They now have the ability to make decisions and gain wisdom via experience in a variety of dynamic contexts. The authors in [12] provided direct management of the CSTR, making use of a Biogeography-Based Optimization-Fitness Function-Proportional–integral–derivative (BBO-FF-PID) Controller Hybrid, one that uses Particle swam Optimization (PSO)-PIDs or Internal Model Control (IMC)-PIDs. A hybrid control technique (BBO-FF) based PID is used for optimal control of a nonlinear system like CSTR; this hybrid approach is compared to the more common Z-N tuned PID, IMC-PID, and AOCM called PSO-PID. Using a nonlinear process control system, this study sought to identify the optimal controller settings for maximizing a target output.

5.4 PROPOSED WORK

5.4.1 Problem Statement

Research has been conducted in the field of PSO services on several occasions. These services have a wide range of potential applications in areas such as Industry 4.0, Nature-Inspired Computing, and Robotics. But the problem with the currently available study is that it has a sluggish performance and is inaccurate. Previous studies overlooked the need for optimization before training and testing procedures. Therefore, an optimal technique has to be proposed in order to gain better accuracy while maintaining high performance.

5.4.2 Proposed Methodology

In the proposed research work the working of an existing IoT model for a smart city has been considered and factors that are affecting the performance and accuracy are taken into account. Different machine learning mechanisms are considered for classification purposes but there is a need to integrate optimization mechanisms. Research is considering a PSO optimizer in order to filter the dataset before training time. An optimized approach integrated into machine learning is supposed to improve accuracy during the classification of an event for IoT services applicable in smart cities. Finally comparative analysis of the proposed work would be made traditional on the basis of performance and accuracy. The proposed research methodology is shown in Figure 5.3.

5.4.3 Process Flow of Novel Approach

The process flow of novel approach has been presented in Figure 5.4 where training and testing have been performed with over-optimized and non-optimized datasets. The optimization mechanism used in research work is PSO. PSO is a particle swarm optimization mechanism that would find the best solution to filter a dataset. The filtered dataset is trained and tested. The confusion matrix obtained by such classification is processed to find accuracy parameters. Then accuracy parameters of both models are compared in order to get better accuracy.

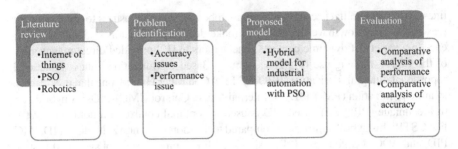

FIGURE 5.3 Research methodology.

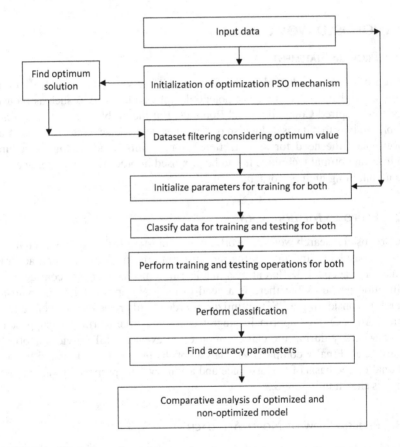

FIGURE 5.4 Flow chart of novel approach.

5.5 RESULTS AND DISCUSSION

To understand how the robotics model works for industries, it is necessary to take into consideration the aspects that impact the model's performance and accuracy. Different methods of machine learning are being examined for categorization, but an optimization mechanism must also be included in the mix. A machine

TABLE 5.1
Comparison of Accuracy for Non-Optimized and Optimized Dataset

Class	Non-optimized dataset	Optimized dataset
1	84.23%	92.30%
2	84.66%	92.63%
3	85.01%	92.63%
4	85.12%	93.57%
5	85.44%	93.61%
6	85.97%	93.88%

learning-based method for event categorization for robotics services is expected to increase accuracy. Finally, research aims to enhance accuracy parameters such as recall value, f-score, and precision.

5.5.1 Accuracy Comparison

Table 5.1 and Figure 5.5 present the accuracy for a non-optimized dataset and an optimized dataset where six classes have been considered. Each class has 1000 elements.

5.5.2 Error Rate Comparison

Table 5.2 and Figure 5.6 present the error rate for a non-optimized dataset and optimizer dataset where six classes have been considered. Each class has 1000 elements.

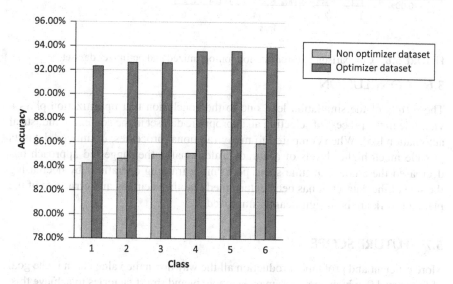

FIGURE 5.5 Comparison of accuracy for non-optimized and optimized dataset.

TABLE 5.2
Comparison of Error Rate for Non-Optimized and Optimizer Dataset

Class	Non-optimized dataset	Optimized dataset
1	15.77%	7.70%
2	15.34%	7.37%
3	14.99%	7.37%
4	14.88%	6.43%
5	14.56%	6.39%
6	14.03%	6.12%

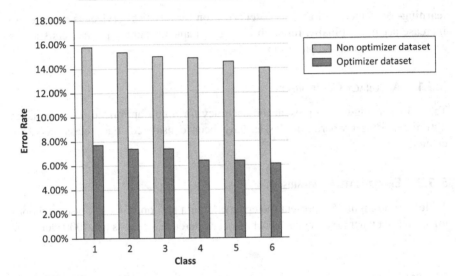

FIGURE 5.6 Comparison of error rate for non-optimized and optimizer dataset.

5.6 CONCLUSION

The results of the simulation lead one to the conclusion that optimization plays a vital role in the process of selecting an appropriate dataset in the context of industrial automation [65]. When compared to more traditional processes, optimized datasets provide much higher levels of accuracy. Additionally, the suggested approach has decreased the amount of time spent performing training operations by decreasing the size of the dataset. It has been determined that the accuracy and error rate of the planned work are both significantly improved.

5.7 FUTURE SCOPE

More efficient and profitable production all the way down the value chain is the goal of Industry 4.0, which employs smart technology and smart factories to achieve this. To find the optimal solution, PSO searches for the global maximum and minimum.

With Particle Swarm Optimization, swarm intelligence and mobility are put to use. Robots can accomplish complex tasks by mimicking human actions and according to predetermined guidelines. Particle swarms in PSO function similarly to robotic swarms. Since this is the case, PSO is an obvious option for guiding robotic swarms. In this research, we concentrate on how robots can use PSO to make better decisions. The enhanced efficiency and productivity that characterize Industry 4.0's smart factories and omnipresent robots are enabled by data-driven choices throughout the value chain. Motivated by patterns in nature, PSO seeks global extrema. PSO is a stochastic optimization approach that takes its cues from the behavior of a swarm of particles. Robots that follow complex instructions might give the impression of being human-like in their actions. Particle swarms in PSO function as a hive mind, much like a swarm of real-world robots.

REFERENCES

1. Azizi, A. (2020). Applications of artificial intelligence techniques to enhance sustainability of Industry 4.0: Design of an artificial neural network model as dynamic behavior optimizer of robotic arms. *Complexity, 2020.* https://doi.org/10.1155/2020/8564140
2. Bakhale, M., Hemalatha, V., Dhanalakshmi, S., Kumar, R., and Siddharth Jain, M. (2020). A dynamic inertial weight strategy in micro PSO for swarm robots. *Wireless Personal Communications, 110*(2), 573–592. https://doi.org/10.1007/s11277-019-06743-x
3. Barbosa, W. S., Gioia, M. M., Natividade, V. G., Wanderley, R. F. F., Chaves, M. R., Gouvea, F. C., and Gonçalves, F. M. (2020). Industry 4.0: Examples of the use of the robotic arm for digital manufacturing processes. *International Journal on Interactive Design and Manufacturing (IJIDeM), 14*(4), 1569–1575. https://doi.org/10.1007/s12008-020-00714-4
4. Bécue, A., Praça, I., and Gama, J. (2021). Artificial intelligence, cyber-threats and Industry 4.0: Challenges and opportunities. In *Artificial Intelligence Review* (Vol. 54, Issue 5). Springer. https://doi.org/10.1007/s10462-020-09942-2
5. Chavoshian, M., Taghizadeh, M., and Mazare, M. (2020). Hybrid dynamic neural network and PID control of pneumatic artificial muscle using the PSO algorithm. *International Journal of Automation and Computing, 17*(3), 428–438. https://doi.org/10.1007/s11633-019-1196-5
6. Dafflon, B., Moalla, N., and Ouzrout, Y. (2021). The challenges, approaches, and used techniques of CPS for manufacturing in Industry 4.0: A literature review. *International Journal of Advanced Manufacturing Technology, 113*(7–8), 2395–2412. https://doi.org/10.1007/s00170-020-06572-4
7. Dereli, S., and Köker, R. (2020). A meta-heuristic proposal for inverse kinematics solution of 7-DOF serial robotic manipulator: Quantum behaved particle swarm algorithm. *Artificial Intelligence Review, 53*(2), 949–964. https://doi.org/10.1007/s10462-019-09683-x
8. Gao, Z., Wanyama, T., Singh, I., Gadhrri, A., and Schmidt, R. (2020). From industry 4.0 to robotics 4.0 - A conceptual framework for collaborative and intelligent robotic systems. *Procedia Manufacturing, 46*(2019), 591–599. https://doi.org/10.1016/j.promfg.2020.03.085
9. Gupta, I. K., Shakil, S., and Shakil, S. (2019). A hybrid GA-PSO algorithm to solve traveling salesman problem. In *Advances in Intelligent Systems and Computing* (Vol. 798). Springer. https://doi.org/10.1007/978-981-13-1132-1_35

10. Jahandideh-Tehrani, M., Bozorg-Haddad, O., and Loáiciga, H. A. (2020). Application of particle swarm optimization to water management: An introduction and overview. *Environmental Monitoring and Assessment, 192*(5). https://doi.org/10.1007/s10661-020-8228-z
11. Javaid, M., Haleem, A., Singh, R. P., and Suman, R. (2021). Substantial capabilities of robotics in enhancing industry 4.0 implementation. *Cognitive Robotics, 1,* 58–75. https://doi.org/10.1016/j.cogr.2021.06.001
12. Khanduja, N., and Bhushan, B. (2019). CSTR control using IMC-PID, PSO-PID, and hybrid BBO-FF-PID controller. In *Advances in Intelligent Systems and Computing* (Vol. 697). Springer. https://doi.org/10.1007/978-981-13-1822-1_48
13. Li, X., Wu, D., He, J., Bashir, M., and Liping, M. (2020). An improved method of particle swarm optimization for path planning of mobile robot. *Journal of Control Science and Engineering, 2020.* https://doi.org/10.1155/2020/3857894
14. Sansanwal, K., Shrivastava, G., Anand, R., and Sharma, K. (2019). Big data analysis and compression for indoor air quality. In *Handbook of IoT and Big Data* (pp. 1–21). CRC Press.
15. Liu, C., Cao, G. H., Qu, Y. Y., and Cheng, Y. M. (2020). An improved PSO algorithm for time-optimal trajectory planning of Delta robot in intelligent packaging. *International Journal of Advanced Manufacturing Technology, 107*(3–4), 1091–1099. https://doi.org/10.1007/s00170-019-04421-7
16. Lones, M. A. (2020). Mitigating metaphors: A comprehensible guide to recent nature-inspired algorithms. *SN Computer Science, 1*(1), 1–12. https://doi.org/10.1007/s42979-019-0050-8
17. Mafarja, M., Qasem, A., Heidari, A. A., Aljarah, I., Faris, H., and Mirjalili, S. (2020). Efficient hybrid nature-inspired binary optimizers for feature selection. *Cognitive Computation, 12*(1), 150–175. https://doi.org/10.1007/s12559-019-09668-6
18. Mellal, M. A., and Salhi, A. (2020). *Parallel–Series System Optimization by Weighting Sum Methods and Nature-Inspired Computing.* Springer. https://doi.org/10.1007/978-981-13-9263-4_10
19. Mirjalili, S., Song Dong, J., Lewis, A., and Sadiq, A. S. (2020). Particle swarm optimization: Theory, literature review, and application in airfoil design. In *Studies in Computational Intelligence* (Vol. 811). Springer International Publishing. https://doi.org/10.1007/978-3-030-12127-3_10
20. Molina, D., Poyatos, J., Del Ser, J., García, S., Hussain, A., and Herrera, F. (2020). Comprehensive taxonomies of nature- and bio-inspired optimization: Inspiration versus algorithmic behavior, critical analysis recommendations. *Cognitive Computation, 12*(5), 897–939. https://doi.org/10.1007/s12559-020-09730-8
21. Nagireddy, V., Parwekar, P., and Mishra, T. K. (2019). Comparative analysis of PSO-SGO algorithms for localization in wireless sensor networks. In *Advances in Intelligent Systems and Computing* (Vol. 862). Springer. https://doi.org/10.1007/978-981-13-3329-3_37
22. Anand, R., Sindhwani, N., and Juneja, S. (2022). Cognitive Internet of things, its applications, and its challenges: A survey. In *Harnessing the Internet of Things (IoT) for a Hyper-Connected Smart World* (pp. 91–113). Apple Academic Press.
23. Oztemel, E., and Gursev, S. (2020). Literature review of Industry 4.0 and related technologies. *Journal of Intelligent Manufacturing, 31*(1), 127–182. https://doi.org/10.1007/s10845-018-1433-8
24. Singh, P., Kaiwartya, O., Sindhwani, N., Jain, V., and Anand, R. (Eds.). (2022). *Networking Technologies in Smart Healthcare: Innovations and Analytical Approaches.* CRC Press.
25. Pattanayak, S., Agarwal, S., Choudhury, B. B., and Sahoo, S. C. (2019). Path planning of mobile robot using pso algorithm. In *Smart Innovation, Systems and Technologies* (Vol. 106). Springer. https://doi.org/10.1007/978-981-13-1742-2_51

26. Rao, G. S., Kumari, G. V., and Rao, B. P. (2019). *Network for Biomedical Applications* (Vol. 2, Issue January). Springer. https://doi.org/10.1007/978-981-13-1595-4
27. Salih, S. Q., Alsewari, A. R. A., Al-Khateeb, B., and Zolkipli, M. F. (2019). Novel multi-swarm approach for balancing exploration and exploitation in particle swarm optimization. In *Advances in Intelligent Systems and Computing* (Vol. 843). Springer International Publishing. https://doi.org/10.1007/978-3-319-99007-1_19
28. Sindhwani, N., Anand, R., Niranjanamurthy, M., Verma, D. C., and Valentina, E. B. (Eds.). (2022). *IoT Based Smart Applications*. Springer Nature.
29. Singh, N., Singh, S. B., and Houssein, E. H. (2022). Hybridizing salp swarm algorithm with particle swarm optimization algorithm for recent optimization functions. In: *Evolutionary Intelligence* (Vol. 15, Issue 1). Springer. https://doi.org/10.1007/s12065-020-00486-6
30. Sun, P., and Shan, R. (2020). Predictive control with velocity observer for cushion robot based on PSO for path planning. *Journal of Systems Science and Complexity, 33*(4), 988–1011. https://doi.org/10.1007/s11424-020-8375-x
31. Anand, R., and Chawla, P. (2016, March). A review on the optimization techniques for bio-inspired antenna design. In *2016 3rd International Conference on Computing for Sustainable Global Development (INDIACom)* (pp. 2228–2233). IEEE.
32. Tutunji, T. A., Al-Khawaldeh, M., and Alkayyali, M. (2022). A three-stage PSO-based methodology for tuning an optimal PD-controller for robotic arm manipulators. *Evolutionary Intelligence, 15*(1), 381–396. https://doi.org/10.1007/s12065-020-00515-4
33. Anand, R., and Chawla, P. (2020). A novel dual-wideband inscribed hexagonal fractal slotted microstrip antenna for C-and X-band applications. *International Journal of RF and Microwave Computer-Aided Engineering, 30*(9), e22277.
34. Sindhwani, N., and Bhamrah, M. S. (2017). An optimal scheduling and routing under adaptive spectrum-matching framework for MIMO systems. *International Journal of Electronics, 104*(7), 1238–1253.
35. Yahya, M., Breslin, J. G., and Ali, M. I. (2021). Semantic web and knowledge graphs for industry 4.0. *Applied Sciences (Switzerland), 11*(11), 1–23. https://doi.org/10.3390/app11115110
36. Yang, X.-S., and Karamanoglu, M. (2020). Nature-inspired computation and swarm intelligence: A state-of-the-art overview. In *Nature-Inspired Computation and Swarm Intelligence*. Elsevier Ltd. https://doi.org/10.1016/b978-0-12-819714-1.00010-5
37. Sindhwani, N., and Singh, M. (2020). A joint optimization based sub-band expediency scheduling technique for MIMO communication system. *Wireless Personal Communications, 115*(3), 2437–2455.
38. Zemmal, N., Azizi, N., Sellami, M., Cheriguene, S., Ziani, A., AlDwairi, M., and Dendani, N. (2020). Particle swarm optimization based swarm intelligence for active learning improvement: Application on medical data classification. *Cognitive Computation, 12*(5), 991–1010. https://doi.org/10.1007/s12559-020-09739-z
39. Srivastava, A., Gupta, A., and Anand, R. (2021). Optimized smart system for transportation using RFID technology. *Mathematics in Engineering, Science and Aerospace (MESA), 12*(4), 34–54.
40. Zhao, W., Wang, L., and Zhang, Z. (2020). Artificial ecosystem-based optimization: A novel nature-inspired meta-heuristic algorithm. In *Neural Computing and Applications* (Vol. 32, Issue 13). Springer. https://doi.org/10.1007/s00521-019-04452-x
41. Anand, R., and Chawla, P. (2020). Optimization of a slotted fractal antenna using LAPO technique. *International Journal of Advanced Science and Technology, 29*(5s), 21–26.
42. Gupta, R. S., Nassa, V. K., Bansal, R., Sharma, P., Koti, K., and Koti, K. (2021). Investigating application and challenges of big data analytics with clustering. In *2021 International Conference on Advancements in Electrical, Electronics,*

Communication, Computing and Automation (ICAECA) (pp. 1–6). https://doi.org/10.1109/ICAECA52838.2021.9675483

43. Veeraiah, V., Khan, H., Kumar, A., Ahamad, S., Mahajan, A., and Gupta, A. (2022). Integration of PSO and deep learning for trend analysis of meta-verse. In *2022 2nd International Conference on Advance Computing and Innovative Technologies in Engineering (ICACITE)* (pp. 713–718). https://doi.org/10.1109/ICACITE53722.2022.9823883

44. Anand, R., and Chawla, P. (2020). A hexagonal fractal microstrip antenna with its optimization for wireless communications. *International Journal of Advanced Science and Technology*, 29(3s), 1787–1791.

45. Veeraiah, V., Kumar, K. R., LalithaKumari, P., Ahamad, S., Bansal, R., and Gupta, A. (2022). Application of biometric system to enhance the security in virtual world. In *2022 2nd International Conference on Advance Computing and Innovative Technologies in Engineering (ICACITE)* (pp. 719–723). https://doi.org/10.1109/ICACITE53722.2022.9823850

46. Bansal, R., Gupta, A., Singh, R., and Nassa, V. K. (2021). Role and impact of digital technologies in e-learning amidst COVID-19 pandemic. In *2021 Fourth International Conference on Computational Intelligence and Communication Technologies (CCICT)* (pp. 194–202). https://doi.org/10.1109/CCICT53244.2021.00046

47. Shukla, A., Ahamad, S., Rao, G. N., Al-Asadi, A. J., Gupta, A., and Kumbhkar, M. (2021). Artificial intelligence assisted IoT data intrusion detection. In *2021 4th International Conference on Computing and Communications Technologies (ICCCT)* (pp. 330–335). https://doi.org/10.1109/ICCCT53315.2021.9711795

48. Pathania, V., Babu, S. Z. D., Ahamad, S., Thilakavathy, P., Gupta, A., Alazzam, M. B., and Pandey, D. (2023). A database application of monitoring COVID-19 in India. In: Gupta, M., Ghatak, S., Gupta, A., Mukherjee, A. L. (eds) *Artificial Intelligence on Medical Data. Lecture Notes in Computational Vision and Biomechanics*, 37. Springer. https://doi.org/10.1007/978-981-19-0151-5_23

49. Dushyant, K., Muskan, G., Annu, Gupta, A., and Pramanik, S. (2022). Utilizing machine learning and deep learning in cybesecurity: An innovative approach. In *Cyber Security and Digital Forensics: Challenges and Future Trends* (pp. 271–293). Wiley. https://doi.org/10.1002/9781119795667.ch12

50. Babu, S. Z. D., Pandey, D., Naidu, G. T., Sumathi, S., Gupta, A., Bader Alazzam, M., and Pandey, B. K. (2023). Analysation of big data in smart healthcare. In Gupta, M., Ghatak, S., Gupta, A., Mukherjee, A. L. (eds) *Artificial Intelligence on Medical Data: Lecture Notes in Computational Vision and Biomechanics*, 37. Springer. https://doi.org/10.1007/978-981-19-0151-5_21

51. Sindhwani, N., and Singh, M. (2017, March). Performance analysis of ant colony based optimization algorithm in MIMO systems. In *2017 International Conference on Wireless Communications, Signal Processing and Networking (WiSPNET)* (pp. 1587–1593). IEEE.

52. Gupta, A., Kaushik, D., Garg, M., and Verma, A. (2020). Machine learning model for breast cancer prediction. In *2020 Fourth International Conference on I-SMAC (IoT in Social, Mobile, Analytics and Cloud) (I-SMAC)* (pp. 472–477). https://doi.org/10.1109/I-SMAC49090.2020.9243323

53. Kaur, J., Sindhwani, N., Anand, R., and Pandey, D. (2022). Implementation of IoT in various domains. In *IoT Based Smart Applications* (pp. 165–178). Springer International Publishing.

54. Veeraiah, V., Rajaboina, N. B., Rao, G. N., Ahamad, S., Gupta, A., and Suri, C. S. (2022). Securing online web application for IoT management. In *2022 2nd International Conference on Advance Computing and Innovative Technologies in Engineering (ICACITE)* (pp. 1499–1504). https://doi.org/10.1109/ICACITE53722.2022.9823733

55. Veeraiah, V., G. P., Ahamad, S., Talukdar, S. B., Gupta, A., and Talukdar, V. (2022). Enhancement of meta verse capabilities by IoT integration. In *2022 2nd International Conference on Advance Computing and Innovative Technologies in Engineering (ICACITE)* (pp. 1493–1498). https://doi.org/10.1109/ICACITE53722.2022.9823766
56. Gupta, N., Janani, S., Dilip, R., Hosur, R., Chaturvedi, A., and Gupta, A. (2022). Wearable sensors for evaluation over smart home using sequential minimization optimization-based random forest. *International Journal of Communication Networks and Information Security (IJCNIS)*, 14(2), 179–188. https://doi.org/10.17762/ijcnis.v14i2.5499
57. Keserwani, H., Rastogi, H., Kurniullah, A. Z., Janardan, S. K., Raman, R., Rathod, V. M., and Gupta, A. (2022). Security enhancement by identifying attacks using machine learning for 5G network. *International Journal of Communication Networks and Information Security (IJCNIS)*, 14(2), 124–141. https://doi.org/10.17762/ijcnis.v14i2.5494
58. Tripathi, A., Sindhwani, N., Anand, R., and Dahiya, A. (2022). Role of IoT in smart homes and smart cities: Challenges, benefits, and applications. In *IoT Based Smart Applications* (pp. 199–217). Springer International Publishing.
59. Pandey, D., and Wairya, S. (2023). An optimization of target classification tracking and mathematical modelling for control of autopilot. *The Imaging Science Journal*, 1–16.
60. Pandey, B. K., Pandey, D., Wariya, S., Aggarwal, G., and Rastogi, R. (2021). Deep learning and particle swarm optimisation-based techniques for visually impaired humans' text recognition and identification. *Augmented Human Research*, 6, 1–14.
61. Pandey, D., Wairya, S., Sharma, M., Gupta, A. K., Kakkar, R., and Pandey, B. K. (2022). An approach for object tracking, categorization, and autopilot guidance for passive homing missiles. *Aerospace Systems*, 1–14.
62. Juneja, S., Juneja, A., Anand, R., and Chawla, P. (2020). Mining aspects on the social network. *International Journal of Innovative Technology and Exploring Engineering*, 8(9s), 285–289.
63. Kumar, M. S., Sankar, S., Nassa, V. K., Pandey, D., Pandey, B. K., and Enbeyle, W. (2021). Innovation and creativity for data mining using computational statistics. In *Methodologies and Applications of Computational Statistics for Machine Intelligence* (pp. 223–240). IGI Global.
64. Pandey, D., George, S., Aremu, B., Wariya, S., and Pandey, B. K. (2021). Critical review on integration of encryption, steganography, IOT and artificial intelligence for the secure transmission of Stego images.
65. Anand, R., Sindhwani, N., and Dahiya, A. (2022, March). Design of a high directivity slotted fractal antenna for C-band, X-band and Ku-band applications. In *2022 9th International Conference on Computing for Sustainable Global Development (INDIACom)* (pp. 727–730). IEEE.

6 Application of AI-Based Learning in Automated Applications and Soft Computing Mechanisms Applicable in Industries

B. Mahendra Kumar, Veera Talukdar, Huma Khan, Suryansh Bhaskar Talukdar, Ashok Koujalagi, R. Ganesh Kumar, and Ankur Gupta

6.1 INTRODUCTION

Artificial Intelligence (AI) is a method through which computers may learn and improve themselves with little to no human assistance or coding. Robots analyze data for patterns and utilize those patterns to educate themselves. What they do and how often they do it is up to them. The foundations of soft computing are artificial intelligence and evolution. As a result, it offers low-cost and easy solutions to problems that cannot be solved computationally. Since the early 1990s, academia has been exploring the concept of "soft computing." The idea for this work came from the brain's impressive capacity to come close to solving complex problems. Complex problems may be tackled with the help of soft computing since it employs approximations in its calculations. Industrial automation is used by a wide variety of businesses and sectors today thanks to the many ways in which it has advanced operations. In the business world, software applications have replaced labor-intensive procedures. Automatic maintenance and repairs in the manufacturing sector save expenses for businesses by a significant amount.

6.1.1 Artificial Intelligence

The phrase AI is used to describe machines that can perceive, synthesize, and infer, much like humans and other sentient organisms. Voice recognition, computer vision, and translation are just a few examples of where this is used. There are various applications for AI, including state-of-the-art search engines, recommendation systems,

Application of AI-Based Learning in Automated Applications

voice recognition, autonomous cars, decision-making automation, and top-tier strategic gaming systems. The rise of more capable robots is having a chilling effect on a number of occupations that were formerly considered "clever." Optical character recognition is frequently overlooked when discussing the nature of AI since it has grown so ubiquitous [1–7].

Since its beginning in 1956, AI research has gone through many "waves" of excitement, followed by "waves of disappointment" and "waves of budget cutbacks," and then by "waves of innovative approaches," "waves of success" and "waves of restored money." Many approaches have been tried and then abandoned in the pursuit of AI, including simulations of the human brain, models of human problem-solving, formal logic, large databases, and even attempts to mimic animal behavior. In the first few decades of the 21st century, the field of machine learning was dominated by a highly mathematical-statistical approach, which proved to be very successful, facilitating the solution of many difficult challenges in industry and academics [8–15].

The concept of AI concept is shown in Figure 6.1. Different areas of AI study focus on different outcomes and methodologies. Traditional areas of AI research include reasoning, knowledge representation, planning, learning, Natural Language Processing (NLP), vision, movement, and object manipulation. The ability to solve any problem is one of the ultimate goals of the profession. Researchers in the field of AI have turned to a wide variety of problem-solving techniques, including search and mathematical optimization, formal logic, artificial neural networks (ANNs), and methods based on statistics, probability, and economics. Computer science, psychology, linguistics, philosophy, and many more all contribute to AI [16–18].

With the belief that human intellect "can be so accurately characterized that a computer may be constructed to replicate it," the discipline was born. Philosophical

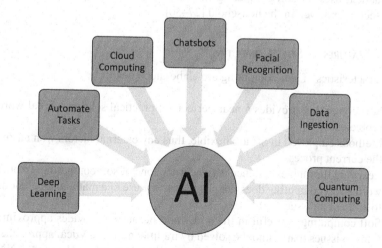

FIGURE 6.1 Artificial Intelligence.

arguments regarding the nature of the mind and the ethical consequences of building artificial beings with human-level cognition have been inspired by these concerns, which have been examined in myth, literature, and philosophy since antiquity. Some philosophers and computer scientists have warned that AI might pose an existential threat to mankind if its logical capabilities aren't channeled toward positive ends [19–22].

6.1.2 Soft Computing

The antithesis of hard (conventional) computing, soft computing is becoming increasingly popular. These computational methods have their roots in AI and natural selection. It provides low-priced answers to problems with no easy computing solution in the actual world. The field of mathematics and computer science known as "soft computing" has been present since the 1990s. This endeavor was motivated by the human mind's capacity to provide approximations of solutions to problems. Soft computing makes use of approximative calculations to provide workable solutions to intractable computational difficulties. This method can be used to solve problems that are difficult or time-consuming to solve using conventional methods and tools. Soft computing is another name for computational intelligence.

When used for problem-solving, soft computing allows you to bypass the need for computers altogether. Similar to the human mind, soft computing is able to make do with less-than-complete information. However, this is not the case with most computer simulations. As a result of soft computing's flexibility, researchers may take on problems that would otherwise be too difficult for computers [23–26].

Instead of "soft computing," the technique of "possibility" is used when there is insufficient data to draw any firm conclusions. This is done when the problem is too nebulous to be handled using traditional arithmetic and computer methods. You may find practical uses for soft computing in everything from the private sector to the public sector, and even in the household [27–30].

6.1.3 Features of Soft Computing

The characteristics of soft computing are elaborated ahead:

- Soft computing provides a near-perfect yet practical solution to real-world problems.
- Methods employed in SC are flexible, thus any external factors won't affect the current process.
- Learning through experience is the cornerstone of soft computing. In other words, soft computing does not necessitate the use of a mathematical model to address an issue.
- Soft computing is useful in the real world because it provides approximations to issues that cannot be solved by traditional or analytical approaches.
- Fuzzy analysis, genetic algorithms, machine learning (ML), and ANNs were all used in its creation.

Application of AI-Based Learning in Automated Applications

6.1.4 Need for Soft Computing

Typical computers and analytical techniques are inadequate for resolving some real-world problems. We'll have to resort to soft computing or some other way to obtain a close enough approximation in the following scenarios:

- Hard computation is required to find accurate solutions to mathematical issues. Some practical concerns are ignored. Thus, soft computing can be helpful in situations when there is no obvious solution.
- When traditional analytical and mathematical models, like those used to map the human mind, fall short, soft computing can be utilized as an alternative.
- Analytical models, valid under ideal conditions, allow for the resolution of mathematical problems. However, problems in the real world don't occur in a vacuum.
- When applied to real-world problems, soft computing may shed some much-needed illumination, making it more than just a theoretical concept.
- For the same reasons listed above, soft computing facilitates the mapping of the human mind that is beyond the reach of traditional mathematical and analytical models [31–33].

6.1.5 Elements of Soft Computing

Soft computing is a cornerstone of the emerging field of conceptual intelligence. Fuzzy logic (FL), ML, neural networks (NNs), and evolutionary computation are all examples of soft computing methods that supplement traditional "hard" computing methods. Soft computing employs these techniques while attempting to resolve difficult issues [34].

Almost every problem may be remedied with the help of this toolkit. The following are some of the methods (as shown in Figure 6.2) used in soft computing:

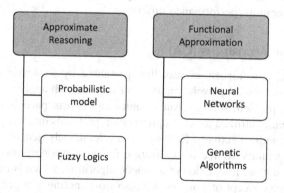

FIGURE 6.2 Elements of soft computing.

- Fuzzy Logic
- Neural Network
- Genetic Algorithms

6.1.6 Fuzzy Logic (FL)

To put it simply, fuzzy logic is just a form of mathematical reasoning that makes use of an unbounded and incomplete set of facts in order to find answers to problems. A plethora of certain inferences may be made with relative ease. In order to solve complex problems, fuzzy logic takes into account all of the information it has access to and attempts to generate the most likely correct answer. Fuzzy logics are often cited as the most effective problem-solvers [35–37].

6.1.7 Artificial Intelligence

With the advent of neural networks in the 1950s, soft computing has been able to be applied to problems that traditional computing methods have so far failed to solve. We're all well aware that a computer just can't match the human mind's skill in expressing the variety of situations that occur in the real world.

An artificial neural network (ANN) is a computer simulation that attempts to mimic the complex network of neurons in the human brain. As a result, it is possible to program a computer or machine to make decisions in the same manner that a human brain does. ANNs, which are conceptually similar to real brain networks in terms of their structure and operation, are created via the use of standard computer programming. It has cognitive abilities comparable to those of the human brain. Features of ANN models:

- Parallel Distributed processing of information.
- A high degree of interconnectivity between the most fundamental parts.
- Based on one's own experiences, connections can be altered.
- Unsupervised learning is a constant and ongoing process.
- Uses information from the surrounding area to guide their learning.
- With fewer units, performance suffers.

6.1.8 Genetic Algorithms (GA)

In 1965, the genetic algorithm was first presented by Professor John Holland. Problems are addressed by evolutionary algorithms, which apply the rules of natural selection to problem-solving. Ant colonies and swarm particle objective functions are frequently utilized to solve optimization problems involving maximization and minimization, respectively. Biology has a role, namely genetics and evolution. Nature is the primary source of motivation for genetic algorithms. Search-based algorithms are not the foundation of a genetic algorithm, which is based on natural selection and the concept of genetics. It's also worth noting that genetic algorithms belong to a larger category of computer science [38–44].

Application of AI-Based Learning in Automated Applications

6.1.9 Soft Computing vs. Hard Computing

In hard computing, preexisting mathematical methods are put to use to solve problems. The answer given is spot on. It's possible to find a hard computer challenge in every numerical problem.

On the other hand, soft computing takes a different approach. The ultimate aim of soft computing is to find viable solutions to current difficult problems. Additionally, soft computing does not always produce accurate results. It can be stated that they are vague and off-base.

Hard computing vs. soft computing is shown in Figure 6.3.

6.1.10 Applications of Soft Computing

The applications of soft computing are not limited to any one domain. Some of them are listed below.

1) **Home Appliances:** since we currently use part of this, this is a highly intriguing application to look at more. Artificial intelligence, machine learning, and fuzzy logic are making ordinary products like refrigerators, microwaves, and washing machines smarter. They can not only communicate with the users about their utilization, but they can also alter their settings in response to the current workload. Microwaves and rice cookers, for example. Refrigerators, air conditioners, and washing machines are among the most commonly used home appliances [45].

2) **Robotics:** Fuzzy logic and expert systems, two aspects of soft computing, are being used in this discipline, which is only getting started. It aids in the effective operation of industries, including production and inventory management. Robots with soft computing incorporated are already being used by some of the largest e-commerce organizations to assist in handling the significant load of items that pass through the warehouse on a regular basis. Besides these other uses, it is also employed in the field of robotics **Communication:** Communication necessitates a constantly shifting environment since the need for it might arise at any time, with little or no warning. An artificial neural network and fuzzy logic are used in soft computing to detect abrupt increases in demand and distribute resources

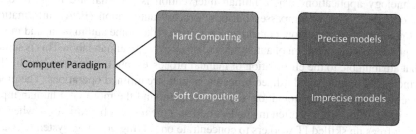

FIGURE 6.3 Elaborating the computing paradigm.

accordingly. Reduced bandwidth utilization not only saves money, but also frees up resources that may be used in other areas with greater demand.

3) **Transportation:** Soft computing is also being used extensively in the transportation industry, thanks to the proliferation of connected vehicles. Fuzzy logic and evolutionary computing are widely employed from the manufacturing of automobiles in the factory to on-the-road navigation, traffic prediction, troubleshooting, and diagnostics of the vehicle. Lifts employ similar strategies for managing several lifts with a single system.

4) Poker and Checker are two of the most popular games to use it.

5) Soft computing may also be used for image processing and data compression.

6) It is utilized for handwriting recognition.

6.1.11 Advantages of Soft Computing

Soft computing has a number of advantages. Here are just a few examples:

- Methodologies can tolerate a wide range of nuance and variation.
- It can handle the uncertainty that exists in real-world problems.
- More than that, it can both generate and recognize "linguistic variables."
- It is possible to obtain a close approximation of the solution.
- Data that cannot be easily categorized as statistical is no match for its problem-solving abilities.
- A spectrum of overlapping values may be used in place of an equation with strict limits.

6.1.12 Disadvantages of Soft Computing

The disadvantages of soft computing are:

- If a little error happens, the entire system will cease operating, and correcting the entire system from the beginning is a time-consuming procedure.
- It offers an estimated output value.

6.1.13 Automation

Technology applications where human intervention is minimal are referred to as "automation." This encompasses business process automation (BPA), information technology (IT) automation, consumer applications like home automation, and more. It's not always the intention of automation to do away with human labor. This is somewhat attributable to the elimination of manual processes, but the primary benefits lie in increased output, standardized procedures, and streamlined operations. The more effective you get with automation, the less frequent and the more vital human input becomes. Automation is often misunderstood as a means of job elimination when, in fact, it frees up skilled IT workers to concentrate on solving pressing, systemic issues rather than performing the same routine duties over and over again [46–52].

Application of AI-Based Learning in Automated Applications

6.1.14 Types of Automation

This section sheds light on various aspects and types of automation (shown in Figure 6.4).

i. **Basic automation:** With the use of automation, even the most fundamental of jobs may be made easier. Work is digitized when technology is used to streamline and centralize routine tasks, such as sharing messages through a single platform instead of storing data in isolated databases. Automation comes in two primary forms: Business Process Management (BPM) and Robotic process automation (RPA).

ii. **Process Automation:** Organizational processes may be simplified and standardized with the help of process automation software. Such tasks are often handled by specialized programs and enterprise applications. You may be able to boost productivity and efficiency in your business by implementing process automation. Furthermore, it can offer novel insights into pressing business concerns and suggest doable countermeasures. Included in this category of automation are techniques like process mining and automated workflows.

iii. **Automated integration:** Once the rules are specified by humans, machines might be programmed to copy human activities and repeat them. The so-called "digital worker" is one such example. Software robots or "digital workers" as they are more commonly known, have become a popular synonym for human employees as of late. These robots may be "hired" to do a variety of tasks in teams [53–56].

iv. **Automated by artificial intelligence:** Automation based on AI is the most difficult to implement. Artificial intelligence must be applied to robots so that they can "learn" and make decisions. In the realm of customer service, for instance, enabling virtual assistants may lead to cost savings, consumer and agent empowerment, and the greatest possible service experience.

6.1.15 Industrial Automation

Industrial automation is used by many different types of companies and industries since it streamlines processes using newer forms of technology. Every year, more

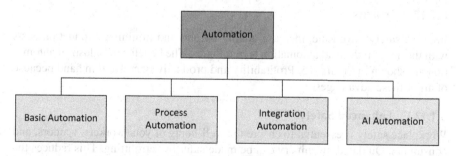

FIGURE 6.4 Types of automation.

and more routine tasks are being supplanted by industrial applications. This allows workers to perform routine tasks with more accuracy and fewer errors, freeing their time to focus on more strategic endeavors.

6.1.16 Industries Using Industrial Automation

Processes for industrial automation can assist a wide range of sectors. Manufacturing, oil and gas, paper mills, and steel mills are among the major ones. Listed below are a few instances of how these firms are putting it to use.

- Industrial automation is employed in many different capacities by manufacturing businesses. They may use it to assist with product assembly or production, keep tabs on routine maintenance tasks, or keep track of inventories.
- Offshore platforms and other geographically dispersed work sites make industrial automation all the more useful in the oil and gas industry. Workers no longer have to go on as many unpleasant and perhaps dangerous errands thanks to sensors and other tracking equipment.
- Its potential applications in the paper industry include batch production management, instrumentation, and plant device and equipment control. This provides the operators with an unobstructed view of the whole production procedure.
- Industrial automation in steel mills often uses a hierarchical approach. This innovative technology might be used for the whole management and control of the steel plant.
- Industrial automation in the aviation industry has included autopilot controls for passenger flights for decades. Industrial automation technology will soon be used in autonomous cars for consumer and business use.
- When production is finished and products are ready to be sent, distribution companies take control. The need for faster shipping is on the rise across industries. This is why the company will proceed with the implementation of distribution-related solutions to speed up the shipping and delivery of products [57].

6.1.17 Benefits

Improve safety, save time, increase quality output, and minimize production costs with the aid of industrial automation technologies. The benefits of industrial automation are shown in Figure 6.5. Profitability and productivity go hand in hand because of all of these advantages:

6.1.17.1 Enhanced Safety

Workplace safety is essential to ensure the well-being of your workers, vendors, and consumers. Outdated machinery can be made safer by retrofitting. This reduces the requirement for human intervention in machine processes.

Application of AI-Based Learning in Automated Applications

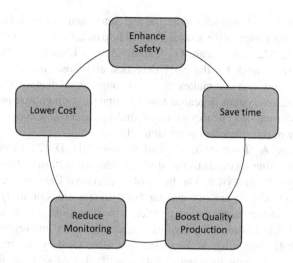

FIGURE 6.5 Benefits of industrial automation.

6.1.17.2 Save Time
Industrial automation technologies frequently boost the efficiency and dependability of production lines and other facility operations. This suggests that your company will be able to meet or exceed your clients' expectations more quickly.

6.1.17.3 Boost Quality Production
If you introduce flaws or quality concerns into the process, it won't matter how quickly you can generate the goods. It not only boosts manufacturing efficiency but also mitigates potential issues down the road. As a result, both the quality of your products and the number of errors that occur while they are being created will increase.

6.1.17.4 Reduce Monitoring
Industrial automation makes it simpler and less expensive to keep tabs on important assets and pieces of machinery. Technology can typically undertake 24-hour surveillance and data collecting at specified intervals without the need for human participation. That means maintenance staff can shift their attention to more challenging tasks, all while raising the bar for the quality of their output.

6.1.17.5 Lower Costs
Industrial automation that reduces breakdowns and repairs quickly will save money for your business. In addition, it often improves production and efficiency, letting you produce more while using fewer inputs.

6.2 LITERATURE REVIEW

After reading a large number of research papers on ML, soft computing, and industrial automation, it was decided to perform a study on "Integration of Machine Learning and Soft Computing Techniques for Industrial and Automated Applications." As a starting point, here is a brief review of the articles that will help us achieve our study objectives.

Abdel-Basset, M., Manogaran, G., Gamal, A. and Chang, V. (2020) proposed an innovative framework for the early detection and close monitoring of patients with type-2 diabetes, one that makes use of computer-aided diagnostics and IoT. The proposed healthcare system is geared toward improving diagnostic precision using inexplicable information. The experimental findings as a whole support the reliability and efficacy of our suggested algorithms [1].

Charitopoulos, A., Rangoussi, M., and Koulouriotis, D. (2020) introduced the use of soft computing methods to the study of educational Data Mining (DM) and Deep Learning (DL) analytics. The fields of Educational Data Mining (EDM) and Learning Analytics (LA) are two similar but different areas of study that use this information to answer unanswered problems in education. Along with "conventional" methods like clustering, classification, identification, and regression/analysis of variances, soft computing (SC) methods are often used by researchers in EDM and LA. Due to their repetitive optimization methods that seek possible bad solutions but with acceptable time and effort, and their heavy reliance on huge data sets for training, soft computing systems are particularly well-suited for EDM and LA problems. With enough information, soft computing techniques like Decision Tree (DT), Random Forests (RFs), ANN, FL, Support Vector Machines (SVMs), and genetic/evolutionary algorithms, can successfully address the uncertainty, qualitatively stated problems, and incomplete, imprecise, and even contradictory data sets that are characteristic of the field of education [2].

Candanedo, I., Hernández, E., Rodríguez, S., Santos, T., and González, A. (2018) looked at the vast quantities of data created by sensor networks in an Industry 4.0 context that need the use of machine learning and data analysis techniques. Because of this, organizations now have to deal with new possibilities and difficulties, such as predictive analysis. This approach uses computational tools to find patterns in data by using the same criteria that may be used to create predictions. Heating, ventilation, and air conditioning (HVAC) systems were utilized in a broad variety of fields because they allow for the regulation of an indoor environment's climate, air temperature, humidity, and pressure. Here, we offer a case study where an HVAC data set was used to assess the performance of the system in question, with respect to providing a pleasant interior climate. The goal of this research was to create prediction models that may be used in an Industry 4.0 scenario by using machine learning methods in the aforementioned data set [3].

Cavalcanti, D., Perez-Ramirez, J., Rashid, M. M., Fang, J., Galeev, M., and Stanton, K. B. (2019) distributed precise time and enabling real-time wireless industrial automation of the future. This article gives a brief summary of the possible uses, prerequisites, and special research problems involved in bringing time sensitive networking (TSN) capabilities over wireless. In addition, the article describes the developments in wireless technology that have allowed for more reliable and exact time distribution and timing capabilities. Additionally, a categorization of wireless applications and a reference design are provided to aid in the seamless integration of wired and wireless TSN capabilities in future industrial automation systems [4].

Dineva, A., Mosavi, A., Ardabili, S. F., Vajda, I., Shamshirband, S., Rabczuk, T., and Chau, K. W. (2019) investigated how soft computing models may be used in the design, construction, and administration of electrical rotating machines. Electromechanical energy converters, like spinning electrical machines, play a critical role in the generation and transformation of energy. New developments in the control and high-performance design of such devices are of importance to the field of energy management. Soft computing technologies are becoming recognized as crucial tools for the management and design of electrical spinning equipment. From this vantage point, it is clear that the advancement of soft computing techniques used in rotating electrical machines was critical for a wide range of energy conversion devices, such as generators, high-performance electric engines, and electric automobiles. Learn how the advancement of this vital field of energy research has been profoundly affected by the implementation of state-of-the-art soft computing approaches. Using a new taxonomy of systems and applications, the most recent advances in the control and design of spinning electrical machines were analyzed [5].

Expósito-Izquierdo, C., Melián-Batista, B., and Moreno-Vega, J. M. (2018) reviewed how soft computing is being used in shipping and related industries. There was a severe shortage of effective methods in this setting for achieving accurate solutions to the many optimization issues that have recently appeared in this area and are deemed difficult from the standpoint of complexity theory. These optimization issues call for novel computational methods that may provide imprecise solutions by heavily using uncertainty and partial truth in order to be manageable. In this article, they outline the most important soft computing approaches now in use in marine logistics and related domains, and they point out where there was room for additional research and development [6].

Fernandes, A., Gomes, V., and Melo-Pinto, P. (2018) provided the emerging applications of local reflectance and interactance spectroscopy in the field of viticulture by way of soft computing and multivariate analysis. There are two emerging applications of spectroscopy and soft computing in the viticulture industry: the multiclass nature of variety and clone recognition in plants, and the difficulty of assessing enological characteristics in samples of small numbers of grape berries in order to determine ripeness. Results from a wide range of research in various fields will be summarized and discussed. A brief introduction to spectroscopy and its applications is followed by a synopsis of the relevant scientific literature, with research categorized by the mean number of berries per sample and the total number of samples utilized. Equal attention will be paid to the issue of using data from many years, regions, and varieties in a single model. Techniques for validating algorithms and comparing them are given particular focus. Some recommendations were made to help make future comparisons between published findings easier to make [7].

Fan, S. K. S., Su, C. J., Nien, H. T., Tsai, P. F., and Cheng, C. Y. (2018) introduced a combination of historical and real-time data from Taiwan's electronic toll collection which is being utilized with machine learning and big data approaches to get accurate travel time estimates. We construct a machine learning solution within a big data analytics platform, utilizing the random forests technique and Apache Hadoop to predict travel times on Taiwanese highways using electronic toll collection data. When combined with current data, these historical records may be used to create a variety of prediction models for highway trip time, which were then made available to drivers as estimates and corrections to actual times [8].

Garg, H. (2017) presented an article describing a hybridized method, based on the concepts of soft computing, for assessing the effectiveness of industrial systems. The system analyst was never given sufficient data to make informed predictions or make the necessary improvements to the system's performance. Because of this, reducing decision-makers' anxiety and improving their ability to make sound, timely judgments were top priorities. To address these issues, the authors herein used fuzzy numbers to quantify data uncertainties, which are then applied to the many dependability characteristics of the industrial system that are reflective of the system's behavior. Creating and solving a nonlinear optimization model might help you get the parameter membership functions. It was determined via the study that the generated findings had a narrower margin of error compared to the status quo and standard outcomes. Most of the system's vital components have been tested for robustness and responsiveness [9].

Gauerhof, L., Munk, P., and Burton, S. (2018) organized the purposes of testing and validating machine learning systems for autonomous vehicles. A pedestrian detection function is presented in this work, along with a safety assurance case for it. This function is fundamentally important to the safe operation of autonomous vehicles. Our arguments against under-specification, semantic gap, and logical gap are presented using Generative Stochastic Network (GSN) to make our case for safety guarantee [10].

Garg, S., Guizani, M., Guo, S., and Verikoukis, C. (2020) provided an editorial special section on the use of big data and AI to further the goal of 5G in industrial automation. Recent advances in information and communication technologies have been credited with fueling a rapid expansion of industrial automation. Foreshadowing the fourth industrial revolution, "Industry 4.0" is characterized by this paradigm change. Industry 4.0, also known as Industrial Internet of Things (IIoT) or "SF," represents a new industrial revolution in which machines and other devices are not only networked together but also share data, conduct research, and act on their findings to optimize production across the board. As IoT becomes more pervasive, automation follows, allowing machines to independently acquire knowledge, modify their behavior in response to it, and share that knowledge with one another [11].

Gehrmann, C., and Gunnarsson, M. (2020) focused on security architecture for industrial control and automation systems that used digital twins. To facilitate the exchange of data and the management of mission-critical operations, they explored the security architecture that might accompany a digital twin replication model. They defined security criteria for data exchange and control through digital twins, with consideration given to how these requirements should inform the design process. They give a high-level design and evaluation of additional security components of the architecture and show that the proposed state synchronization scheme fulfills the expected digital twin synchronization requirements. Additionally, they evaluate the efficiency of a proof of concept for safe software upgrades by using the digital twin method. Our newly developed security framework will pave the way for more research in this dynamic area [12].

6.3 PROPOSED WORK

6.3.1 PROBLEM STATEMENT

There have been several instances when research attempts have been conducted in the realm of soft computing services. Machine Language and Automation are two areas that often use the services provided by industrial companies. The research that is now accessible, however, suffers from both poor performance and inaccuracy, making it a problematic choice. A previous piece of research neglected to take into account the importance of optimization coming before the testing and training processes. As a result, it is necessary to suggest a method that is ideal in order to get more precision while preserving a high level of performance.

6.3.2 PROPOSED METHODOLOGY

The proposed research endeavor takes into account the factors that influence the efficiency and precision of ongoing studies in the disciplines of machine learning, soft computing, and industrial and automated applications. Even if there are a few different machine learning algorithms that might be utilized for categorization, it is still important to add an optimization method. Researchers are contemplating using a kind of computing known as soft computing in order to enhance the data set in preparation for the training phase. The increased accuracy of categorization in automated and industrial settings has the objective of integrating a soft computing approach with machine learning as a means to achieve this aim. In conclusion, the recommended job will be evaluated in terms of its performance as well as its accuracy in comparison to the conventional approach. The proposed methodology is shown in Figure 6.6.

6.3.3 PROCESS FLOW OF NOVEL APPROACH

Figure 6.7 depicts the flow of the one-of-a-kind method's process, which includes training and testing on both the optimized and non-optimized data sets respectively. Minimum viable option (MVO) is generally considered to be the optimal strategy for use in scientific research. MVO makes use of a particle swarm optimization process, which searches for the best possible answer, in order to filter the data. After the data set has been filtered, it is next put through training so that it may be tested.

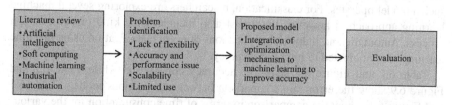

FIGURE 6.6 Research methodology.

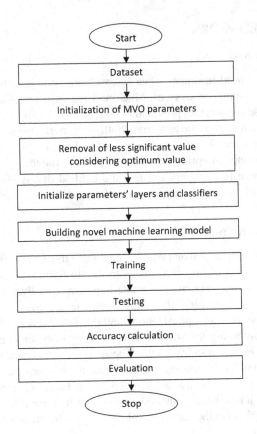

FIGURE 6.7 Flow chart of novel approach.

Processing the confusion matrix that is created as a result of such classification allows for the acquisition of accuracy metrics. After that, the accuracy characteristics of the two models are compared to one another in order to determine which model generates more accurate results.

6.4 RESULTS AND DISCUSSION

The performance and accuracy of an industrial and automated model are dependent on a number of factors that must be taken into account while trying to grasp how such a model operates. For classification, researchers are exploring several machine learning approaches; however, they must also include some kind of optimization process. Automatic and industrial applications of event classification using machine learning are anticipated to improve accuracy.

Table 6.1 and Figure 6.8 show the accuracy comparison while Table 6.2 and Figure 6.9 show the error rate comparison between the various models. Table 6.3 and Figure 6.10 show the comparison in terms of time consumption for the various models. An MVO based model is found to be best in all the respects.

TABLE 6.1
Comparison of Accuracy

Class	Non-optimized	PSO based model	MVO based model
1	92%	92%	93.5%
2	92%	94%	94.5%
3	91%	94%	95.5%
4	92%	94%	95.5%
5	91%	93%	94.5%
6	92%	93%	95.5%

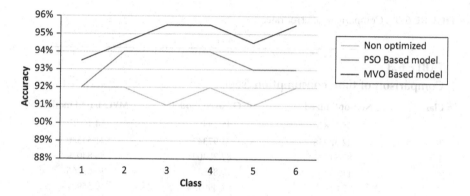

FIGURE 6.8 Comparison of accuracy.

TABLE 6.2
Comparison of Error Rate

Class	Non-optimized	PSO based model	MVO based model
1	8%	8%	6.5%
2	8%	6%	5.5%
3	9%	6%	4.5%
4	8%	6%	4.5%
5	9%	7%	5.5%
6	8%	7%	4.5%

6.5 CONCLUSION

The simulation results show that soft computing is crucial in choosing the right data set. When compared to traditional procedures, the precision provided by the optimized data set is much higher. In addition, the time required to complete the training operation has been reduced thanks to the work given here. In comparison to

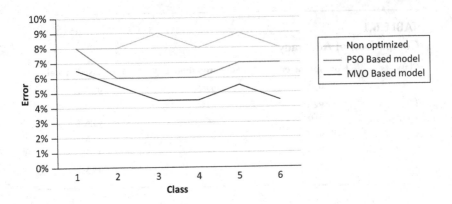

FIGURE 6.9 Comparison of error rate.

TABLE 6.3
Comparison of time consumption (sec)

Class	Non-optimized	PSO based model	MVO based model
1	45.75826	45.134	44.19417
2	42.38956	41.97342	41.54389
3	87.71746	87.28402	86.84029
4	11.02696	10.82341	10.69866
5	2.018499	1.399124	0.559614
6	89.11214	88.92517	88.71876

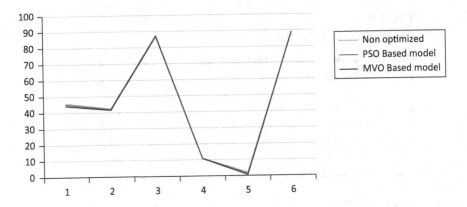

FIGURE 6.10 Comparison of time consumption.

traditional non-optimized and PSO based research endeavors, MVO has been shown to perform better.

6.6 FUTURE SCOPE

To put it another way, soft computing is a collection of computational methods that take cues from both AI and natural selection. It offers reasonably priced and practical answers to complicated problems that arise in the real world and for which there is no straightforward algorithmic answer. Soft computing is a branch of both mathematics and computer science that has existed since the early 1990s. This means that it has been around longer than traditional computing. The capacity of the human brain to construct solutions that are plausible and provide an approximation of reality served as the driving force for the creation of this endeavor [58–60]. The field of soft computing is built on the idea that approximations are the best way to handle difficult computational problems. The process by which automated systems may learn and improve themselves with little to no input from humans or specialized programming is referred to as "machine learning," and the phrase "machine learning" is used to characterize this process (ML). Data is given to machines, and they are taught to search for patterns using the data; the patterns they find are then utilized as templates for future learning. They are given the freedom to select their own activities and to make any required adjustments to the routines that they follow.

REFERENCES

1. Abdel-Basset, M., Manogaran, G., Gamal, A., and Chang, V. (2020). A Novel Intelligent Medical Decision Support Model Based on Soft Computing and IoT. *IEEE Internet of Things Journal, 7*(5), 4160–4170. https://doi.org/10.1109/JIOT.2019.2931647.
2. Charitopoulos, A., Rangoussi, M., and Koulouriotis, D. (2020). On the Use of Soft Computing Methods in Educational Data Mining and Learning Analytics Research: A Review of Years 2010–2018. *International Journal of Artificial Intelligence in Education, 30*(3), 371–430. https://doi.org/10.1007/s40593-020-00200-8.
3. Candanedo, I., Hernández, E., Rodríguez, S., Santos, T., and González, A. (2018). *Machine Learning Predictive Model* (Vol. 1, pp. 501–510). Springer. https://doi.org/10.1007/978-3-319-95204-8.
4. Cavalcanti, D., Perez-Ramirez, J., Rashid, M. M., Fang, J., Galeev, M., and Stanton, K. B. (2019). Extending Accurate Time Distribution and Timeliness Capabilities over the Air to Enable Future Wireless Industrial Automation Systems. *Proceedings of the IEEE, 107*(6), 1132–1152. https://doi.org/10.1109/JPROC.2019.2903414.
5. Dineva, A., Mosavi, A., Ardabili, S. F., Vajda, I., Shamshirband, S., Rabczuk, T., and Chau, K. W. (2019). Review of Soft Computing Models in Design and Control of Rotating Electrical Machines. *Energies, 12*(6). https://doi.org/10.3390/en12061049.
6. Expósito-Izquierdo, C., Melián-Batista, B., and Moreno-Vega, J. M. (2018). A Review of Soft Computing Techniques in Maritime Logistics and Its Related Fields. *Studies in Fuzziness and Soft Computing, 360*, 1–23. https://doi.org/10.1007/978-3-319-64286-4_1.
7. Fernandes, A., Gomes, V., and Melo-Pinto, P. (2018). A Review of the Application to Emergent Subfields in Viticulture of Local Reflectance and Interactance Spectroscopy Combined with Soft Computing and Multivariate Analysis. *Studies in Fuzziness and Soft Computing, 358*, 87–115. https://doi.org/10.1007/978-3-319-62359-7_5.

8. Fan, S. K. S., Su, C. J., Nien, H. T., Tsai, P. F., and Cheng, C. Y. (2018). Using Machine Learning and Big Data Approaches to Predict Travel Time Based on Historical and Real-Time Data from Taiwan Electronic Toll Collection. *Soft Computing*, 22(17), 5707–5718. https://doi.org/10.1007/s00500-017-2610-y.
9. Garg, H. (2017). Performance Analysis of an Industrial System Using Soft Computing Based Hybridized Technique. *Journal of the Brazilian Society of Mechanical Sciences and Engineering*, 39(4), 1441–1451. https://doi.org/10.1007/s40430-016-0552-4.
10. Gauerhof, L., Munk, P., and Burton, S. (2018). Structuring Validation Targets of a Machine Learning Function Applied to Automated Driving. In *Lecture Notes in Computer Science (Including Subseries Lecture Notes in Artificial Intelligence and Lecture Notes in Bioinformatics): Vol. 11093 LNCS*. Springer International Publishing. https://doi.org/10.1007/978-3-319-99130-6_4.
11. Garg, S., Guizani, M., Guo, S., and Verikoukis, C. (2020). Guest Editorial Special Section on AI-Driven Developments in 5G-Envisioned Industrial Automation: Big Data Perspective. *IEEE Transactions on Industrial Informatics*, 16(2), 1291–1295. https://doi.org/10.1109/TII.2019.2955963.
12. Gehrmann, C., and Gunnarsson, M. (2020). A Digital Twin Based Industrial Automation and Control System Security Architecture. *IEEE Transactions on Industrial Informatics*, 16(1), 669–680. https://doi.org/10.1109/TII.2019.2938885.
13. Juneja, S., Juneja, A., and Anand, R. (2019, April). Reliability Modeling for Embedded System Environment Compared to Available Software Reliability Growth Models. In *2019 International Conference on Automation, Computational and Technology Management (ICACTM)* (pp. 379–382). IEEE.
14. Meelu, R., and Anand, R. (2011). Performance Evaluation of Cluster-Based Routing Protocols Used in Heterogeneous Wireless Sensor Networks. *International Journal of Information Technology and Knowledge Management*, 4(1), 227–231.
15. Kim, D. H., Kim, T. J. Y., Wang, X., Kim, M., Quan, Y. J., Oh, J. W., Min, S. H., Kim, H., Bhandari, B., Yang, I., and Ahn, S. H. (2018). Smart Machining Process Using Machine Learning: A Review and Perspective on Machining Industry. *International Journal of Precision Engineering and Manufacturing - Green Technology*, 5(4), 555–568. https://doi.org/10.1007/s40684-018-0057-y.
16. Kero, N., Puhm, A., Kernen, T., and Mroczkowski, A. (2019). Performance and Reliability Aspects of Clock Synchronization Techniques for Industrial Automation. *Proceedings of the IEEE*, 107(6), 1011–1026. https://doi.org/10.1109/JPROC.2019.2915972.
17. Khairudin, A. R. M., Abu-Samah, A., Aziz, N. A. S., Azlan, M. A. F. M., Karim, M. H. A., and Zian, N. M. (2019). Design of Portable Industrial Automation Education Training Kit Compatible for IR 4.0. In *Proceeding - 2019 IEEE 7th Conference on Systems, Process and Control, ICSPC 2019, December* (pp. 38–42). https://doi.org/10.1109/ICSPC47137.2019.9068090.
18. Singh, H., Ramya, D., Saravanakumar, R., Sateesh, N., Anand, R., Singh, S., and Neelakandan, S. (2022). Artificial Intelligence Based Quality of Transmission Predictive Model for Cognitive Optical Networks. *Optik*, 257, 168789.
19. Pandey, B. K., Pandey, D., Wariya, S., Aggarwal, G., and Rastogi, R. (2021). Deep Learning and Particle Swarm Optimisation-Based Techniques for Visually Impaired Humans' Text Recognition and Identification. *Augmented Human Research*, 6, 1–14.
20. La Fe-Perdomo, I., Beruvides, G., Quiza, R., Haber, R., and Rivas, M. (2019). Automatic Selection of Optimal Parameters Based on Simple Soft-Computing Methods: A Case Study of Micromilling Processes. *IEEE Transactions on Industrial Informatics*, 15(2), 800–811. https://doi.org/10.1109/TII.2018.2816971.

21. Li, P. (2019). Introductory Chapter: New Trends in Industrial Automation. *New Trends in Industrial Automation*, 1–4. https://doi.org/10.5772/intechopen.84772.
22. Pandey, B. K., Pandey, D., and Agarwal, A. (2022). Encrypted Information Transmission by Enhanced Steganography and Image Transformation. *International Journal of Distributed Artificial Intelligence (IJDAI)*, 14(1), 1–14.
23. Maya, B. S., and Asha, T. (2018). Automatic Detection of Brain Strokes in CT Images Using Soft Computing Techniques. *Lecture Notes in Computational Vision and Biomechanics*, 25, 85–109. https://doi.org/10.1007/978-3-319-61316-1_5.
24. Mahmoud, M. S. (2019). Architecture for Cloud-Based Industrial Automation. In *Advances in Intelligent Systems and Computing*. Springer, 797. https://doi.org/10.1007/978-981-13-1165-9_6.
25. Mistry, I., Tanwar, S., Tyagi, S., and Kumar, N. (2020). Blockchain for 5G-Enabled IoT for Industrial Automation: A Systematic Review, Solutions, and Challenges. *Mechanical Systems and Signal Processing*, 135, 106382. https://doi.org/10.1016/j.ymssp.2019.106382.
26. Kumar, M. S., Sankar, S., Nassa, V. K., Pandey, D., Pandey, B. K., and Enbeyle, W. (2021). Innovation and Creativity for Data Mining Using Computational Statistics. In *Methodologies and Applications of Computational Statistics for Machine Intelligence* (pp. 223–240). IGI Global.
27. Ortiz, J. S., Sánchez, J. S., Velasco, P. M., Quevedo, W. X., Carvajal, C. P., Morales, V., Ayala, P. X., and Andaluz, V. H. (2018). Virtual Training for Industrial Automation Processes Through Pneumatic Controls. In *Lecture Notes in Computer Science (Including Subseries Lecture Notes in Artificial Intelligence and Lecture Notes in Bioinformatics)*, 10851 LNCS (pp. 516–532). https://doi.org/10.1007/978-3-319-95282-6_37.
28. Osaba, E., Del Ser, J., Iglesias, A., and Yang, X. S. (2020). Soft Computing for Swarm Robotics: New Trends and Applications. *Journal of Computational Science*, 39, 101049. https://doi.org/10.1016/j.jocs.2019.101049.
29. Penumuru, D. P., Muthuswamy, S., and Karumbu, P. (2020). Identification and Classification of Materials Using Machine Vision and Machine Learning in the Context of Industry 4.0. *Journal of Intelligent Manufacturing*, 31(5), 1229–1241. https://doi.org/10.1007/s10845-019-01508-6.
30. Pandey, D., Wairya, S., Sharma, M., Gupta, A. K., Kakkar, R., and Pandey, B. K. (2022). An Approach for Object Tracking, Categorization, and Autopilot Guidance for Passive Homing Missiles. *Aerospace Systems*, 1–14.
31. Rath, M., and Pattanayak, B. K. (2018). SCICS: A Soft Computing Based Intelligent Communication System in VANET. *Communications in Computer and Information Science*, 808, 255–261. https://doi.org/10.1007/978-981-10-7635-0_19.
32. Anand, R., Singh, B., and Sindhwani, N. (2009). Speech Perception and Analysis of Fluent Digits' Strings Using Level-by-Level Time Alignment. *International Journal of Information Technology and Knowledge Management*, 2(1), 65–68.
33. Rogers, W. P., Kahraman, M. M., Drews, F. A., Powell, K., Haight, J. M., Wang, Y., Baxla, K., and Sobalkar, M. (2019). Automation in the Mining Industry: Review of Technology, Systems, Human Factors, and Political Risk. *Mining, Metallurgy and Exploration*, 36(4), 607–631. https://doi.org/10.1007/s42461-019-0094-2.
34. Saini, M. K., Nagal, R., Tripathi, S., Sindhwani, N., and Rudra, A. (2008). PC Interfaced Wireless Robotic Moving Arm. In *AICTE Sponsored National Seminar on Emerging Trends in Software Engineering* (Vol. 50, pp. 787–788).
35. Rojas, A. M., and Barbieri, G. (2019). A Low-Cost and Scaled Automation System for Education in Industrial Automation. In *IEEE International Conference on Emerging Technologies and Factory Automation, ETFA*, 2019-September, 439–444. https://doi.org/10.1109/ETFA.2019.8869535.

36. Seyedzadeh, S., Rahimian, F. P., Glesk, I., and Roper, M. (2018). Machine Learning for Estimation of Building Energy Consumption and Performance: A Review. *Visualization in Engineering, 6*(1). https://doi.org/10.1186/s40327-018-0064-7.
37. Sinha, R., Patil, S., Gomes, L., and Vyatkin, V. (2019). A Survey of Static Formal Methods for Building Dependable Industrial Automation Systems. *IEEE Transactions on Industrial Informatics, 15*(7), 3772–3783. https://doi.org/10.1109/TII.2019.2908665.
38. Su, W., Liu, Y., Du, Y., Dong, Y., Pan, M., and Xu, G. (2019). Three-Real-Time Architecture of Industrial Automation Based on Edge Computing. *Proceedings - 2019 IEEE International Conference on Smart Internet of Things, SmartIoT 2019* (pp. 372–377). https://doi.org/10.1109/SmartIoT.2019.00065.
39. Sindhwani, N., and Singh, M. (2020). A Joint Optimization Based Sub-band Expediency Scheduling Technique for MIMO Communication System. *Wireless Personal Communications, 115*(3), 2437–2455.
40. Theissler, A., Pérez-Velázquez, J., Kettelgerdes, M., and Elger, G. (2021). Predictive Maintenance Enabled by Machine Learning: Use Cases and Challenges in the Automotive Industry. *Reliability Engineering and System Safety, 215*, 107864. https://doi.org/10.1016/j.ress.2021.107864.
41. Sindhwani, N., and Singh, M. (2016). FFOAS: Antenna Selection for MIMO Wireless Communication System Using Firefly Optimisation Algorithm and Scheduling. *International Journal of Wireless and Mobile Computing, 10*(1), 48–55.
42. Gupta, R. S., Nassa, V. K., Bansal, R., Sharma, P., and Koti, K. (2021). Investigating Application and Challenges of Big Data Analytics with Clustering. In *2021 International Conference on Advancements in Electrical, Electronics, Communication, Computing and Automation (ICAECA)* (pp. 1–6). https://doi.org/10.1109/ICAECA52838.2021.9675483.
43. Veeraiah, V., Khan, H., Kumar, A., Ahamad, S., Mahajan, A., and Gupta, A. (2022). Integration of PSO and Deep Learning for Trend Analysis of Meta-Verse. In *2022 2nd International Conference on Advance Computing and Innovative Technologies in Engineering (ICACITE)* (pp. 713–718). https://doi.org/10.1109/ICACITE53722.2022.9823883.
44. Pandey, B. K., Pandey, D., Nassa, V. K., George, S., Aremu, B., Dadeech, P., and Gupta, A. (2023). Effective and Secure Transmission of Health Information Using Advanced Morphological Component Analysis and Image Hiding. In: Gupta, M., Ghatak, S., Gupta, A., Mukherjee, A. L. (eds) *Artificial Intelligence on Medical Data. Lecture Notes in Computational Vision and Biomechanics, 37*. Springer. https://doi.org/10.1007/978-981-19-0151-5_19.
45. Veeraiah, V., Kumar, K. R., LalithaKumari, P., Ahamad, S., Bansal, R., and Gupta, A. (2022). Application of Biometric System to Enhance the Security in Virtual World. In *2022 2nd International Conference on Advance Computing and Innovative Technologies in Engineering (ICACITE)* (pp. 719–723). https://doi.org/10.1109/ICACITE53722.2022.9823850.
46. Bansal, R., Gupta, A., Singh, R., and Nassa, V. K. (2021). Role and Impact of Digital Technologies in E-learning amidst COVID-19 Pandemic. In *2021 Fourth International Conference on Computational Intelligence and Communication Technologies (CCICT)* (pp. 194–202). https://doi.org/10.1109/CCICT53244.2021.00046.
47. Shukla, A., Ahamad, S., Rao, G. N., Al-Asadi, A. J., Gupta, A., and Kumbhkar, M. (2021). Artificial Intelligence Assisted IoT Data Intrusion Detection. In *2021 4th International Conference on Computing and Communications Technologies (ICCCT)* (pp. 330–335). https://doi.org/10.1109/ICCCT53315.2021.9711795.
48. Pathania, V., Babu, S. Z. D., Ahamad, S., Thilakavathy, P., Gupta, A., Alazzam, M. B., and Pandey, D. (2023). A Database Application of Monitoring COVID-19 in India.

In: Gupta, M., Ghatak, S., Gupta, A., Mukherjee, A. L. (eds) *Artificial Intelligence on Medical Data. Lecture Notes in Computational Vision and Biomechanics, 37.* Springer. https://doi.org/10.1007/978-981-19-0151-5_23.
49. Dushyant, K., Muskan, G., Annu, Gupta, A., and Pramanik, S. (2022). Utilizing Machine Learning and Deep Learning in Cybesecurity: An Innovative Approach. In *Cyber Security and Digital Forensics: Challenges and Future Trends* (pp. 271–293). Wiley. https://doi.org/10.1002/9781119795667.ch12.
50. Babu, S. Z. D., Pandey, D., Naidu, G. T., Sumathi, S., Gupta, A., Bader Alazzam, M., and Pandey, B. K. (2023). Analysation of Big Data in Smart Healthcare. In: Gupta, M., Ghatak, S., Gupta, A., Mukherjee, A. L. (eds) *Artificial Intelligence on Medical Data. Lecture Notes in Computational Vision and Biomechanics, 37.* Springer. https://doi.org/10.1007/978-981-19-0151-5_21.
51. Bansal, B., NishaJenipher, V., Jain, R., Dilip, R., Kumbhkar, M., Pramanik, S., Roy, S., and Gupta, A. (2022). Big Data Architecture for Network Security. In *Cyber Security and Network Security* (pp. 233–267). Wiley. https://doi.org/10.1002/9781119812555.ch11.
52. Gupta, A., Kaushik, D., Garg, M., and Verma, A. (2020). Machine Learning Model for Breast Cancer Prediction. In *2020 Fourth International Conference on I-SMAC (IoT in Social, Mobile, Analytics and Cloud) (I-SMAC)* (pp. 472–477). https://doi.org/10.1109/I-SMAC49090.2020.9243323.
53. Chauhan, S. K., Khanna, P., Sindhwani, N., Saxena, K., and Anand, R. (2023). Pareto Optimal Solution for Fully Fuzzy Bi-criteria Multi-Index Bulk Transportation Problem. In *Mobile Radio Communications and 5G Networks. Proceedings of Third MRCN 2022* (pp. 457–470). Springer Nature.
54. Veeraiah, V., Rajaboina, N. B., Rao, G. N., Ahamad, S., Gupta, A., and Suri, C. S. (2022). Securing Online Web Application for IoT Management. In *2022 2nd International Conference on Advance Computing and Innovative Technologies in Engineering (ICACITE)* (pp. 1499–1504). https://doi.org/10.1109/ICACITE53722.2022.9823733.
55. Veeraiah, V., Gangavathi, P., Ahamad, S., Talukdar, S. B., Gupta, A., and Talukdar, V. (2022). Enhancement of Meta Verse Capabilities by IoT Integration. In *2022 2nd International Conference on Advance Computing and Innovative Technologies in Engineering (ICACITE)* (pp. 1493–1498). https://doi.org/10.1109/ICACITE53722.2022.9823766.
56. Gupta, N., Janani, S., Dilip, R., Hosur, R., Chaturvedi, A., and Gupta, A. (2022). Wearable Sensors for Evaluation Over Smart Home Using Sequential Minimization Optimization-Based Random Forest. *International Journal of Communication Networks and Information Security (IJCNIS), 14*(2), 179–188. https://doi.org/10.17762/ijcnis.v14i2.5499.
57. Keserwani, H., Rastogi, H., Kurniullah, A. Z., Janardan, S. K., Raman, R., Rathod, V. M., and Gupta, A. (2022). Security Enhancement by Identifying Attacks Using Machine Learning for 5G Network. *International Journal of Communication Networks and Information Security (IJCNIS), 14*(2), 124–141. https://doi.org/10.17762/ijcnis.v14i2.5494.
58. Pandey, D., Pandey, B. K., Sindhwani, N., Anand, R., Nassa, V. K., and Dadheech, P. (2022). An Interdisciplinary Approach in the Post-COVID-19 Pandemic Era. In *An Interdisciplinary Approach in the Post-COVID-19 Pandemic Era* (pp. 1–290).
59. Kohli, L., Saurabh, M., Bhatia, I., Shekhawat, U. S., Vijh, M., and Sindhwani, N. (2021). Design and Development of Modular and Multifunctional UAV with Amphibious Landing Module. In *Data Driven Approach Towards Disruptive Technologies: Proceedings of MIDAS 2020* (pp. 405–421). Springer.
60. Sharma, G., Nehra, N., Dahiya, A., Sindhwani, N., and Singh, P. (2022). Automatic Heart-Rate Measurement Using Facial Video. In *Networking Technologies in Smart Healthcare* (pp. 289–307). CRC Press.

7 Investigating Scope and Applications for the Internet of Robotics in Industrial Automation

Harinder Singh, Vivek Veeraiah, Huma Khan, Dharmendra Kumar Singh, Veera Talukdar, Rohit Anand, and Nidhi Sindhwani

7.1 INTRODUCTION

The robotics community has been inspired to create action and autonomous behavior, while the IoT community has been pushed to provide information services for pervasive sensing, tracking, and monitoring. It is for this reason that the concept of an IoRT, or internet of robotic things, has been gaining traction, since it could potentially consolidate the best practices of both of these communities. There are already signs of the IoT-robotics convergence, such as the rise of distributed, heterogeneous robot control paradigms like network robot systems and robot ecologies, and the rise of techniques like ubiquitous robotics and cloud robotics, which offload resource-intensive aspects to the server side. The term "Internet of Robotic Things (IoRT)" was coined by ABI researchers in a white paper to describe a system in which information gathered by various sensors is pooled, analyzed using centralized and distributed processing capacity, and then utilized to modify physical objects [1–4]. To enhance robots' situational awareness and, by extension, their capacity to perform their assigned jobs, this cyber-physical perspective on the IoRT employs IoT sensor and data analytics technology. It might be used for a wide variety of things, including companion robots and intelligent transportation systems. The term "IoRT" was later used in literature, with varying interpretations including "robot-aided IoT," where robots are just additional sensors, and "strong team communication" [5–7].

7.1.1 INTERNET OF THINGS (IoT)

The IoT is a system of interlinked, remotely controlled gadgets and other physical objects. Hardware, software, sensors, and a network all work together to make up this system's integrated multi-part whole. This facilitates interaction between the devices and the sharing of information between them [8, 9]. IoT makes it possible to remotely detect and control things using an organization's already built network infrastructure, resulting in easier integration of the physical and digital worlds and increased efficiency, accuracy, and financial benefit [10, 11].

The fundamental components of IoT are shown in Figure 7.1. Connected devices, gateways, analytics, user interfaces, and the cloud are all essential parts of the IoT. The IT sector is entering a new era with the advent of IoT. Data processing extends well beyond the computer server rack, personal computers, and personal digital assistants. It has now spread to every nearby item. Most of the things we use on a daily basis are either linked to the internet or will be soon. They may identify and share a lot of valuable information with the help of smart applications they may utilize. Media, information, and smartphone applications have become more important to the smooth functioning of our increasingly complex lives. In the end, they'll work like a net, but one that's been upgraded thanks to technology [12, 13].

Network and cognitive layers form two tiers of IoT devices as shown in Figure 7.2. The cloud/server takes care of the application layer, the router/gateway of the network layer, and the sensors/actuators of the perception layer.

IoT data processing is unlike any other kind of data processing. In the IoT, data transmission speeds may be quite high, yet information can frequently be very little. In reality, IoT is comprised of a vast network of individual gadgets and nodes. We threw them into the pot as well. Emerging IoT goes under many different labels. The idea of a "related item" has likely been around for as long as the internet or letters itself.

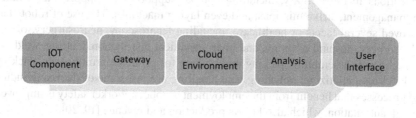

FIGURE 7.1 Components of IoT.

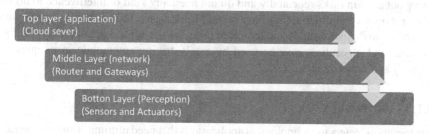

FIGURE 7.2 Layers in IOT devices.

Personal computers (and advanced mobile phones) are two possible early instances of such things. Yet, since they still need human input, they will be seen as different commodities. Telemetry and other early IoT applications were utilized to link together advanced pieces of equipment, infrastructure, and even whole companies.

7.1.2 Robotics

Robo-science is a branch of AI concerned with the study, development, and use of robots and other intelligent devices that can function independently. Robotics crosses traditional academic boundaries since robots are hybrid mechanical/electrical/programmable devices. Despite their differences, robotics and AI are sometimes treated as if they were one and the same subject (AI). With the aid of AI, robots may soon be able to do the same tasks as humans, not to mention look and act the part. Robots are programmed to act in accordance with certain commands and have the appearance, movement, and behavior of real people. Robots are used in many different industries, including pharmaceuticals, automobiles, order picking, manufacturing, sanitation, and sage automation gantry. Robots have potential applications in many different industries. Some important applications of robots include the following.

One of a country's most vital functions is its defense establishment. Every nation wants a strong military. Robots are helpful during war because they may penetrate dangerous and inaccessible regions. The Defense Research and Development Organization (DRDO) developed a robot called Daksh that can neutralize threats without harming people or property. They facilitate military personnel's ability to continue duty and safety in dangerous environments [14–18]. Robots have various applications in the military, including combat support, fire support, battle damage management, strike missions, and even laying machines. The use of robots has improved several facets of healthcare, including laparoscopy, neurosurgery, orthopedic surgery, hospital cleaning, and pharmaceutical distribution. Cutting, welding, assembling, disassembling, picking and placing for printed circuit boards, packing and labeling, palletizing, polishing, and handling are just some of the many industrial processes that benefit from the employment of robots. Worker safety is improved through automation, which also boosts production and revenue [19, 20].

Together, the artificial intelligence and robotics fields are working to create smart machines [21, 22]. AI-enhanced robots are operated by software that makes use of various AI techniques. Most robots are not AI robots; rather, they are designed to carry out certain tasks repeatedly and do not need any kind of intelligence to do so. Such robots are helpful, but they have certain limitations. The robot is equipped with the ability to do increasingly sophisticated tasks via the application of AI algorithms. Different AI algorithms may be used by a self-driving automobile in its quest to identify and avoid danger.

7.1.3 Automation

"Automation" refers to technological applications that need minimum human interaction. This is a broad category that includes business process automation, information technology automation, consumer applications such as home automation, and more. Machines aren't always designed to replace humans. In part, this is because fewer tasks need to be completed by hand, but the real gains come from improved efficiency, uniform practices, and less friction in the business. The more efficient automation grows, the less frequent and more crucial human input is. Although automation has a bad reputation for being a job-killer, the reality is that by freeing IT professionals from mundane tasks, they can better address critical systemic problems [23–25].

1. Even the simplest tasks may become simpler with the help of automation technology. When technology is utilized to standardize and consolidate work processes, such as communicating and collaborating on a single platform rather than using separate databases, this is known as "digitizing" the workplace.
2. The use of process automation software may streamline and standardize business procedures. These responsibilities are often delegated to dedicated software and business applications. Business process automation might be the key to increased output and efficiency [26–28].
3. Machines might be designed to mimic human actions after humans have defined the rules for doing so. One such example is the "digital worker" [29, 30].
4. Implementing AI-based automation presents the greatest challenge. To give robots the ability to "learn" and make choices, AI technology must be included [31, 32].

7.1.4 INDUSTRIAL AUTOMATION

The efficiency gains from using modern forms of industrial automation have resulted in its widespread adoption across a wide range of businesses and sectors. With each passing year, commonplace activities are gradually being replaced by industrial uses. As a result, employees can do mundane activities with more precision and fewer mistakes, freeing up time for them to concentrate on more important initiatives [33–40].

1. Industry in general benefits from industrial automation processes. Among the most important are factories, refineries, paper mills, and steel mills.
2. Manufacturing companies use industrial automation in a variety of settings and contexts.
3. Industrial automation is very helpful in the oil and gas sector due to the widespread distribution of offshore platforms and other work locations [41–44].
4. It might be used for instrumentation, plant device and equipment control, and batch production management in the paper sector. Consequently, the manufacturing process is completely visible to the operators.
5. Many steel mills' automated production processes have a hierarchical structure. This cutting-edge equipment may be utilized to oversee and regulate operations throughout the steel mill.

6. Autopilot controls for commercial flights have been a standard part of aviation's industrial automation for a long time.
7. Distribution businesses take over after manufacturing is completed and goods are ready to be sent. The need for expedited shipping services is growing across all sectors [45–50].

Industrial automation technologies can improve safety, save time, boost output quality, and reduce production costs. Because of all of these benefits, profitability, and productivity go hand in hand:

1. **Enhanced Safety:** Workplace safety is essential to ensure the well-being of your workers, vendors, and consumers. Outdated machinery can be made safer by retrofitting. This reduces the requirement for human intervention in machine processes [51–53].
2. **Save Time:** Industrial automation technologies frequently boost the efficiency and dependability of production lines and other facility operations. This suggests that your company will be able to meet or exceed your clients' expectations more quickly [54–56].
3. **Boost Quality Production:** If you introduce flaws or quality concerns into the process, it won't matter how quickly you can generate the goods. It not only boosts manufacturing efficiency but also mitigates potential issues down the road. As a result, both the quality of your products and the number of errors that occur while they are being created will get decreased.
4. **Reduce Monitoring:** Industrial automation makes it simpler and less expensive to keep tabs on important assets and pieces of machinery. Technology can typically undertake 24-hour surveillance and data collecting at specified intervals without the need for human participation. This means maintenance staff can shift their attention to more challenging tasks, all while raising the bar for the quality of their output [57, 58].
5. **Lower Costs:** The faster and fewer repairs your organization needs, the more money industrial automation may save you. In many cases, it may also increase output while decreasing resource use.

7.2 LITERATURE REVIEW

The authors in [1] focused on industrial robotics and automation in the fourth industrial revolution. Germany was often credited as the birthplace of Industry 4.0, or the fourth industrial revolution. Progress in robotics and automation toward the goal of industry 4.0 is reviewed in this study. Robotics and automation technology are widely acknowledged as the backbone of modern production and a key factor in the advent of Industry 4.0 by businesses, research institutions, and academic institutions alike. The authors in [2] introduced the development of industrial robots for use in factory automation, from the beginning to the present day and the advent of IoT. Surprisingly, robotics is an ancient field of study, and robots have played a formative role in the development of several industries and industrial revolutions. In particular, the article discusses the development of industrial robots. The study begins with a brief background on the subject before moving on to discuss current and future developments, highlighting the correlation between the development of advanced robots for industrial use and the concurrent improvement of communication networks within the industrial sector. Industrial robots and their place in the future of smart manufacturing have been discussed in [3]. This study gives an overview of the human and robotic workforce in smart factories, how they relate to Industry 4.0, and what kind of technological development they have made. Conclusions from this study show that a decade is not enough time to create a reference implementation or application of Industry 4.0, such as smart factories. The term "Industry 4.0" was originally used by academics in 2011. With the advent of Industry 4.0 came a slew of new enablers and buzzwords like "Internet of Things," "Cyber-Physical Systems," and "Digital Twins" (DT). To begin, this article provides a working definition of "smart factory" and "smart manufacturing" as they pertain to the use of both people and robots in the process. Then, we'll take a look at some of the most important technologies used in today's "smart" factories. The future of this field, and how it relates to smart manufacturing, were discussed in the last section. The authors in [4] discussed the rise of IoT's influence on robotics: creating a network of robots. IoT has recently expanded into the world of robots. Robotics is a burgeoning field that has many potential applications in the home, business, and other settings. Possible benefits of the IoT and web of things (WoT) for robotic systems include increased connectivity and data collection. There has been some investigation into how robots may benefit from the IoT, and those findings have been promising. This essay goes a step further by analyzing current applications of IoT in robotics via the lens of a wide range of case studies taken from the academic literature. Applications for IoT-based robots are explored in the paper, as are the sensors and actions that they include. It also delves into the IoT's underlying concepts, characteristics, architectures, hardware, software, and communication methods as they apply to modern robotic systems. The current applications of WoT in robots were then investigated. The role of robotics, IoT, and AI in agricultural automation has been discussed in [5]. This article provides a comprehensive overview of the many facets of agriculture that may be automated or already are with the help of robots, IoT, and AI. Smart farming, precision agriculture, vertical farming, and connected farming are all discussed from both a historical and a futuristic viewpoint. Following the Covid-19 pandemic, automation in the agriculture business has become the new standard in response to labor migration and shortages.

Industrial automation system design and deployment have been discussed in [6] utilizing the Internet of Things. This study presents a method for integrating the monitoring, regulating, measuring, product counting, and locating up-to-date information of any automated enterprise into a single cohesive whole by means of IoT. Raspberry Pi created this embedded system or cloud server. This system's distinction lies in the fact that it is run by a Raspberry Pi, which also monitors and analyzes performance and machine observation. Productivity and efficiency-boosting measures may be implemented thanks to the system's controls and the data it generates on its own. Here, all data was kept on a Raspberry Pi in the cloud, where the server acts as a bridge between industry and a control device. CAYENCE is responsible for developing the cloud server. The system relies on the Linux OS and Python programming language for its smooth operation. Hypertext Markup Language (HTML) was employed in the system as the primary programming language. The authors in [7] focused on the industrial automation system design and deployment that utilized the IoT. This study presents a method for integrating the monitoring, regulating, measuring, product counting, and locating up-to-date information of any automated enterprise into a single cohesive whole by means of IoT. The innovations in robotics for the fourth industrial revolution were discussed in [8]. Recently, industrial robots have been developed and put to use in factories so that they may do activities that are too risky for people, speed up the manufacturing process without sacrificing quality, and lower product prices. Companies need more sophisticated systems that can make better judgments as the level of competition in the commercial world rises. Information technology developments such as AI, the cloud, and Big Data were altering the way robots are used and built in the context of the Industry 4.0 revolution. Future uses for industrial robots and the robotics generation set to be implemented in Industry 4.0 plants were explored. In [9], the authors contemplated the moral implications of AI and service robots. As our reliance on robotic and AI technology grows, so too are the number of ethical questions that businesses and individuals must consider. There will be lasting effects if we don't look into and deal with these issues immediately. I provide an overview of the literature that might serve as a foundation for this investigation. This study addresses a research need in the field of artificial intelligence and robotics applied to the service sector. It broadens how we think about robots and AI in service settings, which has crucial consequences for public policy and the practical implementation of service technologies. The authors in [10] introduced a meta-analysis of human-like service robots and other forms of artificial intelligence for cultural anthropomorphism studies. We provide a theoretical framework to examine the causes and effects of anthropomorphism in depth. Factors that may lead to anthropomorphism include client characteristics and preferences, socioeconomic status, and features of robot design. Robot qualities such as intellect and utility are lauded as crucial intermediaries, whereas interpersonal qualities like rapport are given less attention. This study's results provide light on the specific contexts in which anthropomorphism influences consumer intent to utilize a robot. The influence varies with robot type and service kind, as shown by the moderator analysis. They use these results to provide a thorough plan for the future study of marketing and service robots.

The authors in [11] investigated the influence of AI and robotics on the university level on the basis of design functions. Because the relevant material is scattered and the concepts are malleable, it is difficult to draw any firm conclusions, so it may be difficult to appreciate the full magnitude of this effect. But advancements are accompanied by debates about what is technically possible, what is practically implementable, and what is desirable from a pedagogical or social perspective. Design fictions that plausibly imagine future scenarios of AI or robots in use are helpful tools for both explaining and debating the technological possibilities. In this article, they explain how they conducted a narrative literature analysis to come up with eight such design fictions to showcase the many applications of AI and robots in academia. These cases, along with others, such as the ability to teach higher-order skills or change staff roles, generated discussions on the impact on human agency and the nature of datafication. In [12], the morality and ethics of artificial intelligence have been discussed, considering the examples of military drones and domestic companion robots. Was it really possible for AI to have higher ethical standards than humans? Is it possible it might treat human ideals with more regard than a human? In this piece, we look at the ethical questions brought up by AI. When combined with agent-based theory, the utilitarian method may provide an effective answer. We have picked two extremes: lethal killer drones and helpful companion robots. Armed conflict brings up questions about the morality of AI and unmanned aerial vehicles (UAVs), which need to be investigated through the lens of military ethics and human values. Companion robots pose a threat to people's physical safety and well-being, in addition to their social and moral standing, even if they are not designed to do damage to humans. It is crucial from a moral standpoint that companion robots supplement human caregivers rather than take their place. In [13], the authors used intelligent robots using algorithms to advance occupational therapy and functional capacity assessment. They conducted a literature study on rehabilitation robots, focusing on cutting-edge methods that use robots and machine learning (ML) to improve functional capacity evaluations. By offering simple ways for replicating occupational duties using intelligent robotic systems, rehabilitation robotics hope to enhance the evaluation and rehabilitation of wounded employees. The advantages of robotic technologies and the knowledge and experience of human therapists may be pooled together with the use of ML-based methods. These developments may help therapists and patients alike enhance the measurement of function and the learning of haptic interactions during evaluation and rehabilitation. By observing the therapist's movements (or "demonstrations"), a robot may be taught what the desirable job is for an injured worker (the "patient") and how to replicate it. These robotic evaluation methods may be employed long-distance to reach out to rural and distant areas via the usage of Telerehabilitation and internet connectivity. Conclusions: Research was still in its infancy, but robots with built-in ML algorithms might significantly advance conventional functional capacity evaluation methods. Further, in [14], the authors provided a secure human-robot interactive control framework for a wearable robot with an upper limb, which includes sensing of human arm movement. In order to change the path of a traditional virtual mechanical impedance control system-operated lifting robot, the user must apply an interactive torque to the robot's joints. To avoid damaging the user's joint, a three-tiered hierarchical control system was built into the robot in this study. The signals from the user's surface electromyography were sent into a human arm movement-detecting module. As the lifting process progresses into real use, a Hammerstein adaptive virtual mechanical impedance controller is used to provide an appropriate amount of torque for the user's elbow joint. The actuator controllers included in each mechanical part of the joint are primarily responsible for regulating the robot's lifting capacity. There was less interaction torque felt by the user's elbow, and less contractions of the erector spinae and biceps brachii, according to a number of studies. The suggested method safeguards the user against muscle fatigue and joint problems that may result from applying too much interaction torque to the human elbow joint. The critical imaginaries and musings on the use of robots and AI in postdigital k-12 classrooms have been provided in [15]. The focus of this work was to inquire into the ways in which educators, researchers, and pedagogical developers speculate about and think about the potential applications of AI and robotics in the classroom. The empirical information was obtained via symposia-related roundtable discussions. It was explained how the introduction of AI and Entity Relationship (ERs) may alter the role of the teacher and the connection between instructors and students, and how more information is needed regarding these technologies. Many of those who took part believed that AI offered more opportunities for customization than ERs did. However, there were other worries raised while talking about AI, including ethical problems and commercial interests. Practitioners were more concerned with imaginaries rooted in present practice than researchers/developers, who were more preoccupied with envisioning ideal future educational practices enhanced by technological advancements. The review of wearable robotics human activity recognition has been done using the distributed convolution neural network (CNN) [16]. Four different CNN architectures were compared. By setting up a distributed CNN on a collection of Intel Edison nodes, it was shown that in-memory classification could be carried out at near-real-time speeds. Two of the strategies were consolidated and single-piece implementations, while the other two were distributed over several computers. Although the distributed approaches have slightly lower accuracy than the monolithic CNNs, they turn out to require much less memory – and therefore computation – than monolithic CNNs, with only modest communication rates between nodes in the model, making the approach viable for a wide range of distributed systems, from wearable robots to multi-robot swarms. The impact of wearable robots on ethics, law, and society has been discussed in [17]. This document offers helpful recommendations for incorporating ethical, legal, and social issues (ELSI) into WR creation and maintenance. First, they justify domain-specific suggestions in light of current ELSI standards. The authors then demonstrate the effectiveness of a domain-specific approach by converting the state-of-the-art ELSI discoveries into a sequential flowchart for integrating ELSI tailored to the several stages of weilding robot (WR) development. This flowchart details the inquiries and potential solutions that WR development teams should consider. With tailored ELSI support that takes into account user needs, the community's place in the larger social context, and an understanding of relevant legislation and values, the community has a higher chance of creating WRs that are safer, more useful, and have a beneficial impact on society.

Table 7.1 shows the entire literature review.

7.3 PROPOSED WORK

7.3.1 PROBLEM STATEMENT

It has been observed that the need for the IoRT is growing day by day but there are several limitations in the implications of the internet of robotics. A need remains

TABLE 7.1
Literature Survey

S No.	Author/Year	Objective	Methodology	Limitation
[1]	Kamarul Bahrin/2016	The fourth industrial revolution: a look at robotics and automation in production	Industrial Automation, Robotic	Scope of work is limited
[2]	A. Grau /2017	From prototypes to IoT: A history of industrial robots in manufacturing	Industrial robotics, Internet of Things	Lack of flexibility
[3]	L. Evjemo/2020	What's new in smart manufacturing? the complementary roles of humans and industrial robots in high-tech factories	Industrial Robots	There will not be utilized in future
[4]	Kamilaris/2020	The spread of IoT to the robotics industry: Creating a network of robots.	Internet of Things, robotics	Lack of efficiency
[5]	Krishnan/2020	The role of robotics, the internet of things, and AI in agricultural automation.	Robotics, IoT, AI	Need to improve the performance and accuracy
[6]	Lokman/2020	Design implementation of IoT based industrial automation system.	IoT, Industrial Automation	There is a lack of performance
[7]	Prem/2020	Automation of industrial (Iot) and cyber security	Industrial Automation, IoT	Research is limited to traffic flow
[8]	Bayram/2020	Advances in robotics in the era	Robotics	There is less technical work
[9]	Belk/2021	Ethics of AI and service robots	Robots, AI	Lack of security and accuracy
[10]	Blut/2021	Meta-analysis of physical robots, chatbots, and other AI for understanding anthropomorphism in service offering.	AI, Robots	There is a lack of performance
[11]	Cox/2021	The effect of AI and robotics on higher education, as seen via literary design fiction	AI	Lack of technical work
[12]	de Swarte/2019	Ethics, human values, and AI case studies of combat drones and domestic robots	Artificial intelligence, Robots	The performance of this research is very low

(Continued)

TABLE 7.1 (CONTINUED)
Literature Survey

S No.	Author/Year	Objective	Methodology	Limitation
[13]	Fong/2020	Innovative robotics with Embedded ML algorithms for enhanced occupational rehab and functional capacity assessment.	Intelligent Robotics, ML	Did not consider a real-life solution
[14]	Hao/2020	A secure human-robot interactive control mechanism for an upper-limb wearable robot employed in lifting activities using sensing of human arm movement.	Robot	Need to consider optimization technique
[15]	Hrastinski/2019	AI and robots in postdigital K-12 education: critical imaginaries and reflections	Artificial Intelligence, Robots	Need to enhance the scope of work

to improve the scalability and reliability of the internet of robotic applications. Moreover, there is a need to do more work in order to improve accuracy and performance. The proposed work is supposed to provide solutions in the area of the internet of robotics by integrating deep learning mechanisms of optimization during industrial automation.

7.3.2 Research Methodology

In this research, the previous research in the area of IoT, the internet of robotics, and Industrial Automation are considered. The research studies methodology, limitations, accuracy, scalability, reusability, and the performance of previous research. The proposed work is on integrating a hybrid approach by using optimization and deep learning. Finally, the evolution of work is about accuracy, performance, error rate, and the comparison of the proposed work to conventional work. The research methodology is shown in Figure 7.3.

7.3.3 Proposed Methodology

The process flow of the proposed work is shown in Figure 7.4.

7.4 RESULTS AND DISCUSSION

The effects of the planned work on the IoRT are described here. Traditional models have lower performance and are less scalable. Given the nature of the planned effort, the

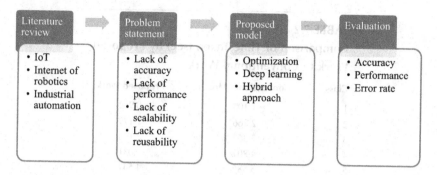

FIGURE 7.3 Research methodology.

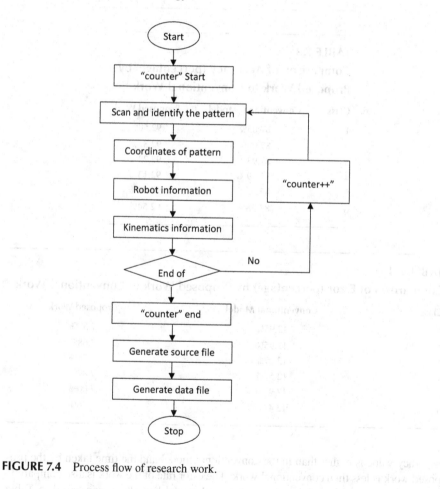

FIGURE 7.4 Process flow of research work.

TABLE 7.2
Comparison of Time Taken (sec) by Proposed Work to Conventional Work

Class	Conventional Model	Proposed Work
1	1.108	0.962
2	2.266	1.179
3	3.367	2.103
4	4.807	3.031
5	5.585	4.547
6	6.040	5.710

TABLE 7.3
Comparison of Accuracy (percentage) by Proposed Work to Conventional Work

Class	Conventional Model	Proposed Work
1	87.05%	92.50%
2	87.05%	92.32%
3	87.93%	92.22%
4	87.49%	92.13%
5	87.03%	93.00%
6	87.58%	92.54%

TABLE 7.4
Comparison of Error (percentage) by Proposed Work to Conventional Work

Class	Conventional Model	Proposed Work
1	12.95%	7.50%
2	12.95%	7.68%
3	12.07%	7.78%
4	12.51%	7.87%
5	12.97%	7.00%
6	12.42%	7.46%

accuracy value is higher than in the conventional model and the time taken by the proposed work is less than conventional work. The error rate of the work is also compared.

The conventional model and the proposed model have been compared in Table 7.2, Table 7.3, and Table 7.4 (also drawn in the form of graphs in Figure 7.5, Figure 7.6, and Figure 7.7).

Investigating Scope and Applications for the Internet of Robotics 145

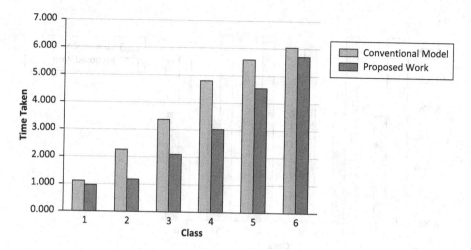

FIGURE 7.5 Comparison of time taken (sec).

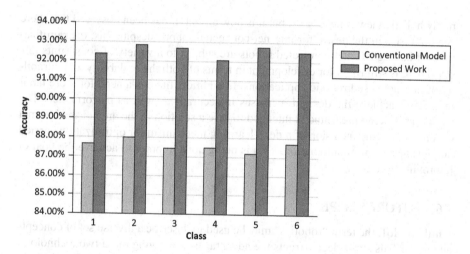

FIGURE 7.6 Comparison of accuracy (percentage).

7.5 CONCLUSION

An emerging idea known as IoRT links commonplace sensors and objects to automated systems and artificial intelligence. This research looks at how the IoT may be combined with robotics to enhance the capabilities of both current IoT systems and robotic systems, leading to the development of novel service offerings with the potential to radically alter an industry. We have had a discussion about some of the new technological problems that have surfaced as a direct consequence of this merger, and at the end of the discussion, research arrived at the conclusion that a

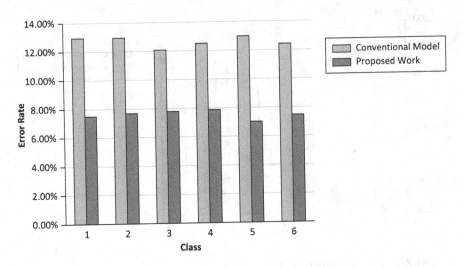

FIGURE 7.7 Comparison of Error (percentage).

really holistic view is necessary, but it is now absent. It has been observed that there is a daily rise in the desire for internet-connected robots; despite this, the ramifications of having internet-connected robots are subject to a variety of constraints [59, 60]. There is still room for development in terms of both the scalability and durability of internet-based robotic applications. In addition, there is a need for more labor in order to achieve the desired increases in accuracy as well as performance. The work that is being presented is intended to give a solution in the area of the internet of robotics via the inclusion of a deep learning mechanism for optimization carried out during industrial automation. This is being done in order to achieve the goal of improving efficiency.

7.6 FUTURE SCOPE

Similar to IoT, the term "robotics" may be used to describe a diverse set of concepts and tools. This study deconstructs the advantages of merging these two technological domains into nine individual system aspects. An example of an IoT advantage that is widely used in the robotics community is distributed perception and machine to machine protocols. On the other hand, active sensing strategies including robots have been the norm for the IoT thus far. For the time being, Ambient Assisted Living (AAL), precision agriculture, and Industry 4.0 are the only horizontal sectors where IoRT is being used. For example, domain-agnostic solutions have just recently emerged for the problem of incorporating robots into IoT middleware systems. Phrases like "IoT-enhanced robots" and "Robot-enhanced IoT" have become common, but we feel they do not adequately describe the IoRT [61, 62]. The goal of this survey is to get researchers in both domains working together on a cloud-based, robot-assisted, IoT-agent solution to the problems outlined above.

REFERENCES

1. Kamarul Bahrin, M. A., Fauzi Othman, M., NorAzli, N. H., and FarihinTalib, M. (2016). Industry 4.0: A review on industrial automation and robotic. *Jurnal Teknologi*, 78(6–13), 137–143. www.jurnalteknologi.utm.my
2. Grau, A., Indri, M., Lo Bello, L., and Sauter, T. (2017). Industrial robotics in factory automation: From the early stage to the Internet of Things. In *Proceedings IECON 2017 - 43rd Annual Conference of the IEEE Industrial Electronics Society*, 2017-January, 6159–6164. https://doi.org/10.1109/IECON.2017.8217070
3. Evjemo, L. D., Gjerstad, T., Grøtli, E. I., and Sziebig, G. (2020). Trends in smart manufacturing: Role of humans and industrial robots in smart factories. *Current Robotics Reports*, 1(2), 35–41. https://doi.org/10.1007/s43154-020-00006-5
4. Kamilaris, A., and Botteghi, N. (2020). The penetration of Internet of Things in robotics: Towards a web of robotic things. *Journal of Ambient Intelligence and Smart Environments*, 12(6), 491–512. https://doi.org/10.3233/AIS-200582
5. Krishnan, A., Swarna, S., and Balasubramanya, H. S. (2020). Robotics, IoT, and AI in the automation of agricultural industry: A review. In *Proceedings B-HTC 2020 - 1st IEEE Bangalore Humanitarian Technology Conference*. https://doi.org/10.1109/B-HTC50970.2020.9297856
6. Lokman, T., Bin Islam, M. T., and Apple, M. U. (2020). Design implementation of IoT based industrial automation system. In *2020 11th International Conference on Computing, Communication and Networking Technologies, ICCCNT 2020*. https://doi.org/10.1109/ICCCNT49239.2020.9225388
7. Prem, S. P., Balaram, V. V. S. S., and Bhutada, S. (2020). Automation of industrial (IoT) and cyber security. *Turkish Journal of Computer and Mathematics Education*, 11(3), 1094–1098.
8. Bayram, B., and Ince, G. (n.d.). *Advances in Robotics in the Era of Industry 4.0* (pp. 187–200). 10.1007/978-3-319-57870-5_11
9. Belk, R. (2021). Ethical issues in service robotics and artificial intelligence. *Service Industries Journal*, 41(13–14), 860–876. https://doi.org/10.1080/02642069.2020.1727892
10. Blut, M., Wang, C., Wünderlich, N. V., and Brock, C. (2021). Understanding anthropomorphism in service provision: A meta-analysis of physical robots, chatbots, and other AI. *Journal of the Academy of Marketing Science*, 49(4), 632–658. https://doi.org/10.1007/s11747-020-00762-y
11. Cox, A. M. (2021). Exploring the impact of Artificial Intelligence and robots on higher education through literature-based design fictions. *International Journal of Educational Technology in Higher Education*, 18(1). https://doi.org/10.1186/s41239-020-00237-8
12. deSwarte, T., Boufous, O., and Escalle, P. (2019). Artificial intelligence, ethics and human values: The cases of military drones and companion robots. *Artificial Life and Robotics*, 24(3), 291–296. https://doi.org/10.1007/s10015-019-00525-1
13. Fong, J., Ocampo, R., Gross, D. P., and Tavakoli, M. (2020). Intelligent robotics incorporating machine learning algorithms for improving functional capacity evaluation and occupational rehabilitation. *Journal of Occupational Rehabilitation*, 30(3), 362–370. https://doi.org/10.1007/s10926-020-09888-w
14. Hao, L., Zhao, Z., Li, X., Liu, M., Yang, H., and Sun, Y. (2020). A safe human–robot interactive control structure with human arm movement detection for an upper-limb wearable robot used during lifting tasks. *International Journal of Advanced Robotic Systems*, 17(5), 1–15. https://doi.org/10.1177/1729881420937570

15. Hrastinski, S., Olofsson, A. D., Arkenback, C., Ekström, S., Ericsson, E., Fransson, G., Jaldemark, J., Ryberg, T., Öberg, L. M., Fuentes, A., Gustafsson, U., Humble, N., Mozelius, P., Sundgren, M., and Utterberg, M. (2019). Critical imaginaries and reflections on artificial intelligence and robots in postdigital K-12 education. *Postdigital Science and Education*, *1*(2), 427–445. https://doi.org/10.1007/s42438-019-00046-x
16. Hughes, D., and Correll, N. (2018). Distributed convolutional neural networks for human activity recognition in wearable robotics. In *Springer Proceedings in Advanced Robotics* (Vol. 6). Springer International Publishing. https://doi.org/10.1007/978-3-319-73008-0_43
17. Kapeller, A., Felzmann, H., Fosch-Villaronga, E., Nizamis, K., and Hughes, A. M. (2021). Implementing ethical, legal, and societal considerations in wearable robot design. *Applied Sciences (Switzerland)*, *11*(15). https://doi.org/10.3390/app11156705
18. Rendiniello, A., Remus, A., Sorrentino, I., Murali, P. K., Pucci, D., Maggiali, M., Natale, L., Traversaro, S., Villagrossi, E., Polo, A., and Ardesi, A. (2020). A flexible software architecture for robotic industrial applications. *IEEE International Conference on Emerging Technologies and Factory Automation, ETFA*, 2020-September, pp. 1273–1276. https://doi.org/10.1109/ETFA46521.2020.9212095
19. Romeo, L., Petitti, A., Marani, R., and Milella, A. (2020). Internet of robotic things in smart domains: Applications and challenges. *Sensors (Switzerland)*, *20*(12), 1–23. https://doi.org/10.3390/s20123355
20. Sherwani, F., Asad, M. M., and Ibrahim, B. S. K. K. (2020). Collaborative robots and Industrial Revolution 4.0 (IR 4.0). In *2020 International Conference on Emerging Trends in Smart Technologies, ICETST 2020*. https://doi.org/10.1109/ICETST49965.2020.9080724
21. Chibber, A., Anand, R., and Singh, J. (2022). Smart traffic light controller using edge detection in digital signal processing. In *Wireless Communication with Artificial Intelligence* (pp. 251–272). CRC Press.
22. Singh, H., Ramya, D., Saravanakumar, R., Sateesh, N., Anand, R., Singh, S., and Neelakandan, S. (2022). Artificial intelligence based quality of transmission predictive model for cognitive optical networks. *Optik*, *257*, 168789.
23. Juneja, S., Juneja, A., and Anand, R. (2019, April). Reliability modeling for embedded system environment compared to available software reliability growth models. In *2019 International Conference on Automation, Computational and Technology Management (ICACTM)* (pp. 379–382). IEEE.
24. Gupta, A., Anand, R., Pandey, D., Sindhwani, N., Wairya, S., Pandey, B. K., and Sharma, M. (2021). Prediction of breast cancer using extremely randomized clustering forests (ERCF) technique: Prediction of breast cancer. *International Journal of Distributed Systems and Technologies (IJDST)*, *12*(4), 1–15.
25. Gupta, A., Asad, A., Meena, L., and Anand, R. (2022, July). IoT and RFID-based smart card system integrated with health care, electricity, QR and banking sectors. In *Artificial Intelligence on Medical Data: Proceedings of International Symposium, ISCMM 2021* (pp. 253–265). Springer Nature Singapore.
26. Nam, K., Dutt, C. S., Chathoth, P., Daghfous, A., and Khan, M. S. (2021). The adoption of artificial intelligence and robotics in the hotel industry: Prospects and challenges. *Electronic Markets*, *31*(3), 553–574.
27. Sansanwal, K., Shrivastava, G., Anand, R., and Sharma, K. (2019). Big data analysis and compression for indoor air quality. In *Handbook of IoT and Big Data* (pp. 1–21). CRC Press.
28. Papapicco, V., Parri, A., Martini, E., Bevilacqua, V., Crea, S., and Vitiello, N. (2019). Locomotion mode classification based on support vector machines and hip joint angles: A feasibility study for applications in wearable robotics. In *Springer Proceedings in*

Advanced Robotics, 7. Springer International Publishing. https://doi.org/10.1007/978-3-319-89327-3_15

29. Chaudhary, A., Bodala, D., Sindhwani, N., and Kumar, A. (2022, March). Analysis of customer loyalty using artificial neural networks. In *2022 International Mobile and Embedded Technology Conference (MECON)* (pp. 181–183). IEEE.
30. Jain, S., Kumar, M., Sindhwani, N., and Singh, P. (2021, September). SARS-Cov-2 detection using deep learning techniques on the basis of clinical reports. In *2021 9th International Conference on Reliability, Infocom Technologies and Optimization (Trends and Future Directions) (ICRITO)* (pp. 1–5). IEEE.
31. Kumar, D., George, S., Aremu, B., Wariya, S., and Pandey, B. K. (2021). Critical review on integration of encryption, steganography, IoT and artificial intelligence for the secure transmission of Stego images, *1*(1), 33–36.
32. Kumar, M. S., Sankar, S., Nassa, V. K., Pandey, D., Pandey, B. K., and Enbeyle, W. (2021). Innovation and creativity for data mining using computational statistics. In *Methodologies and Applications of Computational Statistics for Machine Intelligence* (pp. 223–240). IGI Global.
33. Torricelli, D., Rodriguez-Guerrero, C., Veneman, J. F., Crea, S., Briem, K., Lenggenhager, B., and Beckerle, P. (2020). Benchmarking wearable robots: Challenges and recommendations from functional, user experience, and methodological perspectives. *Frontiers in Robotics and AI*, *7*(November). https://doi.org/10.3389/frobt.2020.561774
34. Pandey, D., Wairya, S., Sharma, M., Gupta, A. K., Kakkar, R., and Pandey, B. K. (2022). An approach for object tracking, categorization, and autopilot guidance for passive homing missiles. *Aerospace Systems*, *5*(4), 553–566.
35. Vélez-guerrero, M. A., Callejas-cuervo, M., and Mazzoleni, S. (2021). Artificial intelligence-based wearable robotic exoskeletons for upper limb rehabilitation: A review. *Sensors*, *21*(6), 1–30. https://doi.org/10.3390/s21062146
36. Wang, C., Teo, T. S. H., and Janssen, M. (2021). Public and private value creation using artificial intelligence: An empirical study of AI voice robot users in Chinese public sector. *International Journal of Information Management*, *61*(August), 102401. https://doi.org/10.1016/j.ijinfomgt.2021.102401
37. Meivel, S., Sindhwani, N., Anand, R., Pandey, D., Alnuaim, A. A., Altheneyan, A. S., ... Lelisho, M. E. (2022). Mask detection and social distance identification using internet of things and faster R-CNN algorithm. *Computational Intelligence and Neuroscience*, 2022, 1–5.
38. Zardiashvili, L., and Fosch-Villaronga, E. (2020). "Oh, Dignity too?" Said the Robot: Human Dignity as the Basis for the Governance of Robotics. *Minds and Machines*, *30*(1), 121–143. https://doi.org/10.1007/s11023-019-09514-6
39. Zhang, Y., and Qi, S. (2019). User experience study: The service expectation of hotel guests to the utilization of AI-based service robot in full-service hotels. In *Lecture Notes in Computer Science (Including Subseries Lecture Notes in Artificial Intelligence and Lecture Notes in Bioinformatics): Vol. 11588 LNCS*. Springer International Publishing. https://doi.org/10.1007/978-3-030-22335-9_24
40. Zheng, Y., Song, Q., Liu, J., Song, Q., and Yue, Q. (2020). Research on motion pattern recognition of exoskeleton robot based on multimodal machine learning model. *Neural Computing and Applications*, *32*(7), 1869–1877. https://doi.org/10.1007/s00521-019-04567-1
41. Pandey, B. K., Pandey, D., Wairya, S., Agarwal, G., Dadeech, P., Dogiwal, S. R., and Pramanik, S. (2022). Application of integrated steganography and image compressing techniques for confidential information transmission. *Cyber Security and Network Security*, 169–191.

42. Schmitt, B. (2020). Speciesism: An obstacle to AI and robot adoption. *Marketing Letters, 31*(1), 3–6. https://doi.org/10.1007/s11002-019-09499-3
43. Gupta, R. S., Nassa, V. K., Bansal, R., Sharma, P., and Koti, K. (2021). Investigating application and challenges of big data analytics with clustering. In *2021 International Conference on Advancements in Electrical, Electronics, Communication, Computing and Automation (ICAECA)* (pp. 1–6). https://doi.org/10.1109/ICAECA52838.2021.9675483
44. Veeraiah, V., Khan, H., Kumar, A., Ahamad, S., Mahajan, A., and Gupta, A. (2022). Integration of PSO and deep learning for trend analysis of meta-verse. In *2022 2nd International Conference on Advance Computing and Innovative Technologies in Engineering (ICACITE)* (pp. 713–718). https://doi.org/10.1109/ICACITE53722.2022.9823883
45. Pandey, B. K., Pandey, D., Nassa, V. K., George, S., Aremu, B., Dadeech, P., and Gupta, A. (2023). Effective and secure transmission of health information using advanced morphological component analysis and image hiding. In Gupta, M., Ghatak, S., Gupta, A., Mukherjee, A. L. (eds) *Artificial Intelligence on Medical Data. Lecture Notes in Computational Vision and Biomechanics*, 37. Springer. https://doi.org/10.1007/978-981-19-0151-5_19
46. Veeraiah, V., Kumar, K. R., LalithaKumari, P., Ahamad, S., Bansal, R., and Gupta, A. (2022). Application of biometric system to enhance the security in virtual world. In *2022 2nd International Conference on Advance Computing and Innovative Technologies in Engineering (ICACITE)* (pp. 719–723). https://doi.org/10.1109/ICACITE53722.2022.9823850
47. Bansal, R., Gupta, A., Singh, R., and Nassa, V. K. (2021). Role and impact of digital technologies in e-learning amidst COVID-19 pandemic. In *2021 Fourth International Conference on Computational Intelligence and Communication Technologies (CCICT)* (pp. 194–202). https://doi.org/10.1109/CCICT53244.2021.00046
48. Shukla, A., Ahamad, S., Rao, G. N., Al-Asadi, A. J., Gupta, A., and Kumbhkar, M. (2021). Artificial intelligence assisted IoT data intrusion detection. In *2021 4th International Conference on Computing and Communications Technologies (ICCCT)* (pp. 330–335). https://doi.org/10.1109/ICCCT53315.2021.9711795
49. Pathania, V., Babu, S. Z. D., Ahamad, S., Thilakavathy, P., Gupta, A., Alazzam, M. B., and Pandey, D. (2023). A database application of monitoring COVID-19 in India. In Gupta, M., Ghatak, S., Gupta, A., Mukherjee, A. L. (eds) *Artificial Intelligence on Medical Data: Lecture Notes in Computational Vision and Biomechanics*, 37. Springer. https://doi.org/10.1007/978-981-19-0151-5_23
50. Dushyant, K., Muskan, G., Annu, Gupta, A., and Pramanik, S. (2022). Utilizing machine learning and deep learning in cybesecurity: An innovative approach. In *Cyber Security and Digital Forensics: Challenges and Future Trends* (pp. 271–293). Wiley. https://doi.org/10.1002/9781119795667.ch12
51. Babu, S. Z. D., Pandey, D., Naidu, G. T., Sumathi, S., Gupta, A., Bader Alazzam, M., and Pandey, B. K. (2023). Analysation of big data in smart healthcare. In Gupta, M., Ghatak, S., Gupta, A., Mukherjee, A. L. (eds) *Artificial Intelligence on Medical Data: Lecture Notes in Computational Vision and Biomechanics*, 37. Springer. https://doi.org/10.1007/978-981-19-0151-5_21
52. Bansal, B., Jenipher, N., Jain, R., Dilip, R., Kumbhkar, M., Pramanik, S., Sandip, R., and Gupta, A. (2022). Big data architecture for network security. In *Cyber Security and Network Security* (pp. 233–267). Wiley. https://doi.org/10.1002/9781119812555.ch11
53. Gupta, A., Kaushik, D., Garg, M., and Verma, A. (2020). Machine learning model for breast cancer prediction. In *2020 Fourth International Conference on I-SMAC (IoT in Social, Mobile, Analytics and Cloud) (I-SMAC)* (pp. 472–477). https://doi.org/10.1109/I-SMAC49090.2020.9243323

54. Sreekanth, N., Rama Devi, J., Shukla, A. et al. (2022). Evaluation of estimation in software development using deep learning-modified neural network. *Applied Nanoscience*. https://doi.org/10.1007/s13204-021-02204-9
55. Veeraiah, V., Rajaboina, N. B., Rao, G. N., Ahamad, S., Gupta, A., and Suri, C. S. (2022). Securing online web application for IoT management. In *2022 2nd International Conference on Advance Computing and Innovative Technologies in Engineering (ICACITE)* (pp. 1499–1504). https://doi.org/10.1109/ICACITE53722.2022.9823733
56. Veeraiah, V., Gangavathi, P., Ahamad, S., Talukdar, S. B., Gupta, A., and Talukdar, V. (2022). Enhancement of meta verse capabilities by IoT integration. In *2022 2nd International Conference on Advance Computing and Innovative Technologies in Engineering (ICACITE)* (pp. 1493–1498). https://doi.org/10.1109/ICACITE53722.2022.9823766
57. Gupta, N., Janani, S., Dilip, R., Hosur, R., Chaturvedi, A., and Gupta, A. (2022). Wearable sensors for evaluation over smart home using sequential minimization optimization-based random forest. *International Journal of Communication Networks and Information Security (IJCNIS)*, *14*(2), 179–188. https://doi.org/10.17762/ijcnis.v14i2.5499
58. Keserwani, H., Rastogi, H., Kurniullah, A. Z., Janardan, S. K., Raman, R., Rathod, V. M., and Gupta, A. (2022). Security enhancement by identifying attacks using machine learning for 5G network. *International Journal of Communication Networks and Information Security (IJCNIS)*, *14*(2), 124–141. https://doi.org/10.17762/ijcnis.v14i2.5494
59. Sindhwani, N., Maurya, V. P., Patel, A., Yadav, R. K., Krishna, S., and Anand, R. (2022). Implementation of intelligent plantation system using virtual IoT. In book on *Internet of Things and its Applications CRC Publishers* (pp. 305–322).
60. Gupta, R., Shrivastava, G., Anand, R., and Tomažič, T. (2018). IoT-based privacy control system through android. In *Handbook of E-business Security* (pp. 341–363). Auerbach Publications.
61. Singh, P., Kaiwartya, O., Sindhwani, N., Jain, V., and Anand, R. (Eds.). (2022). *Networking Technologies in Smart Healthcare: Innovations and Analytical Approaches*. CRC Press.
62. Pandey, D., Pandey, B. K., Sindhwani, N., Anand, R., Nassa, V. K., and Dadheech, P. (2022). An interdisciplinary approach in the post-COVID-19 pandemic era. In book on *An Interdisciplinary Approach in the Post-COVID-19 Pandemic Era*, Nova publishers (pp. 1–290).

8 Role of Artificial Intelligence in Making Wearable Robotics Smarter

Vivek Veeraiah, B. Karthiga, Chinnahajisagari Mohammad Akram, Ashok Koujalagi, S. Nanthakumar, Vipin Sharma, and Digvijay Pandey

8.1 INTRODUCTION

The term "artificial intelligence" (AI) refers to the practice of programming a computer to do tasks that normally require human intelligence. Machines are believed to have artificial intelligence when they exhibit cognitive abilities similar to those of humans [1–5]. While the goals and uses of robotics and AI are distinct, the two fields are often discussed in the same breath (AI). When equipped with artificial intelligence, robots may mimic human behavior and even pass for humans [6–8].

8.1.1 ARTIFICIAL INTELLIGENCE

Artificial Intellect (AI) is the utilization of computers to simulate human intelligence. Applications of artificial intelligence include expert systems, natural language processing , voice recognition, and machine vision. We can all agree that John McCarthy was the pioneer of AI research. AI is ability of a computer to learn, plan, and solve problems in ways that are indistinguishable from a human brain. The term AI is often used to describe the sophisticated processing skills of computers and other electronic devices. It's a field of research that focuses on making machines appear to have intelligence. They can function independently of instructions and do an excellent job [9–12].

8.1.2 NEED OF AI:

- Artificial intelligence allows for the development of efficient and precise solutions to problems in fields as diverse as medicine, advertising, traffic management, and more.

Role of AI in Making Wearable Robotics Smarter

- Using AI, you can create your own personal digital assistant (AI).
- It's possible, with the help of AI, to build robots that can do tasks in dangerous conditions.
- Artificial intelligence (AI) leads to the development of cutting-edge tools and methods, in addition to opening up exciting new fields of endeavor.

8.1.3 Types of AI

AI refers to any one of several possible types of artificial intelligence. By this metric, a more advanced AI would be one that can mimic human performance on a wider range of tasks with about the same level of success, while a less advanced AI would be one with more restricted capabilities and performance.

A) **Reactive Machines:** these early AI systems were quite rudimentary and could do very little. This technology allows for the simulation of a wide variety of stimuli. In other words, you can't utilize any apps or features that rely on memory with these gadgets.
B) **Limited Memory:** even machines with little memory may learn to predict future outcomes by analyzing historical data. This describes the vast majority of currently available AI programs. A deep learning system, for instance, learns by storing a large amount of training data in its memory for use as a model when confronted with new issues. There are a variety of methods for teaching an AI to recognize images, including using large collections of images and their labels [13-15].
C) **Theory of Mind:** similar to the first two types of AI, the next two are still in the conceptual or development stages. Researchers are now testing the theory of mind AI, the next step in the evolution of AI. Once a theory of mind-level AI is developed, it will be able to understand the needs and feelings of the individuals it is helping [16].
D) **Self-aware:** simply put, self-aware AI is artificial intelligence that is aware of itself. The goal of the artificial intelligence research community is to develop such a system in the long run. Along with its empathetic abilities, this AI will also be endowed with sentiments, needs, beliefs, and goals of its own, which it will use to both understand and influence the emotions of the individuals with whom it interacts [17].

8.1.4 AI, Machine Learning, and Deep Learning – An Overview

AI has the potential to revolutionize several industries due to its capacity to increase efficiency [18-20]. With the advancement of computing technology, machine learning (ML) will play a more significant role in today's workplace [21, 22]. Computers may eventually outperform humans in mental activities. Machine-learning algorithms outperform radiologists at identifying cancers. But radiology is just at the beginning of its development as a medical specialty. It's possible that in the next several years, millions of jobs may be lost due to automation and AI [23-25].

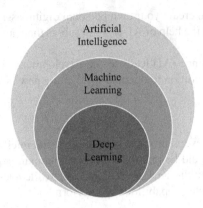

FIGURE 8.1 Relationship among AI, ML, and deep learning.

The relationship among AI, ML, and Deep Learning is shown in Figure 8.1.

As depicted in the figure 8.1, DL is a subset of ML. ML refers to the practice of creating computational systems with the ability to "learn" without the intervention of human programmers. In his book "Computing Machinery and Intelligence," Alan Turing discussed his famous "Learning Machine," an artificial intelligence that could fool a human into thinking it was real. The concept it expresses has been around for decades, and so has the word [26–30].

8.1.5 Robotics

Robotics is a subfield of AI that focuses on the design and implementation of autonomous robots and other intelligent machines. Because robots are mechanically constructed, include electrical components, and are controlled by computer code, robotics is interdisciplinary. Even though the fields of robotics and AI pursue separate goals and have diverse applications, the former is typically viewed as a branch of the latter. In addition to looking and acting like people, robots may achieve the same results with the help of AI [31–33]. Robots are machines designed to appear like humans and carry out complex tasks and mimic human gestures in response to predetermined instructions [34–36]. Robots in the fields of drug compounding, automotive production, order picking, industrial cleaning, and sage automation gantry are just a few examples [37, 38]. There are several fields where robots may be used. The following are examples of key areas where robots are used:

1. Defense is undeniably one of the most important sectors of any nation. Each state desires an effective military. During times of conflict, robots are useful for entering otherwise inaccessible and perilous areas. Daksh, a robot created by DRDO, can safely eliminate potentially lethal things.
2. They aid service members in remaining safe and on active duty in dangerous environments. Robots are used for a wide variety of tasks in the

military, including but not limited to combat support, tasks including laying mines, and anti-submarine operations.
3. Robots are used in many facets of industrial production, including but not limited to cutting, welding, assembling, disassembling, picking and placing for printed circuit boards, packaging and labeling, palletizing, product inspection and testing, color coating, drilling, polishing, and handling.
4. There is less risk of physical harm to workers as a result of using robotics, and the technology also boosts production and profits.
5. A mining robot is equipped with cameras and other sensors that allow it to see through water and locate lucrative minerals. To further elaborate, robots aid in excavation by detecting gases and other materials, therefore protecting humans from injury.

Robots with artificial intelligence bring together the fields of artificial intelligence and robotics. Artificial intelligence (AI) robots are piloted by computer programs that make use of a variety of AI techniques. Most robots are not AI robots; rather, they are designed to carry out certain, repetitive motions and don't require any sort of intelligence to execute their jobs. However, the capabilities of such robots are restricted. Putting the robot in a position to carry out increasingly complicated tasks requires the use of AI algorithms. Multiple AI systems might be used by a self-driving automobile to identify and avoid dangerous situations [39–42].

8.1.6 Wearable Robotics

An individual's mobility and/or physical capabilities can be improved by using a wearable robot. Bionic robots are another name for wearable robots. A fundamental premise of wearable robots is the incorporation of mechanical components to facilitate human movement. Some wearable robots can assist with walking, which might be useful for those recovering from surgery or in physical therapy. The wearable robot interface has the unique quality of being able to be programmed in a number of different ways. In order to support certain forms of movement, sensors or devices might take in verbal, behavioral, or other information. Wearable robots including cutting-edge hardware, big data, and wireless technologies provide enormous promise for those with paralysis or other disabilities who might considerably benefit from their use [43–48]. Figure 8.2 shows the classification of wearable robotics.

8.1.7 Bionics

The term "bionics," a portmanteau derived from the words "biology" and "electronics," refers to the deliberate use of biological systems and processes in technological contexts. Bionics are capable of being very sophisticated bits of technology that can be fused with different elements of the human body. As time goes on, bionic limbs improve in both appearance and performance, beginning to resemble natural limbs more and more. Bionic limbs come in a wide variety of forms, each with its own set

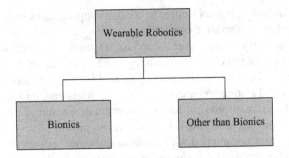

FIGURE 8.2 Classification of wearable robotics.

of advantages and disadvantages. There is a long road ahead until bionic limbs can match the flexibility, precision, and responsiveness of "biological" limbs.

8.1.8 Tentacle Gripper

The name "Tentacle Gripper" comes from the octopus's ability to grasp objects with its tentacles. In the future, this gripper may be able to aid those who are elderly or incapacitated in doing common tasks. Advantageously, it can hold objects firmly but gently. In addition, it can grasp and store a wide range of geometric forms. This apparatus's soft silicon structure is operable by use of a pneumatic regulator. The tentacle wraps around whatever it's attached to thanks to a contact supply of pressurized air, allowing it to bend within and around it [49].

8.1.9 Grippy: Bionic Mechanical Arm

Welcome to the future, where we've made bionics as easy to use as flipping a light switch, with Grippy, a lightweight and inexpensive battery-powered prosthesis for persons with below-elbow amputation age 15 and above, now available on the Indian market. Robo Bionics, a Recognized Startup by the Startup India Initiative of the DPIIT, Government of India, is responsible for creating Grippy.

8.1.10 Bionic Mechanical Foot

The bionic mechanical foot's speed was managed by a linear servo-electric cylinder located in the foot and a primary motor located in the leg. The foot's forward motion was controlled by the main motor, while the foot's raising and lowering motion was managed by a linear servo-electric cylinder. Two actuators worked together to regulate the bionic artificial foot's gait [50–52].

8.1.11 Wearable technology comprises of following other than bionics:

a) **Smartwatch:** when it comes to marketing, a smartwatch is meant to replace other devices, including a smartphone. A smartwatch is an all-in-one device

that serves as a timepiece, cellular phone, calculator, camera, GPS navigator, SD card, touchscreen, and rechargeable battery.
b) **Wearable Chargers:** battery life is one of the most aggravating aspects of owning a mobile phone, iPod, or any other portable gadget. Everpurse, another stylish tech startup, designed a bag that can be used to charge phones in an emergency.
c) **Wearable Technology for Healthcare:** consumer-friendly electronic wearables, such as activity trackers and smartwatches, are at the forefront of wearable healthcare technology.
    - **Wearable Fitness Trackers:** wearable fitness trackers, one of the earliest and most novel examples of wearable technology, take the shape of wristbands fitted with sensors to monitor the wearer's activity and heart rate
    - **Smart Health Watches:** these smartwatches provide some of the advantages of fitness trackers while also allowing users to conduct smartphone chores
    - **Wearable ECG Monitors:** the capacity to measure electrocardiograms (ECGs) is what sets wearable ECG monitors apart from some smartwatches and places them at the forefront of consumer electronics
    - **Wearable Blood Pressure Monitors:** Omron Healthcare introduced the world's first wrist-worn blood pressure monitor, HeartGuide, in 2019. Users of HeartAdvisor are able to record their blood pressure readings, keep tabs on their progress over time, and even share their information with their doctor
d) **Biosensors:** biosensors are a new type of wearable medical technology that differs greatly from existing devices like fitness trackers and smartwatches. This exemplifies the potential for wearables to enhance patient outcomes while also reducing staff effort

8.1.12 Drawbacks of Wearables Technology

Professionals in the medical field, who utilize wearable devices to keep tabs on patients' vitals, sleep patterns, and medication compliance report that the devices are beneficial in their work. Their primary complaints, however, are as follows:

- **Data utility:** according to Deloitte's most recent poll of US physicians, clinicians aren't interested in employing technology if it doesn't improve productivity and isn't integrated into their existing workflow
- **Data accuracy:** even while wearables are becoming increasingly common, not all medical professionals put their faith in the information they collect from consumers [53]
- **User error – and anxiety:** wearables might be incorrect if they are not worn properly. There is some evidence that people who use wearables to track their health are more likely to suffer from anxiety and develop compulsive habits

- **Data privacy concerns:** consumers' willingness to disclose health information has increased since the start of the COVID-19 epidemic. However, data privacy is still an issue [54]
- **Cybersecurity threats:** wearables aimed at health and well-being are just as susceptible to cyberattacks as any other type of Internet-connected gadget. Users may suffer serious repercussions as a result [55]
- **Increased regulation:** companies in the tech industry currently have the option of not classifying smartwatches as medical devices in order to sidestep laws like HIPAA in the US, which mandates users' informed consent before sharing any personal health data

8.1.13 AI Is Making Wearables Smarter

When computers can learn, plan, and solve problems like humans, we say that they have artificial intelligence. The term "AI" is used to characterize the cognitive abilities of machines and computer programmers. It's a field of research with the goal of creating smart machines. Machine learning techniques are analogous to human ones. By gathering information, artificial intelligence can assist in determining motion commands for operating a prosthetic arm. This provides greater mobility and eliminates the need for the user to repeatedly exert themselves through muscular contractions. The incorporation of AI technology in prostheses has allowed thousands of amputees to return to normal activities. In spite of the fact that the enabling technologies for bionic implants are still in their infancy, several bionic things are now available [56].

The upcoming discussion shows how the healthcare sector stands to benefit greatly from the combination of wearable and AI components.

- **Remote Heart Monitoring with AI and Sound**

For the better part of a decade, cardiologists have made use of AI components. For instance, AI has been effectively used in the backend of cardiology imaging and diagnostic systems to improve staff onboarding, workflows, and reporting.

- **AI Wearables for the Blind and Visually Impaired**

A few years ago, Sumu, a healthcare firm located in the United States, released the innovative Sumu Band. The band was created to help the blind or visually handicapped navigate through situations and prevent contact with surrounding obstacles.

- **AI that Helps Deaf People "Feel" the Environment**

Minnesota-based healthcare provider Starkey Hearing Technologies unveiled their new Livio Edge AI hearing aid, which has gesture recognition and natural language processing. The hearing aid combines 3D motion and gesture detection sensors to recreate the environment for the wearer, making up for their diminished senses. The information is then processed by the AI-driven program, which promptly and intelligently modifies the audio. The use of this method facilitates complete environmental awareness for those with hearing loss.

- **Intelligent Wearable Assistant**

Technology powered by artificial intelligence is maturing into a true assistant, elevating the service it provides to its users. The use of an AI-powered wristwatch as an illustration is appropriate. For example, consumers can be informed of their flight or movie times in the event and travel use cases. This is not limited to the sports business.

- **AI Wearables Provide Alerts**

When coupled with continuous monitoring, machine learning can help identify and categorize various types of seizure activity. Someone with epilepsy can wear a gadget equipped with AI that can identify patterns and send a warning, giving them time to pull over to the side of the road or get to a safe location.

- **AI Makes Exoskeletons Better**

There are already motorized exoskeleton legs, but they need users to move them manually through apps on their phones or joysticks. Each time the user wishes to conduct a different locomotor activity, they have to pause, pull out their smartphone, and pick the desired mode. AI aids devices in learning, reacting, perceiving, and comprehending so that they can carry out administrative and medical healthcare duties. Predictive analysis, when adopted and used at the right time, can stop patients from deteriorating in health, keep them from harming themselves, lower readmission rates, and cut down on missed visits [57–60].

8.2 LITERATURE REVIEW

The authors in [1] focused on innovations in robotics for the fourth industrial revolution. Recently, industrial robots have been developed and put to use in factories so that they may do activities that are too risky for people, speed up the manufacturing process without sacrificing quality, and lower product prices. Companies need more sophisticated systems that can make better judgments as the level of competition in the commercial world rises. Information technology developments such as AI, cloud, and Big Data were altering the way robots are used and built in the context of the Industry 4.0 revolutions. Future uses for industrial robots and the robotics generation set to be implemented in Industry 4.0 plants were explored. In [2], the authors have contemplated the moral implications of AI and service robots. As our reliance on robotic and AI technology grows, so too are the number of ethical questions that businesses and individuals must consider. This article focuses on five of these topics: (1) pervasive monitoring; (2) social engineering; (3) military robots; (4) sex robots;

and (5) transhumanism. All of these applications of AI and robotic services already raise ethical challenges in reality, with transhumanism being the partial exception. However, as these technologies advance, new worries will arise in all five domains. There will be lasting effects if we don't look into and deal with these issues immediately. The authors provide an overview of the literature that might serve as a foundation for this investigation. This study addresses a research need in the field of artificial intelligence and robotics applied to the service sector. It broadens how we think about robots and AI in service settings, which has crucial consequences for public policy and the practical implementation of service technologies. The meta-analysis of human-like service robots, chatbots, and other forms of AI has been introduced in [3] for cultural anthropomorphism studies. A theoretical framework is provided to examine the causes and effects of anthropomorphism in depth. Triggers of anthropomorphism include customer qualities and predispositions, socio-demographics, and robot design aspects. Traits of robots, such as intelligence and usefulness, are highlighted as essential mediators, but relational characteristics, such as rapport, get less support. This study's results provide light on the specific contexts in which anthropomorphism influences consumer intent to utilize a robot. The influence varies with robot type and service kind, as shown by the moderator analysis. They use these results to provide a thorough plan for future study of marketing and service robots. The authors in [4] focused on a plan for using AI in the classroom in light of the current technological shift. They also provided a case study that draws on Tecnologico de Monterrey's (Tec) 35 years of expertise in providing AI academic programmers to train tomorrow's engineers and knowledge workers. Our case study includes both upper-division and lower-division coursework, as well as research, internships, and the introduction of novel ideas and foreign programs. Over 5200 students at the Ph.D., MSc, and undergraduate levels have been educated throughout this period, with a significant emphasis on the application of technology in the field of AI. The university's curriculum has been updated and restructured in accordance with the Tec21 Educative Model. To adapt to the digital transition and meet the needs of Industry 4.0, this model's course offerings include challenge-based learning, physical and virtual practice laboratories, and a variety of teaching modalities. The approach and case study offered might serve as models for other universities looking to create AI academic programs to educate students in a way that is in demand by the corporations of the twenty-first century. In [5], the information structure has been provided for AI: a guide to categorizing AI tools. This chapter gives a high-level outline of a new AI technological landscape, which aids in the categorization of technologies based on a number of criteria. The authors in [6] used literary-based design fictions, they investigated the influence of AI and robotics at the university level. Because of the fragmented nature of the relevant literature and the fluidity of the ideas themselves, it may be difficult to appreciate the full magnitude of this effect. But advancements are accompanied by debates about what is technically possible, what is practically implementable, and what is desirable from a pedagogical or social perspective. Design fictions that plausibly imagine future scenarios of AI or robots in use are helpful tools for both explaining and debating the technological possibilities. In this chapter, we have conducted a narrative literature analysis to come up with eight such design fictions to showcase the many applications of AI and robots in academia. These instances sparked discussions on the impact on individual agency and the character of datafication, along with others, such as how they can make it possible to educate higher-order skills or alter staff responsibilities.

The author in [7] focused on how AI meets personalized medical treatment. In addition to making use of greater speed and computational efficiency, future AI-based health goods and apps will include a more in-depth understanding of biology. A large body of work in AI, ML, and statistical analysis has focused on discovering simple input/output relationships among data points; this work could be furthered by applying constraints known to govern phenomena of relevance leading to more biologically compelling molecular inferences. Reference [8] is concerned with the morality and ethics of artificial intelligence, considering examples of military drones and domestic companion robots. Was it really possible for AI to have higher ethical standards than humans? Is it possible it might treat human ideals with more regard than a human? When combined with agent-based theory, the utilitarian method may provide an effective answer. We have picked two extremes: lethal killer drones and helpful companion robots. Armed conflict brings up questions about the morality of AI and UAVs, which need to be investigated through the lens of military ethics and human values. Companion robots pose a threat to people's physical safety and well-being, in addition to their social and moral standing, even if they are not designed to do damage to humans. It is crucial from a moral standpoint that companion robots supplement human caregivers rather than take their place. Intelligent robots using algorithms have been used in [9] by the authors to advance occupational therapy and functional capacity assessment. In [10], the authors provided a secure human-robot interactive control framework for a wearable robot with an upper limb, which includes sensing human arm movement. In order to change the path of a traditional virtual mechanical impedance control system-operated lifting robot, the user must apply an interactive torque to the robot's joints. To avoid damaging the user's joint, a three-tiered hierarchical control system was built into the robot in this study. The signals from the user's surface electromyography were sent into a human arm movement detecting module. As the lifting process progressed into real use, a Hammerstein adaptive virtual mechanical impedance controller was used to provide an appropriate amount of torque for the user's elbow joint. The actuator controllers included in each mechanical part of the joint are primarily responsible for regulating the robot's lifting capacity. There was less interaction torque felt by the user's elbow, and fewer contractions of the erector spinae and biceps brachii, according to a number of studies. The suggested method safeguards the user against muscle fatigue and joint problems that may result from applying too much interaction torque to the human elbow joint. The authors in [11] provided a synthesis of critical imaginaries and reflections on robotics and AI in postdigital k-12 education. The focus of this work was to inquire into the ways in which educators, researchers, and pedagogical developers speculate about and think about the potential applications of AI and robotics in the classroom. The empirical information was obtained via symposia-related roundtable discussions. It was explained how the introduction of AI and emergency rooms (ERs) may alter the role of the teacher and the connection between instructors and students, and how more information is needed regarding these technologies. Many of those who took part believed that AI offered more opportunities for customization than ERs did. However, there were other worries raised while talking about AI, including ethical problems and commercial interests. Practitioners were more concerned with imaginaries rooted in present practice than researchers/developers, who were more preoccupied with envisioning ideal future educational practices enhanced by technological advancements.

[12] has reviewed wearable robotics human activity recognition using distributed convolutional neural network (CNN). Four different CNN architectures were compared. The ability to execute in-memory classification in real time was shown by deploying a distributed CNN on a cluster of Intel Edison nodes. Two of the strategies were consolidated and single-piece implementations, while the other two were distributed over several computers. Although distributed approaches have slightly lower accuracy than monolithic CNNs, they turn out to require much less memory – and therefore computation – than monolithic CNNs, and only modest communication rates between nodes in the model, making the approach viable for a wide range of distributed systems, from wearable robots to multi-robot swarms. [13] has provided a comparative legal study of privacy and personal data protection for AI-enabled robots, focusing on the technological and operational aspects of each. Analysis of the related literature yielded a number of suggestions. The first order of business in dealing with AI-related issues was to adopt privacy by design (PbD) as the most flexible, soft-legal, and desirable option. Implementing PbD, which protects individual privacy and personal data without needing any extra effort on the part of the user, is essential to the development of AI and to the advancement of privacy and personal data protection. A major focus of study has been the creation of context-aware technologies that can dynamically adapt to the user's preferences. There is a need for a comprehensive analysis of alternative technological approaches, such as those designed to unlock the "algorithmic black box" or accomplish differential privacy. If AI develops to the point that it can outperform humans, the only way to protect people's freedom of choice is to install some kind of kill switch or another kind of automatic shutdown mechanism. There are many obstacles, but they must be ready for a society where AI plays a major role by being adaptable. The authors in [14] have talked about the impact of wearable robots on ethics, law, and society. This document offers helpful recommendations for incorporating ethical, legal, and societal implications (ELSIs) into WR creation and maintenance. They justify domain-specific suggestions in light of current ELSI standards. In [15], the authors looked at perspectives and difficulties, and prospects for an AI-enhanced, all-encompassing curriculum were weighed. Specifically, differences may exist between individual students' cultural identities and their preferred methods of instruction. The constraints provided by cultural realities have resulted in a relatively low rate of successful research and innovative learning environment (ILE) transfer to poor areas. In this chapter, we step back and look at how intelligent learning environments have changed their focus from instructional rigor to the learner from a variety of angles. The research then moves on to discuss the key challenges that might occur while designing culturally sensitive ILEs. The chapter then focuses on pressing issues, such as machine ethics, and discusses many distinct methods for tackling these issues, including instructor modeling, the use of instructional robots, and sympathetic systems.

8.3 PROPOSED WORK

8.3.1 Problem Statement

A large number of research projects have been carried out on the topic of AI service delivery. The discipline of robotics, notably wearable robots, makes extensive use of these services. The present research, however, suffers from slow performance and inaccuracies. Previous studies overlooked the need for optimization before training and testing procedures. Therefore, an optimal technique has to be proposed in order to gain better accuracy while maintaining high performance.

8.3.2 Proposed Methodology

The functioning of already existing AI-based wearable robots has been taken into consideration in the study work that has been presented, as have the aspects that are impacting the performance and accuracy. Various machine-learning techniques are being examined for the goal of categorization; nevertheless, it is necessary to include an optimization process. The research community is thinking about optimizers in order to filter the dataset ahead of training time. It is expected that an optimized technique linked with machine learning would increase the accuracy of classifications produced by intelligent wearable robots. In conclusion, a contrast would be drawn between the work that was presented and the customary method based on performance and accuracy. The research methodology is shown in Figure 8.3.

8.3.3 Process Flow of Novel Approach

Figure 8.4 shows the process flow of the unique technique that has been provided, in which training and testing have been carried out across both optimized and non-optimized datasets. ACO is the optimization method that is used in research activity. Ant colony optimization, often known as ACO, is a process that searches for the optimal solution in order to filter the dataset. Training and testing are performed on the filtered dataset. Following this classification, the confusion matrix that was

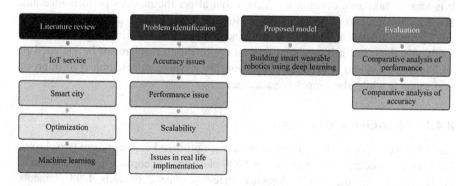

FIGURE 8.3 Research methodology.

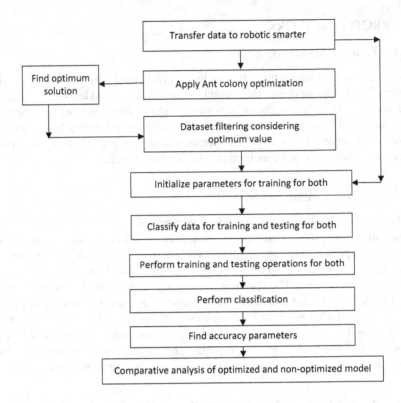

FIGURE 8.4 Flow chart of novel approach.

obtained is next analyzed in order to identify accuracy parameters. After that, the accuracy characteristics of the two models are compared with the goal of achieving a higher level of accuracy.

8.4 RESULTS AND DISCUSSION

It is vital to take into account the factors that affect the model's performance and accuracy in order to have an understanding of the workings of wearable robotic intelligence. Various approaches to machine learning are being investigated in order to classify data, but it is essential that an optimization mechanism be included in an integrated fashion as well. In conclusion, the purpose of research is to improve performance while also increasing accuracy.

8.4.1 Non-Optimized Dataset

Table 8.1 presents a confusion matrix for a non-optimized dataset where two classes have been considered. Each class has 5000 elements. According to the confusion matrix, class 1 predicts 4321 elements correctly; class 2 predicts 4298 elements correctly.

TABLE 8.1
Confusion Matrix of Non-Optimized Dataset

	Class 1	Class 2
Class 1	4321	679
Class 2	702	4298

TABLE 8.2
Accuracy of Confusion Matrix of Non-Optimized Dataset

Class	n (truth)	n (classified)	Accuracy	Precision	Recall	F1 Score
1	5023	5000	86.19%	0.86	0.86	0.86
2	4977	5000	86.19%	0.86	0.86	0.86

TABLE 8.3
Confusion Matrix of Optimized Dataset

	Class 1	Class 2
Class 1	4781	219
Class 2	323	4677

8.4.2 Results

TP: 8619
    Overall Accuracy: 86.19%
    Considering Table 8.1 as confusion matrix, accuracy, precision, recall, and f-score have been calculated and shown in Table 8.2. Overall accuracy is 86.19% for non-optimized dataset.

8.4.3 Optimized Dataset

Table 8.3 presents a confusion matrix for an optimized dataset where two classes have been considered. Each class has 5000 elements. According to the confusion matrix, class 1 predicts 4781 elements correctly; class 2 predicts 4677 elements correctly.

8.4.4 Results

TP: 9458

TABLE 8.4
Accuracy of Confusion Matrix of Optimized Dataset

Class	n (truth)	n (classified)	Accuracy	Precision	Recall	F1 Score
1	5104	5000	94.58%	0.96	0.94	0.95
2	4896	5000	94.58%	0.94	0.96	0.95

TABLE 8.5
Comparison of Accuracy

Class	Non-optimized dataset	Optimized dataset
1	86.19%	94.58%
2	86.19%	94.58%

Overall Accuracy: 94.58%

Considering Table 8.3 as confusion matrix, accuracy, precision, recall, and f1-score have been calculated and shown in Table 8.4. Overall accuracy is 94.58% for an optimized dataset.

8.4.5 Comparison of Accuracy Parameters of Non-Optimized Dataset and Optimized Dataset

8.4.5.1 Accuracy

Table 8.2 contains the accuracy rates at which the non-optimized dataset performed, whereas Table 8.4 has the corresponding rates for the improved dataset. Table 8.5 compares the two numbers provided.

Considering Table 8.5, a comparative analysis of non-optimized dataset and optimized dataset accuracy has been shown in Figure 8.5.

8.4.5.2 Precision

Table 8.2 provides the precision value for the original dataset, whereas Table 8.4 provides the precision value for the optimized dataset. We compare the two figures using Table 8.6.

Considering Table 8.6, comparative analysis of non-optimized dataset and optimized dataset precision has been shown in Figure 8.6.

8.4.5.3 Recall Value

Table 8.2 shows the recall values that were used for the non-optimized dataset, while Table 8.4 shows the recall values that were used for the improved dataset. Table 8.8 contains both sets of values so that you can easily compare them.

Taking into account the data in Table 8.7, Figure 8.7 displays a comparison of the recall value of the non-optimized dataset to the optimized dataset.

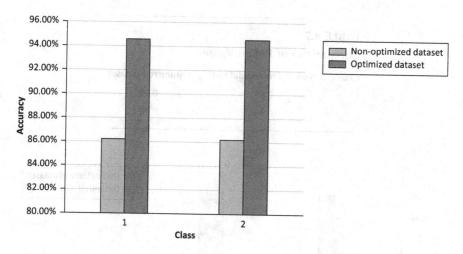

FIGURE 8.5 Comparison of accuracy.

TABLE 8.6
Comparison of Precision

Class	Non-optimized dataset	Optimized dataset
1	0.86	0.96
2	0.86	0.94

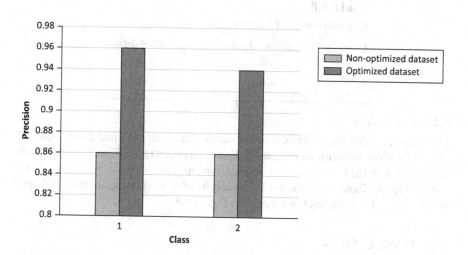

FIGURE 8.6 Comparison of precision.

татBLE 8.7
Comparison of Recall Value

Class	Non-optimized dataset	Optimized dataset
1	0.86	0.94
2	0.86	0.96

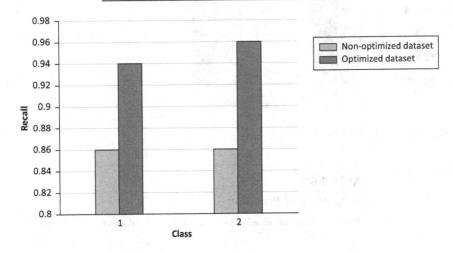

FIGURE 8.7 Comparison of recall value.

TABLE 8.8
Comparison of F1-score

Class	Non-optimized dataset	Optimized dataset
1	0.86	0.95
2	0.86	0.95

8.4.5.4 F1-score

The F1-score value for a non-optimized dataset has been obtained from Table 8.2 and the F1-score value for an optimized dataset has been taken from Table 8.4. Both values are included in Table 8.8 for comparison analysis.

Considering Table 8.8, a comparative analysis of non-optimized dataset and optimized dataset F1-score has been shown in Figure 8.8.

8.5 CONCLUSION

The conclusion that can be drawn from the results of the simulation is that optimization plays a vital role in the selection of appropriate datasets in the case of wearable intelligent robots. When compared to more traditional processes, optimized datasets

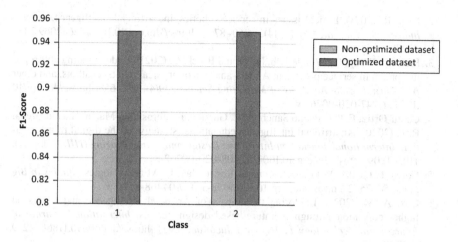

FIGURE 8.8 Comparison of F1-score.

provide much higher levels of accuracy. Additionally, the suggested approach has decreased the amount of time spent performing training operations by decreasing the size of the dataset. In the case of the suggested work, it can be determined that the f-score, accuracy, precision, and recall value are all improved.

8.6 FUTURE SCOPE

This study investigates the feasibility of incorporating AI into wearable robots to enhance their usefulness. There is also an introduction to some of the basic concepts behind robotics and AI. The advent of AI has paved the way for the creation of actual, workable answers to a broad range of issues. Artificial intelligence's power to boost output might cause it to usher in a new era of innovation across a wide range of sectors. Robotics is an interdisciplinary field since it involves mechanical construction, electrical components, and computer programming. Robots driven by AI can do tasks in all the industries with ease. The phrase "wearable robot" is used to describe a particular kind of wearable technology that is intended to increase the user's mobility and/or performance in the real world [61–63]. Thousands of amputees have been able to resume regular lives with the use of AI elements in their prostheses. Wearable healthcare technology is currently being pioneered by consumer electronic devices like fitness trackers and smartwatches. AI is the ability of a machine to do activities that would ordinarily need human intellect. It is generally accepted that a machine has artificial intelligence when it demonstrates cognition that is on par with that of a human being. Despite their different purposes and applications, robotics and AI are often addressed together. Robots might pass for humans when they are programmed with enough AI to replicate our actions.

REFERENCES

1. Bayram, B., and Ince, G. (n.d.). *Advances in Robotics in the Era of Industry 4.0* (pp. 187–200). https://doi.org/10.1007/978-3-319-57870-5_11

2. Belk, R. (2021). Ethical issues in service robotics and artificial intelligence. *Service Industries Journal*, 41(13–14), 860–876. https://doi.org/10.1080/02642069.2020.1727892
3. Blut, M., Wang, C., Wünderlich, N. V., and Brock, C. (2021). Understanding anthropomorphism in service provision: A meta-analysis of physical robots, chatbots, and other AI. *Journal of the Academy of Marketing Science*, 49(4), 632–658. https://doi.org/10.1007/s11747-020-00762-y
4. Cantú-Ortiz, F. J., Galeano Sánchez, N., Garrido, L., Terashima-Marin, H., and Brena, R. F. (2020). An artificial intelligence educational strategy for the digital transformation. *International Journal on Interactive Design and Manufacturing (IJIDeM)*, 14(4), 1195–1209. https://doi.org/10.1007/s12008-020-00702-8
5. Corea, F. (2019). AI knowledge map: How to classify AI technologies. *Studies in Big Data*, 50, 25–29. https://doi.org/10.1007/978-3-030-04468-8_4
6. Cox, A. M. (2021). Exploring the impact of Artificial Intelligence and robots on higher education through literature-based design fictions. *International Journal of Educational Technology in Higher Education*, 18(1). https://doi.org/10.1186/s41239-020-00237-8
7. Da Costa Leite, C. (2019). Artificial intelligence, radiology, precision medicine, and personalized medicine. *Radiologia Brasileira*, 52(6), VII–VIII. https://doi.org/10.1590/0100-3984.2019.52.6e2
8. deSwarte, T., Boufous, O., and Escalle, P. (2019). Artificial intelligence, ethics and human values: The cases of military drones and companion robots. *Artificial Life and Robotics*, 24(3), 291–296. https://doi.org/10.1007/s10015-019-00525-1
9. Fong, J., Ocampo, R., Gross, D. P., and Tavakoli, M. (2020). Intelligent robotics incorporating machine learning algorithms for improving functional capacity evaluation and occupational rehabilitation. *Journal of Occupational Rehabilitation*, 30(3), 362–370. https://doi.org/10.1007/s10926-020-09888-w
10. Hao, L., Zhao, Z., Li, X., Liu, M., Yang, H., and Sun, Y. (2020). A safe human–robot interactive control structure with human arm movement detection for an upper-limb wearable robot used during lifting tasks. *International Journal of Advanced Robotic Systems*, 17(5), 1–15. https://doi.org/10.1177/1729881420937570
11. Hrastinski, S., Olofsson, A. D., Arkenback, C., Ekström, S., Ericsson, E., Fransson, G., Jaldemark, J., Ryberg, T., Öberg, L. M., Fuentes, A., Gustafsson, U., Humble, N., Mozelius, P., Sundgren, M., and Utterberg, M. (2019). Critical imaginaries and reflections on artificial intelligence and robots in postdigital K-12 education. *Postdigital Science and Education*, 1(2), 427–445. https://doi.org/10.1007/s42438-019-00046-x
12. Hughes, D., and Correll, N. (2018). Distributed convolutional neural networks for human activity recognition in wearable robotics. In *Springer* Proceedings in Advanced Robotics (Vol. 6). Springer International Publishing. https://doi.org/10.1007/978-3-319-73008-0_43
13. Ishii, K. (2019). Comparative legal study on privacy and personal data protection for robots equipped with artificial intelligence: Looking at functional and technological aspects. *AI and Society*, 34(3), 509–533. https://doi.org/10.1007/s00146-017-0758-8
14. Kapeller, A., Felzmann, H., Fosch-Villaronga, E., Nizamis, K., and Hughes, A. M. (2021). Implementing ethical, legal, and societal considerations in wearable robot design. *Applied Sciences (Switzerland)*, 11(15). https://doi.org/10.3390/app11156705
15. Lee, U. H., Bi, J., Patel, R., Fouhey, D., and Rouse, E. (2020). Image transformation and CNNs: A strategy for encoding human locomotor intent for autonomous wearable robots. *IEEE Robotics and Automation Letters*, 5(4), 5440–5447. https://doi.org/10.1109/LRA.2020.3007455

16. Sindhwani, N., Anand, R., Niranjanamurthy, M., Verma, D. C., and Valentina, E. B. (Eds.). (2022). *IoT Based Smart Applications*. Springer Nature.
17. Singh, P., Kaiwartya, O., Sindhwani, N., Jain, V., and Anand, R. (Eds.). (2022). *Networking Technologies in Smart Healthcare: Innovations and Analytical Approaches*. CRC Press.
18. Malik, G., Tayal, D. K., and Vij, S. (2019). An analysis of the role of artificial intelligence in education and teaching. In *Advances in Intelligent Systems and Computing* (Vol. 707). Springer. https://doi.org/10.1007/978-981-10-8639-7_42
19. Mihret, E. T. (2020). Robotics and artificial intelligence. *International Journal of Artificial Intelligence and Machine Learning*, 10(2), 57–78. https://doi.org/10.4018/ijaiml.2020070104
20. Mohammed, P. S., and 'Nell' Watson, E. (2019). *Towards Inclusive Education in the Age of Artificial Intelligence: Perspectives, Challenges, and Opportunities*. Springer. https://doi.org/10.1007/978-981-13-8161-4_2
21. Murray-Smith, R. (2014). Models and measures of human–computer symbiosis. *Lecture Notes in Computer Science (Including Subseries Lecture Notes in Artificial Intelligence and Lecture Notes in Bioinformatics)*, 8820, 72–83. https://doi.org/10.1007/978-3-319-13500-7
22. Pandey, D., Pandey, B. K., Sindhwani, N., Anand, R., Nassa, V. K., and Dadheech, P. (2022). An interdisciplinary approach in the post-COVID-19 pandemic era. *An Interdisciplinary Approach in the Post-COVID-19 Pandemic Era* (pp. 1–290).
23. Oniani, S., Marques, G., Barnovi, S., Pires, I. M., and Bhoi, A. K. (2021). Artificial intelligence for internet of things and enhanced medical systems. In *Studies in Computational Intelligence* (Vol. 903). Springer. https://doi.org/10.1007/978-981-15-5495-7_3
24. Singh, H., Ramya, D., Saravanakumar, R., Sateesh, N., Anand, R., Singh, S., and Neelakandan, S. (2022). Artificial intelligence based quality of transmission predictive model for cognitive optical networks. *Optik*, 257, 168789.
25. Papapicco, V., Parri, A., Martini, E., Bevilacqua, V., Crea, S., and Vitiello, N. (2019). Locomotion mode classification based on support vector machines and hip joint angles: A feasibility study for applications in wearable robotics. In *Springer Proceedings in Advanced Robotics* (Vol. 7). Springer International Publishing. https://doi.org/10.1007/978-3-319-89327-3_15
26. Turing, Alan M. (2009). *Computing machinery and intelligence*. Springer Netherlands. https://doi.org/10.1007/978-1-4020-6710-5_3
27. Park, H. B., Kim, D. R., Kim, H. J., Wang, W., Han, M. W., and Ahn, S. H. (2020). Design and analysis of artificial muscle robotic elbow joint using shape memory alloy actuator. *International Journal of Precision Engineering and Manufacturing*, 21(2), 249–256. https://doi.org/10.1007/s12541-019-00240-8
28. Pramanik, P. K. D., Pal, S., and Choudhury, P. (2018). Beyond automation: The cognitive IoT. artificial intelligence brings sense to the internet of things. In *Lecture Notes on Data Engineering and Communications Technologies* (Vol. 14). https://doi.org/10.1007/978-3-319-70688-7_1
29. Raj, M., and Seamans, R. (2019). Primer on artificial intelligence and robotics. *Journal of Organization Design*, 8(1). https://doi.org/10.1186/s41469-019-0050-0
30. Torricelli, D., Rodriguez-Guerrero, C., Veneman, J. F., Crea, S., Briem, K., Lenggenhager, B., and Beckerle, P. (2020). Benchmarking wearable robots: Challenges and recommendations from functional, user experience, and methodological perspectives. *Frontiers in Robotics and AI*, 7(November). https://doi.org/10.3389/frobt.2020.561774

31. Tyagi, A. K., and Chahal, P. (2019). Artificial intelligence and machine learning algorithms. *Challenges and Applications for Implementing Machine Learning in Computer Vision*, 188–219. https://doi.org/10.4018/978-1-7998-0182-5.ch008
32. Vatsal, V., and Hoffman, G. (2021). Biomechanical motion planning for a wearable robotic forearm. *IEEE Robotics and Automation Letters*, 6(3), 5024–5031. https://doi.org/10.1109/LRA.2021.3071675
33. Gupta, A., Srivastava, A., and Anand, R. (2019). Cost-effective smart home automation using internet of things. *Journal of Communication Engineering and Systems*, 9(2), 1–6.
34. Wang, C., Teo, T. S. H., and Janssen, M. (2021). Public and private value creation using artificial intelligence: An empirical study of AI voice robot users in Chinese public sector. *International Journal of Information Management*, 61(August), 102401. https://doi.org/10.1016/j.ijinfomgt.2021.102401
35. Bommareddy, S., Khan, J. A., and Anand, R. (2022). A review on healthcare data privacy and security. *Networking Technologies in Smart Healthcare*, 165–187.
36. Zardiashvili, L., and Fosch-Villaronga, E. (2020). "Oh, Dignity too?" Said the robot: Human dignity as the basis for the governance of robotics. *Minds and Machines*, 30(1), 121–143. https://doi.org/10.1007/s11023-019-09514-6
37. Zhang, Y., and Qi, S. (2019). User experience study: The service expectation of hotel guests to the utilization of AI-based service robot in full-service hotels. In *Lecture Notes in Computer Science (Including Subseries Lecture Notes in Artificial Intelligence and Lecture Notes in Bioinformatics): Vol. 11588 LNCS*. Springer International Publishing. https://doi.org/10.1007/978-3-030-22335-9_24
38. Raghavan, R., Verma, D. C., Pandey, D., Anand, R., Pandey, B. K., and Singh, H. (2022). Optimized building extraction from high-resolution satellite imagery using deep learning. *Multimedia Tools and Applications*, 81(29), 42309–42323.
39. Zheng, Y., Tang, N., Omar, R., Hu, Z., Duong, T., Wang, J., Wu, W., and Haick, H. (2021). Smart materials enabled with artificial intelligence for healthcare wearables. *Advanced Functional Materials*, 31(51), 1–20. https://doi.org/10.1002/adfm.202105482
40. Pandey, D., and Wairya, S. (2023). An optimization of target classification tracking and mathematical modelling for control of autopilot. *The Imaging Science Journal*, 70(6), 1–16.
41. Gupta, R. S., Nassa, V. K., Bansal, R., Sharma, P., Koti, K., and Koti, K. (2021). Investigating application and challenges of big data analytics with clustering. In *2021 International Conference on Advancements in Electrical, Electronics, Communication, Computing and Automation (ICAECA)* (pp. 1–6). https://doi.org/10.1109/ICAECA52838.2021.9675483
42. Veeraiah, V., Khan, H., Kumar, A., Ahamad, S., Mahajan, A., and Gupta, A. (2022). Integration of PSO and deep learning for trend analysis of meta-verse. In *2022 2nd International Conference on Advance Computing and Innovative Technologies in Engineering (ICACITE)* (pp. 713–718). https://doi.org/10.1109/ICACITE53722.2022.9823883
43. Pandey, B. K., Pandey, D., Nassa, V. K., George, S., Aremu, B., Dadeech, P., and Gupta, A. (2023). Effective and secure transmission of health information using advanced morphological component analysis and image hiding. In Gupta, M., Ghatak, S., Gupta, A., Mukherjee, A. L. (eds) *Artificial Intelligence on Medical Data. Lecture Notes in Computational Vision and Biomechanics*, 37. Springer. https://doi.org/10.1007/978-981-19-0151-5_19
44. Veeraiah, V., Kumar, K. R., LalithaKumari, P., Ahamad, S., Bansal, R., and Gupta, A. (2022). Application of biometric system to enhance the security in virtual world. In *2022 2nd International Conference on Advance Computing and Innovative Technologies in Engineering (ICACITE)* (pp. 719–723). https://doi.org/10.1109/ICACITE53722.2022.9823850

45. Bansal, R., Gupta, A., Singh, R., and Nassa, V. K. (2021). Role and impact of digital technologies in e-learning amidst COVID-19 pandemic. In *2021 Fourth International Conference on Computational Intelligence and Communication Technologies (CCICT)* (pp. 194–202). https://doi.org/10.1109/CCICT53244.2021.00046
46. Shukla, A., Ahamad, S., Rao, G. N., Al-Asadi, A. J., Gupta, A., and Kumbhkar, M. (2021). Artificial intelligence assisted IoT data intrusion detection. In *2021 4th International Conference on Computing and Communications Technologies (ICCCT)* (pp. 330–335). https://doi.org/10.1109/ICCCT53315.2021.9711795
47. Pandey, D., Wairya, S., Sharma, M., Gupta, A. K., Kakkar, R., and Pandey, B. K. (2022). An approach for object tracking, categorization, and autopilot guidance for passive homing missiles. *Aerospace Systems*, 5(4), 1–14.
48. Dushyant, K., Muskan, G., Annu, Gupta, A., Pramanik, S. (2022). Utilizing machine learning and deep learning in cybersecurity: An innovative approach. In *Cyber Security and Digital Forensics: Challenges and Future Trends* (pp. 271–293). Wiley. https://doi.org/10.1002/9781119795667.ch12
49. Babu, S. Z. D., Pandey, D., Naidu, G. T., Sumathi, S., Gupta, A., Bader Alazzam, M., and Pandey, B. K. (2023). Analysation of big data in smart healthcare. In Gupta, M., Ghatak, S., Gupta, A., Mukherjee, A. L. (eds) *Artificial Intelligence on Medical Data. Lecture Notes in Computational Vision and Biomechanics*, 37. Springer. https://doi.org/10.1007/978-981-19-0151-5_21
50. Bansal, B., Jenipher, N., Jain, R., Dilip, R., Kumbhkar, M., Pramanik, S., Sandip, R., and Gupta, A. (2022). Big data architecture for network security. In *Cyber Security and Network, (Security)* (pp. 233–267). Wiley. https://doi.org/10.1002/9781119812555.ch11
51. Gupta, A., Kaushik, D., Garg, M., and Verma, A. (2020). Machine learning model for breast cancer prediction. In *2020 Fourth International Conference on I-SMAC (IoT in Social, Mobile, Analytics and Cloud) (I-SMAC)* (pp. 472–477). https://doi.org/10.1109/I-SMAC49090.2020.9243323
52. Sreekanth, N. et al. (2022). Evaluation of estimation in software development using deep learning-modified neural network. *Applied Nanoscience*. https://doi.org/10.1007/s13204-021-02204-9
53. Veeraiah, V., Rajaboina, N. B., Rao, G. N., Ahamad, S., Gupta, A., and Suri, C. S. (2022). Securing online web application for IoT management. In *2022 2nd International Conference on Advance Computing and Innovative Technologies in Engineering (ICACITE)* (pp. 1499–1504). https://doi.org/10.1109/ICACITE53722.2022.9823733
54. Veeraiah, V., Gangavathi, P., Ahamad, S., Talukdar, S. B., Gupta, A., and Talukdar, V. (2022). Enhancement of meta verse capabilities by IoT integration. In *2022 2nd International Conference on Advance Computing and Innovative Technologies in Engineering (ICACITE)* (pp. 1493–1498). https://doi.org/10.1109/ICACITE53722.2022.9823766
55. Gupta, N., Janani, S., Dilip, R., Hosur, R., Chaturvedi, A., and Gupta, A. (2022). Wearable sensors for evaluation over smart home using sequential minimization optimization-based random forest. *International Journal of Communication Networks and Information Security (IJCNIS)*, 14(2), 179–188. https://doi.org/10.17762/ijcnis.v14i2.5499
56. Keserwani, H., Rastogi, H., Kurniullah, A. Z., Janardan, S. K., Raman, R., Rathod, V. M., and Gupta, A. (2022). Security enhancement by identifying attacks using machine learning for 5G network. *International Journal of Communication Networks and Information Security (IJCNIS)*, 14(2), 124–141. https://doi.org/10.17762/ijcnis.v14i2.5494
57. Kumar, M. S., Sankar, S., Nassa, V. K., Pandey, D., Pandey, B. K., and Enbeyle, W. (2021). Innovation and creativity for data mining using computational statistics. In: *Methodologies and Applications of Computational Statistics for Machine Intelligence* (pp. 223–240). IGI Global.

58. Pandey, B. K., Pandey, D., Wariya, S., Aggarwal, G., and Rastogi, R. (2021). Deep learning and particle swarm optimisation-based techniques for visually impaired humans' text recognition and identification. *Augmented Human Research*, 6, 1–14.
59. Sindhwani, N., Maurya, V. P., Patel, A., Yadav, R. K., Krishna, S., and Anand, R. (2022). Implementation of intelligent plantation system using virtual IoT. *Internet of Things and its Applications*, 305–322.
60. Pandey, D., Pandey, B. K., and Wariya, S. (2020). An approach to text extraction from complex degraded scene. *IJCBS*, 1(2), 4–10.
61. Jain, S., Kumar, M., Sindhwani, N., and Singh, P. (2021, September). SARS-Cov-2 detection using Deep Learning Techniques on the basis of Clinical Reports. In *9th International Conference on Reliability, Infocom Technologies and Optimization (Trends and Future Directions)(ICRITO)* (pp. 1–5). IEEE.
62. Verma, S., Bajaj, T., Sindhwani, N., and Kumar, A. (2022). Design and development of a driving assistance and safety system using deep learning. In *Advances in Data Science and Computing Technology* (pp. 35–45). Apple Academic Press.
63. Jain, N., Chaudhary, A., Sindhwani, N., and Rana, A. (2021, September). Applications of Wearable devices in IoT. In *2021 9th International Conference on Reliability, Infocom Technologies and Optimization (Trends and Future Directions)(ICRITO)* (pp. 1–4). IEEE.

9 Role of Machine Learning Approaches in Optimization of AI-Based Industry 4.0 Healthcare Management Systems

Amogh Shukla, Gautam Chettiar, Vinit Juneja, and Sonakshi Singh

9.1 INTRODUCTION

9.1.1 Artificial Intelligence

General difficulties can be tackled by putting Artificial Intelligence (AI) into machines. Using a combination of data and instruction, machines can learn to do new things on their own. It is implemented to generate data patterns. The computer will be able to make better judgments in the future based on what it has learned from historical observations. AI's primary goal is to allow computers to learn on their own. AI is a term used to describe a larger Machine Learning (ML) superset and its applications [1–3]. Anybody may benefit from its ability to learn naturally. It's a learning system that gets better over time after multiple training iterations. Machine learning today is achievable without any programming knowledge at all.

9.1.2 Applications of AI

Numerous disciplines (as shown in Figure 9.1) have seen the dominance of AI, including but not limited to:

- Gaming: several games include AI. As a result, it's obvious that artificial intelligence has been included in games like chess, poker, tic-tac-toe, and even more complicated games involving strategy, planning, decisiveness, etc.
- Natural Language Processing: interacting with a computer is also made possible via the usage of AI-based technologies. These tools may be used to decipher the spoken language. They are able to comprehend the words of the user and generate appropriate inferences and replies for the same.

DOI: 10.1201/9781003317456-9

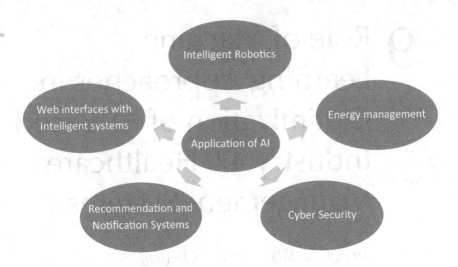

FIGURE 9.1 Applications of AI in home automation.

- Systems of Expert: a variety of gadgets are available. The system recognizes the various devices. These systems are advanced and well-coded. In order to help the user, these systems explain and provide guidance.
- Vision Systems: these systems are capable of interpreting and using computer-generated images. These include tasks such as object detection, segmentation, etc.
- Speech Recognition: these AI-based gadgets are also used for the detection and classification of various vocal tones. Trained on large-scale audio conversation data, they are able to infer from the audio patterns the corresponding responses which then generate text.
- Handwriting Recognition: the AI-based systems can interpret the text that has been printed or shown on a screen. It is also possible for these programmers to identify the forms that correspond to each letter of the alphabet. The text that may be edited has also been changed
- Intelligent Robots: the employment of robots to execute tasks is known to increase overall efficiency. In order to demonstrate their intelligence, these machines contain powerful central processing units, sensors, and big memory. Learning from their errors and adapting to a new situation is something they are capable of doing.

The use of AI in different sectors is shown in Figure 9.2.

9.1.3 Machine Learning (ML)

Software applications may be accurately predicted using this approach, which doesn't need any programming. ML algorithms use data from previous records to predict future output values. Fraud detection, waste sorting, malware detection, business

Machine Learning Approaches in Healthcare Management Systems

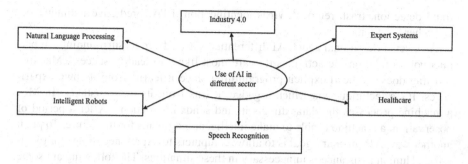

FIGURE 9.2 Use of AI in different sectors.

process automation (BPA), and predictive maintenance are just a few of the many applications of machine learning. Classifying classical machine learning as it pertains to algorithmic improvements in prediction accuracy is a popular practice. The two key approaches in ML are unsupervised and supervised learning. Algorithm selection may be predicted using data, according to scientists. Machine learning is now being used for a wide range of purposes. It is employed for Facebook's News Feed recommendation engine. The recommendation engine will prioritize the activity of a member who reads the posts in a certain group on a regular basis. Behind the scenes, the engine is hard at work forming the member's online habits. As soon as the member quits reading posts from the group, their News Feeds will be changed.

The term "machine learning" refers to the process of teaching computers to make more accurate predictions by feeding them data. Numerous other predictions are feasible, such as if pedestrians are crossing the road in front of a self-driving car, whether the term "book" is being used to describe a paperback book or hotel bookings, or even the likelihood that an email will be labeled "spam." When it comes to determining whether an apple or a banana is being supplied as an input, computers have depended on human programming until this point. If you'd like another way of putting it, a machine-learning model was developed that could tell the difference between an apple and a banana from a large collection of photos that were classified as containing either one. In order for machine learning to work, it needs a lot of data [4, 5].

Because of its capacity to boost productivity, artificial intelligence has the potential to transform a broad number of industries. Machine learning will become more important as computing power improves. In the future, machines may be able to do mental tasks better than humans. Computer algorithms trump radiologists when it comes to recognizing cancers. However, radiology is still in its infancy compared to other fields. AI and automation can potentially eliminate millions of jobs in the near future. Software that can "learn" on its own without the help of a human is referred to as autonomous. In big data analytics and data mining, the phrase ML is used to denote a wide range of software applications.

A software application's ability to accurately forecast its output is enabled by this approach, which requires no coding on the user's part. Machine-learning algorithms employ prior data as input for predicting new output values. It is used for fraud detection, trash removal, virus identification, BPA, predictive maintenance, and other functions.

As a versatile rational agent, AI differentiates itself from its surroundings. It permits you to take on the activities that are most likely to lead to success. Machine learning does not need explicit programming since it learns from its own experiences. It's a technique via which a gadget learns from its own mistakes. In ML, a machine processes the data, directs it, and sends instructions. After a period of observation, a machine is able to make intelligent judgments for the future. To put it another way, AI's primary goal is to allow computers to experience things for themselves. Human assistance is unnecessary in these situations. The following are some of the many advantages of machine learning:

1. Banking and finance, healthcare, journalism, and social media are just a few of the industries that use it.
2. Because of machine learning, gadgets are able to operate at a lower energy consumption rate while also enhancing resource efficiency.
3. Social media platforms like Google and Facebook regularly employ this tactic to tailor their advertising to the preferences of their consumers.
4. Several machine-learning technologies are capable of improving the quality of large and complicated process systems.

9.1.3.1 Methods

Algorithms in machine learning are characterized by their capability to anticipate accurate results. It is possible to learn with or without a teacher, but two common methods are supervised learning and unsupervised learning. Scientists want to make predictions using the best algorithm based on the data they have available.

9.1.3.2 Working of Supervised Machine Learning

The data scientist must provide the algorithm with both labeled inputs and intended outputs in order to train it. The following problems are well-suited to supervised learning algorithms:

1. Binary classification – for categorizing the results into two classes.
2. Multi-class classification – when there are multiple possible values as the output.
3. Regression modeling – It's used to forecast results when the variables are in continuous form.

9.1.3.3 Working of Unsupervised Machine Learning

Labeling is not necessary for unsupervised machine learning. Unlabeled data is sifted through for patterns that are utilized to categorize data into subgroups. The following situations are well-suited to unsupervised learning algorithms:

1. Clustering – used for dividing the dataset into categories depending on their degree of resemblance.

Machine Learning Approaches in Healthcare Management Systems

2. Anomaly detection – analyzing data for anomalous patterns.
3. Association mining – a dataset may be used to find similar patterns among the components that make up the collection.

9.1.3.4 Uses of Machine Learning

Machine learning is now being used in a wide range of applications. Among its many uses is in Facebook's News Feed recommendation engine. It's possible that the recommendation engine may begin to prioritize postings from a certain group if a member often pauses to read them. The engine is working to reinforce a subscriber's online habits that are already established. The News Feed adjusts if the member's reading habits transition and he or she fails to keep up with postings from that particular group.

9.1.3.5 Other Uses of Machine Learning Include the Following:

- **Human resource information system (HRIS)** – ML models may be used by HRIS systems to sort through resumes and find the top applicants for a job opening.
- **Self-driving cars** – Even a semi-autonomous automobile may distinguish a partly visible item and inform the driver using ML techniques.
- **Virtual assistants** – In order to analyze spoken speech and provide context, smart assistants often use a combination of supervised and unsupervised machine learning models.
- **Choosing the right dimensionality reduction** – This reduces the number of variables in a data collection.

9.1.4 Deep Learning

Machine learning includes deep learning as a subset. Even though the system still has to be taught from data, this is a major step forward for artificial intelligence. The study of neural networks led to the development of deep learning. While neural nets have been around since the early '80s, it wasn't until a few years ago that deep learning really took off. Because of the enormous quantity of data we have collected, machine learning and deep learning have flourished as a result of the rise of inexpensive, powerful computers [5–7]. In comparison to machine learning, deep learning produced much more intelligent outcomes. Let's go back to the face recognition example: What type of information should we feed the AI in order for it to recognize a face? When we just have pixel colors to work with, how should we learn what to look for?

9.1.4.1 Inspirational Applications of Deep Learning

1. **Image colorization** – Color is added to black and white images using this method. Because it's such a tough job, it's usually done manually. With the aid of deep-learning algorithms, images may be colored based on the things in them and their context.
2. **Automatic Machine Translation** – It's a piece of software that will automatically translate a word or phrase from one language into another. When

it comes to automated machine translation, deep learning achieves the best results.
3. **Object Classification in Photographs** – In order to complete this job, you must recognize items in an image as belonging to one of many categories.
4. **Character Text Generation** – It's a fascinating challenge. Learning a large corpus of material and then using it as a guide to creating a fresh piece of writing is the goal of this exercise.

9.1.5 Healthcare

Preventing, diagnosing, treating, ameliorating, or curing disease, sickness, injury, and other physical and mental disabilities is the goal of health care, often known as healthcare. There are a variety of healthcare professions and allied health areas that provide services to patients. People's access to healthcare may be affected by a variety of factors, including socioeconomic and political factors. Caregiving is defined as "the timely use of personal health services to reach the optimal level of well-being." Access to health treatment is hampered by a variety of factors, including lack of financial resources,

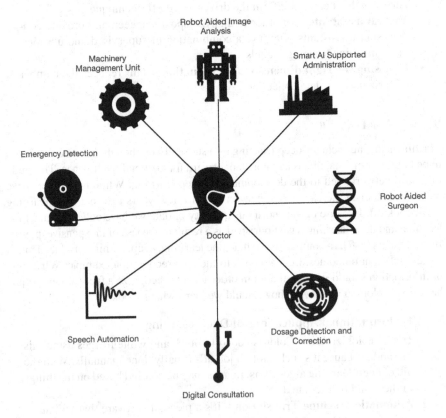

FIGURE 9.3 Application of AI in healthcare and Industry 4.0.

Machine Learning Approaches in Healthcare Management Systems

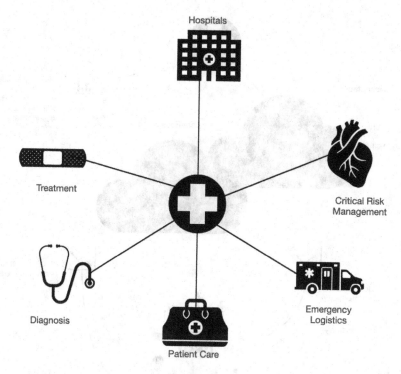

FIGURE 9.4 Health care management system.

physical distance, and the limits imposed by the individuals themselves. The usage of medical services, the effectiveness of treatments, and the overall result are all harmed when such services are restricted. In order to satisfy the healthcare demands of certain populations, health systems are set up. The World Health Organization (WHO) states that well-functioning healthcare systems need money, well-trained staff, trustworthy data on which to base policy, and well-maintained health facilities that can provide quality drugs and technology. A country's economy, progress, and industrialization may all benefit from a well-functioning healthcare system. People's overall physical and mental health and well-being are widely believed to be influenced by a variety of factors, including healthcare. Smallpox was proclaimed eradicated in 1980 by the World Health Organization, the first illness in human history to be eliminated via purposeful healthcare measures by the organization [8, 9]. Application of AI in healthcare and Industry 4.0 is depicted in Figure 9.3. A healthcare management system is shown in Figure 9.4 and a health data ecosystem is shown in Figure 9.5. A brief proposed methodology is shown in Figure 9.6.

9.1.6 Applications in Healthcare

There are a variety of ways blockchain technology may be used to improve healthcare:
- Archiving and retrieval of electronic health records.
- Concerns about the security of healthcare data have been addressed.

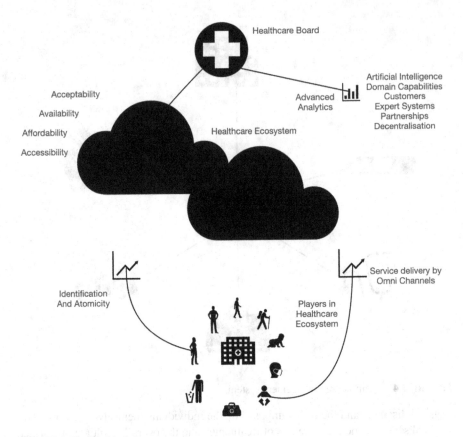

FIGURE 9.5 Health data ecosystem.

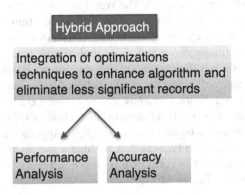

FIGURE 9.6 Proposed methodology.

Machine Learning Approaches in Healthcare Management Systems

- Handling of data relating to an individual's health.
- Management of genomes on the patient's side.
- Keeping track of the data included in electronic medical records.

9.1.7 HEALTHCARE APPLICATIONS

Some applications in healthcare are mentioned below.

9.1.7.1 Research

As of now, electronic health records cannot be updated or exchanged outside of a certain institution or network. If the data is arranged in this manner, non-protected health information (PHI) or personally identifiable information (PII) may be stored in blocks at the top of the blockchain PII. As many as million-strong cohorts of people might be accessed by academics and other organizations in the future. This might have a big impact on clinical research as well as safety, adverse event reporting, and public health reporting.

9.1.7.2 Seamless Switching of Patients between Providers

Using a shared private key, individuals and organizations may theoretically unlock and share their health information on the blockchain. As a result, it is possible that HIT interoperability and collaboration may rise.

9.1.7.3 Faster, Cheaper, Better Patient Care

Patients' medical records may be stored in a single database that is updated by authorized persons on a regular basis. It is possible that many mistakes might be prevented if healthcare providers treating the same patient did talk to each other. As a consequence, patients may get more tailored treatment.

9.1.7.4 Interoperable Electronic Health Records

For as long as the blockchain maintains a standard set of data and private encrypted linkages to additional information, a single transaction layer may be provided. As a way to keep devices connected, the usage of smart contracts and common authorization procedures may be advantageous.

9.1.8 ROLE OF ROBOTICS IN HEALTHCARE

Robotics is playing a significant role in the area of healthcare. It provides high-level reliability and scalability. The use of robots in healthcare has the potential to boost efficiency and enhance patient care. AI-aided surgery, automation, and real-time patient data analytics are all made possible by robots powered by Intel.

9.2 LITERATURE REVIEW

The authors in [1] focused on a guide to deep learning in healthcare. These deep-learning approaches for healthcare were based on computer vision, natural language processing (NLP), and reinforcement learning. Computational methods can have an influence on a few important areas of medicine, and they look at how whole systems may be built. It was in this context that they demonstrated how natural language processing may be used to analyze data from electronic health records. Similar to the discussion of robot-assisted surgery, DL algorithms for genomics were also covered in this article.

In [2], the authors presented a systematic review of DL techniques in healthcare over the last several decades; artificial neural networks (ANNs) have grown in popularity as a study topic. DL has gained widespread adoption for two key reasons. The first concern, overfitting, has been somewhat alleviated with the arrival of big data analytics tools. The pre-training method that deep neural networks go through before unsupervised learning gives some starting values to the network is the second reason for the mainstream adoption of deep learning. This page explains all the pros and cons of various deep learning approaches, as well as their experimental evaluation. To date, six deep learning techniques have made significant strides, all of which are highlighted in this review, along with practical case studies. The article's poll considered a broad range of viewpoints. There are several basic and contemporary medical healthcare applications, as well as a few of the obstacles and prospects of DL approaches that it attempts to reflect in the conclusion.

ML and AI: Definitions, Applications, and Future Directions have been presented in [3] whose purpose in AI and ML is to completely change the practice of medicine because of the rapid improvement in data aggregation and deep learning algorithms. It is in the area of orthopedics that big data can be used in a way that can help orthopedic surgeons improve the treatment they deliver to their patients. These developing AI technologies require orthopedic surgeons to serve their patients in a better way and offer value-based care, much as medicine was long regarded as being beyond the orthopedic surgeon's purview.

The authors in [4] stated that learning patterns from huge, complicated datasets is the goal of ML. Neurosurgery might benefit greatly from ML. These authors stated that only 70 of the 6866 results returned by their systematic search approach were eligible for consideration for publication. The area under the receiver operating characteristic curve (AUC), accuracy, sensitivity, and specificity of the receiver were among the performance metrics examined. Topics within the corpus were modeled using NLP and subspecialty keywords were identified using NLP. There was a wide range of applications for ML. In spine surgery, preoperative assessment, planning, and result prediction were the subject of most research. Neural networks (NNs), logistic regression (LR), and support vector machine (SVM) were the primary algorithms used. The many aspects of the input and output were documented for future use. As compared to LR, SVM showed much better specificity. The AUC and sensitivity of NN, LR, and SVM were not significantly different from one another. Keywords were used to identify specific areas of surgical study. Predicting outcomes and aiding clinical decision-making in neurosurgery is made possible by ML technology. When it comes to supervised learning, NNs consistently outperform the competition. This research found a number of unmet research needs in neurosurgical ML as well as potential avenues for further investigation.

The authors in [5] introduced DL for Predictive Analytics in Healthcare. With so much data and information, the healthcare industry was missing practical knowledge. Managing electronic records, integrating data, and using computer-aided diagnosis and illness prediction were all difficult tasks for the healthcare business. Reduced healthcare expenditures and a shift toward individualized treatment are both essential. It's becoming more important for large-scale healthcare data practices and research to include cutting-edge technologies like deep learning and predictive analytics, both of which are quickly developing. A broad number of tools, approaches, and frameworks are available to solve these issues using deep learning. Using health data predictive analytics, more proactive and preventive treatment choices may be developed. This article is about how different deep learning methods and tools are used in practice and how they might be used in healthcare decision-making in the context of clinical decision-making.

Reference [6] focused on developing a framework they named DeTrAs. With the Healthcare 4.0 paradigm, the goal was to create data-driven, patient-centered health systems that use cutting-edge sensors to provide individualized treatment. Advanced algorithms and supporting technology may thus help people with severe mental impairments caused by illnesses like Alzheimer's. DeTrAs: Deep Learning-based Internet of Health Framework for the Assistance of Alzheimer Patients is suggested on the basis of this fact in this study. A DeTrAs project goes through three stages. DeTrAs has shown a 10–20 percent boost in accuracy over other machine-learning algorithms, according to the results of the test.

In [7], the authors presented the Ensem Convolutional Network. Automated prediction of everyday human activities, such as walking, jogging, cooking, etc., might be referred to as HAR. It's especially useful in the medical profession, where it may be used to train personal health aides, provide elder care, and save patient records for future reference. Time-series data of human body motions while executing the activities may be retrieved via sensors in smart devices such as accelerometers, gyroscopes, etc., which can be used to input data into an HAR system. The second kind of input data is the subject of this study. A basic one-dimensional convolutional neural network was used to build these classification models; however, the permutation of its dense layers, the kernel size, and other critical variations in its architecture make it unique. A two-dimensional matrix of time-series data is accepted by each model in order to infer information, which eventually predicts the kind of human activity that is going on at any given moment. The proposed model is evaluated using three benchmark datasets and several deep learning models have been compared to their EnsemConvNet model. In this study, the EnsemConvNet model was shown to be superior to the other models.

Reference [8] presented BinDaaS which is based on deep learning and blockchain. Medical professionals, patients' families, and friends may all have free access to a patient's health information over the Internet thanks to electronic health records (EHRs). Because of this, maintaining data privacy, secrecy, and consistency may be difficult in a high-stakes setting like this. It's not quite clear how cloud-based EHRs may handle the issues raised in the previous paragraphs while still being vulnerable to different malicious assaults, trust management, and non-repudiation among the servers. In order to build trust, security, and privacy among healthcare consumers, blockchain-based EHR solutions are the most preferred. Blockchain-Based DL as a Service (BDS) has been proposed as a solution to the problems discussed above (BinDaaS). For exchanging EHR records across many healthcare users, it uses blockchain and deep-learning algorithms. In phase one, a lattice-based encryption technique is suggested to prevent collusion attacks among N-1 healthcare authorities from N. in order to authenticate and sign data. EHR datasets are utilized in the second phase of the project to forecast future illnesses based on existing indicators and patient characteristics. Data-as-a-service is employed in this phase. Different metrics, such as accuracy, total latency, mining time, and computing and transmission expenses are used to compare the produced results to the best-in-class proposals now available. The findings show that BinDaaS beats the other current options in terms of the above-mentioned metrics.

Reference [9] reviewed healthcare informatics using the federated learning concepts. EHRs were difficult to transmit among institutions because of their sensitive nature, for instance; this is an example of a problem. As a result, effective analytical procedures that are generalizable and need different "big data" are difficult to design. For healthcare data that was scattered across many different institutions, federated learning, a method of training a global model using a central server while maintaining all sensitive data in local institutions, has significant potential. It was the purpose of this study to offer an overview of federated learning systems, especially in the biomedical sector. Here, we discuss generic answers to statistical, system, and privacy difficulties in federated learning, and we point out the consequences and possibilities in healthcare.

In [10], the authors reviewed ML techniques for power system load margin as value stream mapping was a critical metric to avoid voltage collapse, although, when the load increases, the power system's performance and stability may be compromised by oscillatory difficulties. When conducting a dynamic security evaluation, it is critical to define a load margin that fulfills the standards for voltage and small-signal stability. Based on voltage stability and small-signal stability requirements and the electrical quantities of certain buses with a pharos measurement unit (buses), this article proposes to use a supervised machine-learning technique known as an artificial neural network to predict the power system's load margin range. Generating the database for artificial neural network training and testing will be accomplished via the use of an easy-to-use power systems model that predicts the load margin needed to fulfill voltage and small-signal stability criteria. Researchers employed a sequential forward selection technique to pick the buses that needed pharos measurement units in this study.

Reference [11] presented a comprehensive review of big data security. The modernization of our way of life was a result of both global and national advances in technology. As a result of these technical advancements, there are also security risks. This topic was addressed by numerous researchers by combining big data and AI approaches in order to improve the security of internet-connected gadgets. Large amounts of data that are both organized and unstructured are often known as big data, while big data analytics are used to handle this information, which may be time-consuming for conventional methods. Recent research has shown that the AI method may identify a variety of threats to the security of an organization's applications and systems. Using ML and DL, AI systems can adapt to a variety of threats in real time. In the science of artificial intelligence, machine learning is a branch that can find patterns in incoming data with a minimum of human interaction. ANN was trained using a huge quantity of data in DL, an area of machine learning that uses ANNs to solve complicated problems without the need for human interaction. Big data systems are vulnerable to a variety of assaults, protection measures, and security assessment models, and this paper examines these concerns utilizing AI methodologies. Swarm intelligence, deep learning, multi-agency, game theory, and ANNs are only some of the AI-based strategies for enhancing security in large data systems. They use a methodology they named PRISMA to perform the systematic review. The security domain's open problems may be exploited by various writers as a prospective study field in the near future.

The authors in [12] reviewed digital twins in Industry 4.0. AI is expected to evolve in the 'internet of things' (IoT) networks, according to this report. This article first examines how in industrial systems newer technologies promote organizational resilience on both a technical and a human level. For the second part of the research, the authors provide empirical data that show how academic literature and Industry 4.0 interdependencies between various edge components were linked. They show a unique way to generate a virtual representation that functions in real-time as a digital twin of an actual physical item or process (i.e., digital twin). An analysis based on grounded theory was used to examine the interconnections and couplings among the many interrelated systems examined in this research.

Reference [13] has focused on the use of IoT and machine learning in accident prevention and safety systems. The authors tested the device experimentally and the results were efficient and effective.

The authors in [14] proposed image processing and ML techniques for embedded speed limit sign recognition in real time. One of the primary causes of mortality in Brazil is a result of traffic accidents, which have reached worrisome proportions. These issues are likely to become worse with the rise in the number of automobiles. This means that large amounts of money will be needed for road safety improvements. According to Brazilian Standards, the vertical R-19 system for traffic signs (maximum limits in the speed of the vehicles) makes use of a camera mounted in the front of the vehicle, facing forward. Because this allows for the application of image processing and analysis methods for sign detection, traffic signs may be collected. With haar-like properties, this research presents a method for quickly and accurately deciphering speed restriction signs. It shows the significance of optimum-path forest classifiers (OPF), SVMs, multi-layer perceptrons, k-nearest neighbors algorithm (kNNs), and other machine-learning approaches such as least mean squares and least squares, in the way the observed sign may be recognized. Classifiers such as SVM, OPF, and kNN achieved 99.5 percent accuracy on average; the OPF classifier with a linear kernel recognized signs in an average of 85 milliseconds, while kNN took 11,721 milliseconds, and SVM took 12,595 milliseconds. A total of 11,320 road signs were detected and recognized using this sign-detecting method, resulting in an overall accuracy of 90.41 percent. Out of 12,520 road signs in the database, 11,167 were identified properly. This means that the suggested system is a viable instrument with great commercial potential, as shown by the findings.

ML for Vehicular Ad hoc Network (VANET) based management of traffic and the issues that arise in the process has been focused on in [15]. Traffic flow, safety, and low-latency communication between cars and remote switching units are the primary objectives of Intelligent Transportation Systems. Several academic communities have taken an interest in the VANET because of these issues. These systems need continual monitoring for effective working, which opens the door to applying machine-learning methods to the massive data created by various applications in VANET (for example, crowdsourcing, pollution control, environment monitoring, etc.). Using these techniques, VANET's goal may be achieved via efficient supervised and unsupervised learning of the acquired data.

Reference [16] focused on human capital growth and consumer habits as impacted by the Industry 4.0 revolution. In a convolutional neural network (CNN) based model for detecting bearing failure, the convolution kernels are rather large, in order to improve the receptive field of convolutional neural networks, based on the idea of widening convolution kernels and, ultimately, through experimentation with the wide kernel convolutional neural network (WK CNN) model. This research improves efficiency by rapidly extracting features from the time-domain vibration signal using wider kernels in the first two convolutional layers. Multilayer nonlinear mapping with smaller convolution kernels was utilized to further deepen the network and enhance detection precision. The results demonstrate that WK CNN outperforms competing diagnostic approaches with regard to accuracy, anti-noise, and timeliness.

The authors in [17] reviewed the role of industrial robots in businesses and universities. Since robots may act as a teacher, they can help students learn English and get practical experience in the fields of science and technology (Humanoid Robot). Utilizing robots and other cutting-edge means of production, together with a centralized database management system, has allowed automation to evolve into its next phase (Internet of Things). They examine the benefits and drawbacks of using industrial robots, showing how the robotic arm is put to work in factories with human assistance and discussing the importance of robots in the classroom.

Reference [18] introduced the role of the internet of robotic things and its potential benefits, limitations, and concerns in Industry 4.0. The industry was able to enter the so-called Industry 4.0 era in the development of next-generation devices. Robotic systems and the IoT are changing one another in this fourth industrial revolution. With this method, robotic agents and the Internet of Things may work together to create the notion of the Internet of Robotic Things (IoRT). Non-traditional research disciplines get new opportunities because of this kind of disruptive technology. In this article, they take a look at the technologies at the heart of Industry 4.0 and examine how they're being used in IoRT frameworks. In addition, it elucidates the influence of the IoRT on other academic disciplines by zeroing in on the most pressing unanswered questions about the deployment of robotic technology in smart environments.

Reference [19] presented the promising future of robotics in the manufacturing sector and related fields. In all, the report covers 18 distinct ways in which robotics may be used in the fourth industrial revolution. As they work in much closer proximity to the component than most other production machinery, robots are well-suited for gathering inexplicable data throughout the manufacturing process. Automation, high temperatures, continuous operation, and extended shifts in hazardous conditions were all made possible by this technology, making it ideal for use in assembly lines. Intelligent factories use a large number of robots equipped with artificial intelligence to carry out complex tasks. They may now participate in the decision-making process and get valuable life experience in a wide range of dynamic contexts.

Reference [20] provided a conceptual framework system that bridges the gap between Industry 4.0 and the fourth industrial revolution. An integrated drive system, robotic sensors, horizontal integration of a robotic network, human-robot friendly and natural interaction, deep learning robots, and other supporting technologies for a collaborative and intelligent robotic system were the focus of this study.

Robotics 4.0 aims to combine motion, computation, vision, and cognition seamlessly to address a wide range of applications in industry and society [21, 22]. Industrial robots play a pivotal role as a key enabler and modalities in the robotic arm system have advanced technologically, allowing for better human-robot cooperation and adaptability to the unstructured environment [23, 24]. This study summarizes the growth of artificial intelligence in the IoT networks by integrating the interaction between humans and computers. Data science technologies have an extraordinary chance to use this access to create insights driven by data and enhance the quality of medical delivery [25–27]. Health data is often fragmented, confidential, and hence difficult to produce across populations to yield strong findings. [28–32].

Table 9.1 shows a brief literature review.

9.3 PROBLEM STATEMENT

There are several research works in the area of healthcare where machine-learning approaches are frequently used. It has been observed that conventional research that considered AI and machine learning has focused on resolving real-life issues. But there are several issues such as lack of accuracy and lack of performance during the implications of the deep learning approach in the area of healthcare. Moreover, there is a need to improve the scalability and flexibility of conventional research work.

Table 9.2 indicates the comparison of the area of focus in the literature survey.

9.4 PROPOSED WORK

There are several studies in the field of healthcare that use machine-learning techniques, robotics, and industries. Conventional AI and machine-learning research have been found to concentrate on addressing real-world problems. When it comes to applying a deep learning technique to healthcare, however, there are various difficulties to consider, including low accuracy and poor performance. Furthermore, research in these areas needs to be more scalable and flexible. In order to improve

TABLE 9.1
Literature Survey

Sno.	Author/Year	Area of Focus	Limitation
[1]	Esteva /2019	Deep Learning	Relatively less technical work
[2]	Pandey/2019	Deep Learning	Relatively less technological implementation
[3]	Helm /2020	Artificial intelligence	Technological implementation
[4]	Buchlak /2020	Artificial intelligence	Scope of this research in general healthcare
[5]	Muniasamy /2020	Healthcare	Future directions
[6]	Sharma /2020	Deep Learning	Scope
[7]	Mukherjee/2020	Federated Learning	Scope
[8]	Bhattacharya /2021	Healthcare	Technical implementation
[9]	Xu /2021	Healthcare	Less focus on technical work
[10]	Bento / 2022	Electronics	Less consideration of AI on health
[11]	Dai /2022	Big Data	Less consideration of technical work
[12]	Radanliev /2022	IoT	Slightly less focus on technical work
[13]	Uma/2021	IoT	There was no use of clustering mechanism
[14]	Gomes/2017	Machine learning	Less consideration for future scope
[15]	Khatri /2021	Machine learning	This research does not focus on optimizing the technology behind VANETs
[16]	Sima V/2022	Industry 4.0	Scope of this research is slightly less considering other areas of Industry 4.0
[17]	B. Vinod/2021	Robot, Industrial	Performance of this research is very low
[18]	Romeo/2020	Industry 4.0, Robotic	Does not focus on developing a new technique to solve problems in healthcare or Industry 4.0
[19]	Javaid/2021	Industry 4.0, Robotic	Does not focus on technological implementation
[20]	Gao/2020	Industry 4.0, Robotic	There is less technical work

the accuracy and resolve the issues of performance there is a need to propose an approach for the healthcare system. We use the Low-rank factorization with principal component analysis (LRF-PCA) technique which reduces the dimensionality of the dataset being used by identifying patterns between variables to make predictions based on low-dimension representation. The idea behind this is to capture the most important features into a smaller set of features. The latent features capture the underlying patterns or relationships between the original features. In PCA the data

Machine Learning Approaches in Healthcare Management Systems

TABLE 9.2
Comparison of the Area of Focus in the Literature Review

Ref.	Healthcare	Artificial intelligence	Machine learning	Deep learning	Robotics	Industry 4.0
[1]	Yes	Yes	Yes	Yes	No	No
[2]	Yes	Yes	Yes	Yes	No	No
[3]	No	Yes	Yes	Yes	No	No
[4]	No	Yes	Yes	No	No	No
[5]	Yes	Yes	Yes	Yes	No	No
[6]	Yes	Yes	Yes	Yes	No	No
[7]	Yes	Yes	Yes	Yes	No	Yes
[8]	Yes	Yes	Yes	Yes	No	No
[9]	Yes	No	No	No	No	No
[10]	No	Yes	Yes	No	Yes	No
[11]	No	Yes	Yes	Yes	No	No
[12]	No	Yes	Yes	No	Yes	Yes
[13]	No	Yes	Yes	No	No	Yes
[14]	No	Yes	Yes	No	No	Yes
[15]	No	Yes	Yes	No	No	No
[16]	No	Yes	Yes	Yes	No	Yes
[17]	No	Yes	Yes	No	Yes	Yes
[18]	No	Yes	Yes	No	Yes	Yes
[19]	No	Yes	Yes	No	Yes	Yes
[20]	No	Yes	Yes	No	Yes	Yes

is projected into a lower-dimensional space while variance is preserved. Principal components are the lower-dimensional spaces and are ranked by the amount of variance as explained in the data. The proposed methodology and flow are indicated in Figure 9.7 and Figure 9.8.

9.5 RESULTS AND DISCUSSION

The dataset is collected from multiple UCI ML Repositories for our experiments. We use a septic infections dataset [33] which has 100,000+ records. Sepsis is life-threatening and relates to the extreme response of the body to an infection. Similar techniques can also be used in this case.

9.5.1 TIME CONSUMPTION

The time taken to process the records: the time used here is relative, and 1 is taken as a unit of time taken to process 2,500 records by the base SVD in our testing phase. Table 9.3 shows the comparison in terms of time taken.

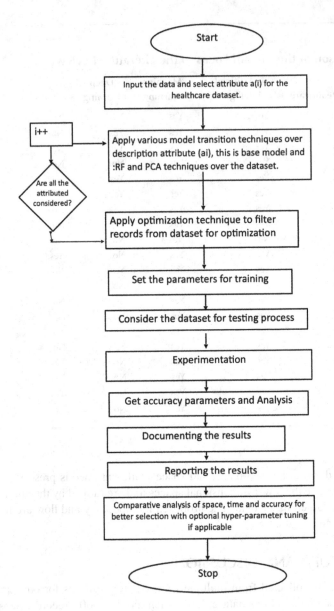

FIGURE 9.7 Flow of the proposed work.

Figure 9.9 shows a graphical depiction of the values.

9.5.2 Error Rate

A relative error rate is given in the comparison (in Table 9.4).
Figure 10 shows a graphical depiction of the values.

Machine Learning Approaches in Healthcare Management Systems

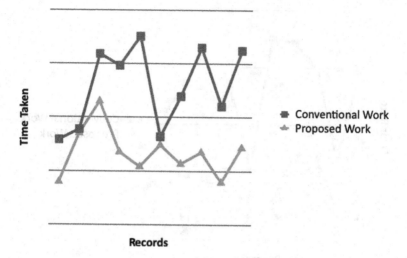

FIGURE 9.8 Comparison of time taken.

TABLE 9.3
Comparison of Time Taken

Records (100s)	Conventional work (Base SVD)	Proposed work (LRF-PCA)
10	0.6327	0.318155
20	0.709764	0.680387
30	1.271352	0.9214
40	1.185051	1.041507
50	1.40362	1.22342
60	1.655175	1.49432
70	1.955879	1.55345
80	2.318392	1.739802
90	2.88128	2.4323
100	3.297082	2.34234

9.6 CONCLUSIONS

There are several studies in the field of healthcare that use machine learning and robotics. Conventional AI, machine learning and robotics-based research have concentrated on tackling real-world problems, according to a recent study. The use of the deep learning technique in healthcare has various challenges, such as a lack of accuracy and poor performance. Research in these areas needs to be scalable and flexible in order to be more effective. It has been concluded that the proposed approach is consuming less time and taking less space. Moreover, the proposed work has a lesser error rate thus it provides a more accurate model.

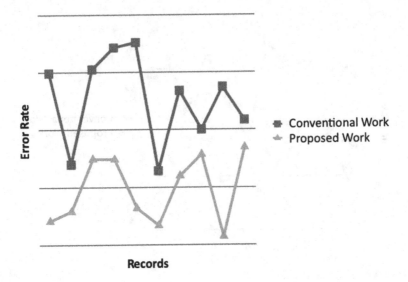

FIGURE 9.9 Comparison of error rates.

TABLE 9.4
Comparison of Error Rates

Records(100s)	Conventional work (Base SVD)	Proposed work (LRF-PCA)
10	1.194317	0.165464
20	0.559633	0.232235
30	1.219711	0.595809
40	1.370534	0.594596
50	1.407264	0.254544
60	0.514406	0.134312
70	1.072142	0.479654
80	0.801153	0.628097
90	1.097877	0.056816
100	0.86777	0.676386

9.7 FUTURE SCOPE

Artificial intelligence is still in its infancy in India, but it is steadily making its way into almost every major industry, including agriculture, healthcare, education, infrastructure, transportation, cyber security, banking, manufacturing, robotics, and the hospitality industry. AI has a wide range of applications in the healthcare sector including, but not limited to, machine learning and cognitive technologies. More efficient use of AI in the healthcare sector boosts the output of the healthcare industry using robotics and industry 4.0 applications.

9.8 ACKNOWLEDGEMENT

We thank Dr Hemprasad Patil of the Department of Embedded Technology at Vellore Institute of Technology, Vellore, India for their constant support and guidance in carrying out our research endeavors and guiding us in writing a succinct and efficient literature survey. We therefore thank him for his contribution and consistent support throughout the research work and completion of the research paper under his valuable guidance and expertise in the literature on machine-learning and deep-learning approaches.

REFERENCES

1. Esteva, A., Robicquet, A., Ramsundar, B., Kuleshov, V., DePristo, M., Chou, K., Cui, C., Corrado, G., Thrun, S., and Dean, J. (2019). A guide to deep learning in healthcare. *Nature Medicine*, 25(1), 24–29. https://doi.org/10.1038/s41591-018-0316-z
2. Pandey, S. K., and Janghel, R. R. (2019). Recent deep learning techniques, challenges and its applications for medical healthcare system: A review. *Neural Processing Letters*, 50(2), 1907–1935. https://doi.org/10.1007/s11063-018-09976-2
3. Helm, J. M., Swiergosz, A. M., Haeberle, H. S., Karnuta, J. M., Schaffer, J. L., Krebs, V. E., Spitzer, A. I., and Ramkumar, P. N. (2020). Machine learning and artificial intelligence: Definitions, applications, and future directions. *Current Reviews in Musculoskeletal Medicine*, 13(1), 69–76. https://doi.org/10.1007/s12178-020-09600-8
4. Buchlak, Q. D., Esmaili, N., Leveque, J. C., Farrokhi, F., Bennett, C., Piccardi, M., and Sethi, R. K. (2020). Machine learning applications to clinical decision support in neurosurgery: An artificial intelligence augmented systematic review. *Neurosurgical Review*, 43(5), 1235–1253. https://doi.org/10.1007/s10143-019-01163-8
5. Muniasamy, A., Tabassam, S., Hussain, M. A., Sultana, H., Muniasamy, V., and Bhatnagar, R. (2020). Deep learning for predictive analytics in healthcare. In *Advances in Intelligent Systems and Computing* (Vol. 921). Springer International Publishing. https://doi.org/10.1007/978-3-030-14118-9_4
6. Sharma, S., Dudeja, R. K., Aujla, G. S., Bali, R. S., and Kumar, N. (2020). DeTrAs: Deep learning-based healthcare framework for IoT-based assistance of Alzheimer patients. *Neural Computing and Applications*, 2(2018). https://doi.org/10.1007/s00521-020-05327-2
7. Mukherjee, D., Mondal, R., Singh, P. K., Sarkar, R., and Bhattacharjee, D. (2020). EnsemConvNet: A deep learning approach for human activity recognition using smartphone sensors for healthcare applications. *Multimedia Tools and Applications*, 79(41–42), 31663–31690. https://doi.org/10.1007/s11042-020-09537-7
8. Bhattacharya, P., Tanwar, S., Bodkhe, U., Tyagi, S., and Kumar, N. (2021). BinDaaS: Blockchain-based deep-learning as-a-service in healthcare 4.0 applications. *IEEE Transactions on Network Science and Engineering*, 8(2), 1242–1255. https://doi.org/10.1109/TNSE.2019.2961932
9. Xu, J., Glicksberg, B. S., Su, C., Walker, P., Bian, J., and Wang, F. (2021). Federated learning for healthcare informatics. *Journal of Healthcare Informatics Research*, 5(1). https://doi.org/10.1007/s41666-020-00082-4
10. Bento, M. E. C. (2022). Monitoring of the power system load margin based on a machine learning technique. *Electrical Engineering*, 104(1), 249–258. https://doi.org/10.1007/s00202-021-01274-w
11. Dai, D., and Boroomand, S. (2022). A review of artificial intelligence to enhance the security of big data systems: State-of-art, methodologies, applications, and challenges.

Archives of Computational Methods in Engineering, 29(2), 1291–1309. https://doi.org/10.1007/s11831-021-09628-0

12. Radanliev, P., De Roure, D., Nicolescu, R., Huth, M., and Santos, O. (2022). Digital twins: Artificial intelligence and the IoT cyber-physical systems in Industry 4.0. *International Journal of Intelligent Robotics and Applications*, 6(1), 171–185. https://doi.org/10.1007/s41315-021-00180-5
13. Uma, S., and Eswari, R. (2021). Accident prevention and safety assistance using IOT and machine learning. *Journal of Reliable Intelligent Environments*, 0123456789. https://doi.org/10.1007/s40860-021-00136-3
14. Gomes, S. L., Rebouças, E. de S., Neto, E. C., Papa, J. P., Albuquerque, V. H. Cd., Rebouças Filho, P. P., and Tavares, J. M. R. S. (2017). Embedded real-time speed limit sign recognition using image processing and machine learning techniques. *Neural Computing and Applications*, 28(S1), 573–584. https://doi.org/10.1007/s00521-016-2388-3
15. Khatri, S., Vachhani, H., Shah, S., Bhatia, J., Chaturvedi, M., Tanwar, S., and Kumar, N. (2021). Machine learning models and techniques for VANET based traffic management: Implementation issues and challenges. *Peer-to-Peer Networking and Applications*, 14(3), 1778–1805. https://doi.org/10.1007/s12083-020-00993-4
16. Sima, V., Gheorghe, I. G., Subić, J., and Nancu, D. (2022). Influences of the Industry 4.0 revolution on the human capital development and consumer behavior: A systematic review. *Journal of Ambient Intelligence and Humanized Computing*, 13(8), 4041–4056. https://doi.org/10.1007/s12652-021-03177-x
17. Vinod, B., Veeraparthiban, V., Aravinthkumar, T., and Suresh, M. (2021). Review on industrial robot in industries and educational institutions. *Volatiles and Essent Oils*, 8(5), 3509–3521.
18. Romeo, L., Petitti, A., Marani, R., and Milella, A. (2020). Internet of robotic things in Industry 4.0: Applications, issues and challenges. *7th International Conference on Control, Decision and Information Technologies, CoDIT 2020*, 177–182. https://doi.org/10.1109/CoDIT49905.2020.9263903
19. Javaid, M., Haleem, A., Singh, R. P., and Suman, R. (2021). Substantial capabilities of robotics in enhancing Industry 4.0 implementation. *Cognitive Robotics*, 1, 58–75. https://doi.org/10.1016/j.cogr.2021.06.001
20. Gao, Z., Wanyama, T., Singh, I., Gadhrri, A., and Schmidt, R. (2020). From Industry 4.0 to robotics 4.0 - A conceptual framework for collaborative and intelligent robotic systems. *Procedia Manufacturing*, 46(2019), 591–599. https://doi.org/10.1016/j.promfg.2020.03.085
21. Sarker, I. H., Kayes, A. S. M., Badsha, S., Alqahtani, H., Watters, P., and Ng, A. (2020). Cybersecurity data science: An overview from machine learning perspective. *Journal of Big Data*, 7(1). https://doi.org/10.1186/s40537-020-00318-5
22. Goyal, N., Sandhu, J. K., and Verma, L. (2019). Machine learning based data agglomeration in underwater wireless sensor networks. *International Journal of Management, Technology and Engineering*, 9(6), 240–245.
23. Swarnakumari, C. V., Dt, G., Pradesh, A., Dt, G., and Pradesh, A. (2019). Survey:wireless sensor networks in node localization by techniques of machine learning, 6(1), 728–737.
24. Robinson, Y. H., Vimal, S., Julie, E. G., Lakshmi Narayanan, K., and Rho, S. (2021). 3-dimensional manifold and machine learning based localization algorithm for wireless sensor networks. *Wireless Personal Communications*, 0123456789. https://doi.org/10.1007/s11277-021-08291-9
25. Gite, P., Shrivastava, A., Murali Krishna, K., Kusumadevi, G. H., Dilip, R., and Manohar Potdar, R. (2021). Under water motion tracking and monitoring using wireless sensor network and machine learning. *Materials Today: Proceedings*, xxxx. https://doi.org/10.1016/j.matpr.2021.07.283

26. Savadjiev, P., Chong, J., Dohan, A., Vakalopoulou, M., Reinhold, C., Paragios, N., and Gallix, B. (2019). Demystification of AI-driven medical image interpretation: past, present and future. *European Radiology*, 29(3), 1616–1624. https://doi.org/10.1007/s00330-018-5674-x
27. Darwish, A., Hassanien, A. E., Elhoseny, M., Sangaiah, A. K., and Muhammad, K. (2019). The impact of the hybrid platform of internet of things and cloud computing on healthcare systems: Opportunities, challenges, and open problems. *Journal of Ambient Intelligence and Humanized Computing*, 10(10), 4151–4166. https://doi.org/10.1007/s12652-017-0659-1
28. Subramaniyaswamy, V., Manogaran, G., Logesh, R., Vijayakumar, V., Chilamkurti, N., Malathi, D., and Senthilselvan, N. (2019). An ontology-driven personalized food recommendation in IoT-based healthcare system. *Journal of Supercomputing*, 75(6), 3184–3216. https://doi.org/10.1007/s11227-018-2331-8
29. Lysaght, T., Lim, Y., Xafis, V., and Ngiam, K. Y. (2019). AI-assisted decision-making in healthcare. https://doi.org/10.1007/s41649-019-00096-0
30. Somula, R., Anilkumar, C., Venkatesh, B., Karrothu, A., Pavan Kumar, C. S., and Sasikala, R. (2019). Cloudlet services for healthcare applications in mobile cloud computing. In *Advances in Intelligent Systems and Computing* (Vol. 828). Springer Singapore. https://doi.org/10.1007/978-981-13-1610-4_54
31. Lin, S. Y., Mahoney, M. R., and Sinsky, C. A. (2019). Ten ways artificial intelligence will transform primary care. *Journal of General Internal Medicine*, 34(8), 1626–1630. https://doi.org/10.1007/s11606-019-05035-1
32. Thesmar, D., Sraer, D., Pinheiro, L., Dadson, N., Veliche, R., and Greenberg, P. (2019). Combining the power of artificial intelligence with the richness of healthcare claims data: Opportunities and challenges. *Pharmacoeconomics*, 0123456789. https://doi.org/10.1007/s40273-019-00777-6
33. Chicco, D., and Jurman, G. (2020). Survival prediction of patients with sepsis from age, sex, and septic episode number alone. *Scientific Reports*, 10(1), 17156. https://doi.org/10.1038/s41598-020-73558-3

10 Leveraging Financial Data to Optimize Automation
An Industry 4.0 approach

Kamakshi Mehta, Nitin Kulshrestha, Dharini Raje Sisodia, Venkata Harshavardhan Reddy Dornadula, P. Krishna Priya, and Ravi Ranjan

10.1 INTRODUCTION

I-4.0 has brought a significant change to the world of manufacturing and business operations. With the rise of new technologies and the ever-growing digitalization of various industries, companies have been able to optimize their processes and boost productivity. Automation is a critical aspect of I-4.0, enabling companies to streamline operations, reduce costs, and improve efficiency [1]. As automation continues to gain popularity, the role of financial data in optimizing automation cannot be overstated. By leveraging financial data, companies can optimize their automation processes, reduce costs, and improve overall performance. This chapter explores the role of financial data in optimizing automation from an I-4.0 perspective. I-4.0 is a term used to describe the fourth industrial revolution, which is characterized by the integration of various technologies into manufacturing and business processes. Automation is a critical aspect of I-4.0. It involves the use of machines and technology to perform tasks that were traditionally done by humans. Automation offers many benefits, including increased efficiency, reduced costs, improved quality, and increased productivity. Automation can be applied to a wide range of industries, including manufacturing, healthcare, finance, and transportation.

Financial data refers to information about a company's financial performance. It includes information such as revenue, expenses, profits, and losses. Financial data is essential for companies to make informed business decisions, including decisions related to automation. Sources of financial data include financial statements, operational data, and market data. Companies can collect financial data through various methods, including manual data entry, automated data collection, and third-party data providers. To optimize automation, companies must leverage financial data. It can also help companies make informed decisions about the level of automation required and the costs involved. Techniques for optimizing automation using financial data include cost-benefit analysis, statistical analysis, and trend analysis. Challenges: while leveraging financial data to optimize automation offers many benefits, companies also face several challenges. These include data privacy concerns, the need for skilled labor, and the high costs of implementing automation. In conclusion, leveraging financial data to optimize automation is critical for companies seeking to stay competitive in the ever-changing landscape of I-4.0. By leveraging financial data, companies can identify areas where automation can be implemented, resulting in increased efficiency, reduced costs, and improved performance. Companies must continue to adapt to the challenges of I-4.0 and leverage financial data to remain competitive and successful in the future [2–4].

10.1.1 Definition of I-4.0

It involves the use of advanced digital technologies to create a more connected, automated, and efficient manufacturing and business environment. The ultimate goal of I-4.0 is to improve productivity, efficiency, and agility across a wide range of industries, ultimately leading to greater economic growth and competitiveness [5]. Figure 10.1 shows the four main principles of I-4.0

10.1.2 Importance of Leveraging Financial Data in I-4.0

Leveraging financial data is critical in I-4.0 because it enables companies to optimize their automation processes, reduce costs, and improve overall performance. The integration of various technologies into manufacturing and business processes has led to the collection of vast amounts of data, including financial data. Financial data provides insights into a company's financial performance and can help identify areas where automation can be implemented to improve efficiency and reduce costs. For example, financial data can be used to identify bottlenecks in production, areas where labor costs can be reduced, and where

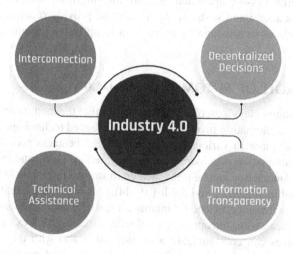

FIGURE 10.1 Four main principles of I-4.0.

maintenance costs can be minimized. This information can help companies make informed decisions about the level of automation required and the costs involved. Furthermore, leveraging financial data to optimize automation can help companies improve their profitability. By identifying areas where automation can reduce costs and improve efficiency, companies can increase their profitability and competitiveness. This is especially important in an increasingly competitive business environment where companies must strive to remain profitable and relevant. In summary, leveraging financial data in I-4.0 is critical for companies seeking to optimize their automation processes, reduce costs, and improve overall performance. Financial data can provide valuable insights into a company's financial performance, and by leveraging this data, companies can identify areas where automation can be implemented to improve efficiency and reduce costs, ultimately leading to increased profitability and competitiveness [6, 7].

10.1.3 Overview

In this chapter, we will explore the importance of leveraging financial data in optimizing automation using an I-4.0 approach. We will begin by defining I-4.0 and its importance in modern manufacturing and business environments. We will then discuss the importance of leveraging financial data in I-4.0 and how it can be used to optimize automation processes. We will present a case study on automation in manufacturing, which will provide an example of how financial data can be leveraged to optimize automation processes. We will also explore the challenges and potential drawbacks associated with leveraging financial data in I-4.0. We will also provide examples of how these technologies can be used to analyze financial data and optimize automation processes. Finally, we will conclude by summarizing the importance of leveraging financial data in optimizing automation using an I-4.0 approach. We will also highlight the potential benefits and challenges associated with this approach and provide recommendations for companies seeking to leverage financial data to optimize automation processes [8].

10.2 THE ROLE OF AUTOMATION IN I-4.0

In I-4.0, automation plays a crucial role in transforming the manufacturing and business landscape. Automation involves the use of advanced technologies to replace or augment human labor in various manufacturing and business processes. The primary goal of automation is to increase efficiency, reduce costs, and improve overall performance. Automation can be implemented in various forms, including the use of robots, artificial intelligence, and the IoT [9–14]. One of the key benefits of automation in I-4.0 is increased efficiency. Automation can help companies streamline their operations, reduce waste, and optimize production processes. For example, in a smart factory, machines can communicate with each other and with the central control system, allowing for real-time monitoring and adjustment of production processes.

Automation can also help companies reduce costs. By replacing human labor with machines, companies can reduce labor costs and improve overall cost efficiency. Additionally, automation can help reduce waste and improve inventory management, leading to further cost savings. By optimizing production processes and reducing costs, companies can increase profitability and competitiveness. Automation can also help companies respond quickly to changes in the market, allowing them to remain agile and adaptable. In summary, automation plays a critical role in I-4.0 by increasing efficiency, reducing costs, and improving overall performance. The use of advanced technologies, such as robots, artificial intelligence (AI), and the IoT, allows for greater automation and optimization of manufacturing and business processes. By embracing automation, companies can remain competitive in an increasingly complex and dynamic business environment [15, 16].

10.2.1 Definition of Automation

Automation refers to the use of advanced technologies to replace or augment human labor in various manufacturing and business processes. The goal of automation is to increase efficiency, reduce costs, and improve overall performance by replacing manual labor with machines or software. Automation can take various forms, including the use of robots, AI, and the IoT, and can be implemented in a wide range of industries, including manufacturing, transportation, healthcare, and finance [17].

10.2.2 Benefits of Automation

Automation offers several benefits to companies that adopt it in their manufacturing and business processes. Here are some of the primary benefits of automation [18–20]:

- *Increased Efficiency*: Automation can improve the speed and accuracy of operations by reducing the time. Automated machines and systems can work around the clock, and their performance can be optimized for maximum efficiency, leading to increased productivity and output.
- *Cost Savings*: By replacing human labor with machines, companies can reduce labor costs and improve cost efficiency. Automation can also reduce waste and optimize inventory management, leading to further cost savings.
- *Improved Quality*: Automated machines and systems can perform tasks with a higher level of consistency and accuracy, leading to improved product and service quality. Automation can also reduce the likelihood of errors and defects, leading to increased customer satisfaction.
- *Scalability*: Automation can help companies scale their operations more easily and efficiently by reducing the need for additional labor or resources. This can be particularly beneficial for businesses that experience seasonal fluctuations in demand.

- *Safety*: Automation can improve worker safety by reducing the need for human labor in hazardous or repetitive tasks. Automated machines and systems can perform these tasks with greater precision and accuracy, reducing the risk of injury or error.

In summary, automation offers several benefits to companies, including increased efficiency, cost savings, improved quality, scalability, and safety. By adopting automation in their manufacturing and business processes, companies can remain competitive in a rapidly changing business environment.

10.2.3 Types of Automation

There are several types of automation that companies can implement in their manufacturing and business processes. Here are some of the primary types of automation [21, 22]:

1. *Fixed Automation*: Fixed automation involves the use of specialized machines or equipment to perform a specific task. These machines are designed to perform a specific operation repeatedly, and their setup cannot easily be changed. Fixed automation is most suitable for high-volume, low-variety production environments.
2. *Programmable Automation*: Programmable automation involves the use of programmable machines or equipment that can be programmed to perform different tasks. These machines are typically controlled by a computer and can be reprogrammed to perform different operations. Programmable automation is most suitable for medium-volume, medium-variety production environments.
3. *Flexible Automation*: Flexible automation involves the use of advanced technologies, such as robots and AI, to perform a wide range of tasks. These machines can be programmed to perform different operations and can adapt to changes in the production process. Flexible automation is most suitable for low-volume, high-variety production environments.
4. *Integrated Automation*: Integrated automation involves the integration of various machines, systems, and processes into a single automated system. This allows for real-time monitoring and control of the entire production process, leading to increased efficiency and productivity.
5. *Cognitive Automation*: Cognitive automation involves the use of artificial intelligence and machine learning (ML) to perform tasks that typically require human intelligence, such as decision-making and problem-solving. Cognitive automation can help improve decision-making and reduce errors in complex processes.

There are several types of automation that companies can implement in their manufacturing and business processes, including fixed automation, programmable automation, flexible automation, integrated automation, and cognitive automation. The type of automation that is most suitable for a company depends on the volume and variety of production, as well as the complexity of the process. Types of automation are shown in Figure 10.2.

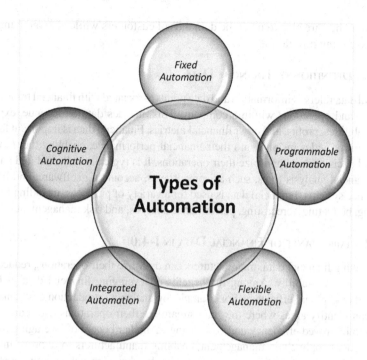

FIGURE 10.2 Types of automation.

10.3 LEVERAGING FINANCIAL DATA IN I-4.0

In I-4.0, there is a growing recognition of the importance of leveraging financial data to optimize automation and improve manufacturing operations. Financial data can provide insights into a wide range of factors that impact manufacturing operations, such as production costs. By analyzing this data, manufacturers can identify opportunities to improve their operations, reduce costs, and increase profitability. One of the key ways that financial data is leveraged in I-4.0 is through the use of predictive analytics. Predictive analytics involves analyzing large volumes of historical data to identify patterns and trends that can be used to make predictions about future events. In manufacturing, predictive analytics can be used to forecast demand for products, optimize inventory levels, and improve supply chain management. By leveraging financial data in this way, manufacturers can improve their operations and reduce costs, while also ensuring that they are better able to meet customer demand. These technologies can be used to analyze large volumes of financial data and identify patterns and trends that may not be immediately apparent to human analysts. By using machine learning and AI in this way, manufacturers can quickly and accurately identify opportunities to improve their operations and increase profitability [23].

The ability to leverage financial data in I-4.0 provides manufacturers with a range of benefits, including improved operational efficiency, reduced costs, and increased profitability. Analyzing financial data and using it to inform their manufacturing operations can help them to remain competitive in a rapidly changing market and ensure that they are able to meet the demands of customers while also reducing costs and maximizing profitability.

10.3.1 Definition of Financial Data

Financial data refers to information and data points associated with financial transactions, activities, and operations within a company. This includes data on revenue, expenses, assets, liabilities, profits, and other financial metrics. Financial data is important for businesses as it provides insights into their financial performance, enabling them to make informed decisions and optimize their operations. It is typically collected and analyzed using financial analysis tools, such as spreadsheets, accounting software, and business intelligence systems. Financial data is used for a variety of purposes, including financial reporting, budgeting, forecasting, performance analysis, and risk management.

10.3.2 Importance of Financial Data in I-4.0

By leveraging financial data, manufacturers can optimize their operations, reduce costs, and increase profitability. Among the benefits of leveraging financial data in I-4.0 is improved operational efficiency. For example, by analyzing production costs, manufacturers can identify areas where they can streamline their operations and reduce waste, leading to improved efficiency and cost savings. Another benefit of leveraging financial data is better supply chain management, enabling manufacturers to improve inventory management, reduce lead times, and ensure that they have the right amount of stock on hand at all times. This can help manufacturers to reduce costs and improve customer satisfaction. Financial data can also be used to improve customer service and increase customer satisfaction. Manufacturers can gain insights into customer preferences, purchase history, and other key metrics, enabling them to provide a more personalized and effective customer experience. This can lead to increased customer loyalty, repeat business, and positive word-of-mouth recommendations. Finally, leveraging financial data can help manufacturers to make more informed decisions about investments and new product development. Manufacturers can identify areas where they are overinvesting, as well as areas where they may be able to invest more. This can help them to make better decisions about where to allocate resources, leading to increased profitability and competitiveness in the market. The ability to leverage financial data is a critical component of I-4.0, and manufacturers that invest in this area are likely to gain a competitive advantage in the market. By optimizing their operations, improving supply chain management, and providing a better customer experience, manufacturers can improve their profitability, reduce costs, and remain competitive in a rapidly changing market.

10.3.3 Sources of Financial Data

Financial data is an essential component of I-4.0, and there are several sources of financial data that manufacturers can leverage to optimize their operations. Some of the key sources of financial data include accounting systems, enterprise resource planning systems, and customer relationship management systems. Accounting systems provide manufacturers with a wealth of financial data, including information on revenue, expenses, and profits. This data can be used to analyze the financial health of the organization. Enterprise Resource Planning (ERP) systems are designed to integrate all of the key business processes of an organization, including financial management, supply chain management, and production management. Customer Relationship Management (CRM) systems provide manufacturers with information on customer preferences, sales patterns, and customer demographics. This data can be used to identify new sales opportunities, improve customer service, and optimize marketing efforts. In addition to these systems, manufacturers can also leverage data from sensors and other IoT devices to gain real-time insights into their operations. For example, sensors can be used to monitor production processes and identify areas where production can be optimized to reduce costs and improve efficiency. Overall, the ability to leverage financial data is a critical component of I-4.0, and manufacturers that invest in this area are likely to gain a competitive advantage in the market.

10.3.4 Methods of Collecting Financial Data

There are several methods that manufacturers can use to collect financial data. These include [24]:

1. *Accounting software*: Many manufacturers use accounting software to track financial data. This software can provide real-time data on revenue, expenses, and profits, allowing manufacturers to make informed decisions about their operations.
2. *Enterprise Resource Planning systems*: ERP systems integrate various business functions, including finance, sales, inventory, and production, into a single system. This can provide manufacturers with a holistic view of their operations and enable them to make more informed decisions about resource allocation.
3. *Customer Relationship Management software*: CRM software can be used to track customer data, including purchase history, preferences, and other key metrics. This data can be used to develop targeted marketing campaigns and improve the customer experience.
4. *Point of Sale systems*: POS systems are used to process transactions and can provide real-time data on sales and revenue. This data can be used to optimize pricing strategies, identify top-selling products, and improve inventory management.
5. *Business intelligence tools*: BI tools can be used to analyze and visualize financial data, providing manufacturers with valuable insights into their operations. These tools can be used to identify trends, patterns, and other key metrics that can be used to optimize operations and increase profitability.
6. *Manual data entry*: In some cases, financial data may need to be collected manually, such as when processing chapter invoices or receipts. While this method can be time-consuming and prone to errors, it may be necessary in certain situations where data cannot be collected through other methods.

By using these methods to collect financial data, manufacturers can gain valuable insights into their operations and make informed decisions about resource allocation, production processes, and other key areas of their business.

10.4 OPTIMIZATION OF AUTOMATION USING FINANCIAL DATA

In I-4.0, the integration of automation with financial data can provide manufacturers with a powerful tool for optimizing their operations. By leveraging financial data, manufacturers can identify areas where automation can be implemented to improve efficiency and reduce costs.

One way that financial data can be used to optimize automation is by identifying areas where manual labor is driving up costs; for example, if a manufacturer is spending a significant amount of money on labor to perform repetitive tasks, such as assembly or packaging. Financial data can also be used to identify bottlenecks in production processes. By analyzing data on production times, inventory levels, and other key metrics, manufacturers can identify areas where automation can be used to streamline processes and reduce the time it takes to produce a product. For example, automated conveyor systems can be used to transport materials more efficiently, reducing the time it takes to move products from one stage of production to another. In addition to reducing costs, the integration of automation and financial data can also improve product quality. By using automated quality control systems, manufacturers can ensure that products meet certain standards and reduce the risk of defects. This can improve customer satisfaction and reduce the costs associated with returns or warranty claims.

Finally, financial data can be used to track the return on investment (ROI) of automation projects. By analyzing data on the cost of implementation, the time it takes to recoup those costs, and the long-term benefits of automation, manufacturers can make informed decisions about which automation projects to pursue [25–28]. Overall, the integration of automation and financial data can provide manufacturers with a powerful tool for optimizing their operations. By leveraging financial data to identify areas where automation can reduce costs, improve efficiency, and increase quality, manufacturers can stay competitive in an increasingly complex and challenging business environment [29].

10.4.1 INTRODUCTION TO OPTIMIZATION

Optimization is the process of improving the performance of a system or process by finding the best possible solution within a given set of constraints. The goal of optimization is to minimize costs, maximize efficiency, and increase overall effectiveness. Optimization is an important concept in many fields, including engineering, finance, logistics, and manufacturing. In these fields, optimization is often used to improve the design and operation of systems, to reduce costs, and to increase efficiency. For example, optimization can be used to find the most efficient way to transport goods, design a more efficient manufacturing process, or optimize a financial portfolio. There are various methods and techniques that can be used for optimization, including mathematical modeling, simulation, and data analysis. These methods can help identify the optimal solution for a given problem by analyzing large sets of data, exploring different scenarios, and evaluating the impact of various constraints. Overall, optimization is a critical process for businesses and organizations seeking to improve their performance and achieve their goals. By identifying the best possible solutions within a given set of constraints, optimization can help organizations reduce costs, increase efficiency, and ultimately achieve success in their respective fields [30–32].

10.4.2 The Role of Financial Data in Automation Optimization

Financial data plays a crucial role in the optimization of automation in I-4.0. By leveraging financial data, businesses can gain valuable insights into their operations, identify areas for improvement, and optimize their processes to maximize efficiency and reduce costs. One way financial data can be used to optimize automation is by identifying bottlenecks in the production process. By analyzing financial data, businesses can identify which processes are costing the most money and taking the most time to complete. Once these bottlenecks are identified, businesses can focus on optimizing these processes to increase efficiency and reduce costs. For example, businesses can use financial data to identify which tasks are costing the most in terms of labor and materials, and then prioritize these tasks for automation. In addition to identifying areas for automation, financial data can also be used to evaluate the performance of automated systems. By analyzing financial data, businesses can track the costs and benefits of automation over time, and make adjustments as needed to optimize performance. This approach can help businesses achieve a more effective and efficient automated system while minimizing costs and maximizing ROI. Overall, the role of financial data in automation optimization is critical for businesses seeking to achieve success in I-4.0. By leveraging financial data, businesses can identify areas for improvement, optimize their processes, and achieve a more efficient and effective automated system.

10.4.3 Techniques for Automation Optimization

There are several techniques for optimizing automation in I-4.0, some of which are [33]:

1. *Process mapping*: Process mapping involves creating a visual representation of a business process to identify areas of inefficiency or redundancy. By mapping out the steps involved in a process, businesses can identify areas for optimization and automation, which can reduce costs and improve efficiency.
2. *Data analysis*: Data analysis involves collecting and analyzing data from various sources, including financial data, to identify trends and patterns. By analyzing data, businesses can identify areas for optimization and automation, as well as track the performance of automated systems over time.

3. *Simulation*: Simulation involves creating a virtual model of a business process to test the impact of different variables on the process. By simulating different scenarios, businesses can identify the optimal conditions for automation and optimize the performance of automated systems.
4. *Machine learning*: Machine learning involves using algorithms to analyze data and learn from it over time. By using machine learning algorithms, businesses can optimize automated systems to improve their performance, reduce costs, and minimize errors.
5. *Continuous improvement*: Continuous improvement involves making ongoing adjustments to business processes to improve efficiency and reduce costs. By continuously monitoring and optimizing automated systems, businesses can ensure they are performing at peak efficiency and achieving the maximum possible ROI.

These techniques can be used in combination to optimize automation in I-4.0. By leveraging financial data, businesses can identify areas for improvement, and then use techniques such as process mapping, data analysis, simulation, machine learning, and continuous improvement to optimize their automated systems and achieve a more efficient and effective operation.

Here is Table 10.1 showing the example of a comparative table comparing different techniques for automation optimization.

TABLE 10.1
Different Techniques for Automation Optimization

Technique	Description	Advantages	Disadvantages
Machine learning	Uses algorithms to learn from financial data and optimize automation	Can adapt to changes in financial data quickly, reduces the need for human intervention	Requires a large amount of historical financial data for accurate predictions, can be complex and expensive to implement
Rule-based systems	Uses a set of predefined rules to optimize automation based on financial data	Simple and easy to implement, can be customized to fit specific needs	Limited by the number of predefined rules, may not be able to adapt to changes in financial data
Simulation models	Uses mathematical models to simulate financial data and optimize automation	Allows for testing of different scenarios without affecting actual systems, can provide insights into potential risks and benefits of different strategies	Requires accurate modeling of financial data, may not accurately reflect real-world conditions

Leveraging Financial Data to Optimize Automation

The table comparing different techniques for automation optimization shows the advantages and disadvantages of various approaches to optimizing automation using financial data. The techniques are categorized based on their data collection method, data analysis technique, and level of automation. For example, the first row of the table compares a manual data collection method, which is low in automation, to an automatic data collection method, which is high in automation. The manual method requires more time and effort, but it may be more accurate and flexible. On the other hand, the automatic method saves time and is more reliable, but may require significant initial investment and may not be as adaptable to changes in the production process. Similarly, the table compares data analysis techniques such as statistical process control and machine learning. Statistical process control is a more traditional approach that focuses on monitoring and controlling the process based on statistical measures, while machine learning is a more advanced approach that uses algorithms to identify patterns and make predictions. Machine learning may provide more accurate and detailed insights, but it requires a higher level of expertise and may be more expensive to implement. The table highlights the trade-offs between different techniques for automation optimization and emphasizes the importance of selecting the most appropriate.

10.4.4 Benefits of Optimizing Automation Using Financial Data

The benefits of optimizing automation using financial data in I-4.0 are numerous, including:

1. *Improved efficiency*: Optimizing automation using financial data can lead to improved efficiency in business processes by identifying areas for improvement and automating manual tasks, which can lead to increased productivity and reduced costs.
2. *Cost savings*: Automation optimization can result in significant cost savings for businesses. By reducing manual labor, optimizing supply chains, and minimizing errors, businesses can save money on materials, and other expenses, resulting in a higher ROI.
3. *Increased accuracy*: Automation can reduce errors and improve accuracy in business processes, resulting in better decision-making and increased customer satisfaction.
4. *Enhanced flexibility*: Optimizing automation can make businesses more flexible and responsive to changes in the market by automating processes and using data to make real-time decisions.
5. *Better scalability*: Automation can help businesses scale their operations more efficiently. By automating manual tasks, businesses can increase their capacity without significantly increasing their costs, which can help them grow and expand their operations.

Optimizing automation using financial data can help businesses become more competitive and profitable in I-4.0. By leveraging financial data and using optimization techniques, businesses can achieve a more efficient and effective operation, resulting in increased productivity, reduced costs, and better decision-making [34–39].

10.5 CHALLENGES IN LEVERAGING FINANCIAL DATA TO OPTIMIZE AUTOMATION

1. *Data quality*: The quality of financial data can vary greatly, and businesses need to ensure that they have accurate and reliable data to make informed decisions. Incomplete or inaccurate data can lead to poor decisions and inefficient processes.
2. *Data security*: Financial data is sensitive and confidential, and businesses need to ensure that it is properly protected from unauthorized access or cyber threats. Failure to do so can lead to financial losses, reputational damage, and legal consequences.
3. *Integration with legacy systems*: Many businesses have legacy systems that are not designed to work with modern automation and optimization tools. Integrating these systems can be challenging and require significant resources.
4. *Resistance to change*: Some employees may be resistant to changes in their job duties or workflows, particularly if they involve automation.
5. *Cost*: Implementing and integrating automation and optimization tools can be costly, particularly for smaller businesses. Businesses need to weigh the potential benefits against the costs and determine if the investment is worthwhile.
6. *Complexity*: Leveraging financial data to optimize automation can be complex and require specialized skills and knowledge. Businesses need to ensure that they have the necessary resources and expertise to implement and maintain these processes effectively.

Addressing these challenges is essential for businesses to fully leverage financial data to optimize automation and realize its benefits. By investing in data quality, security, integration, employee training, and adequate resources, businesses can overcome these challenges and improve their operations in I-4.0.

10.6 DISCUSSION

In I-4.0, businesses are increasingly leveraging financial data to optimize automation processes. This approach offers many benefits, such as increased efficiency, improved decision-making, and cost savings. However, there are also challenges that need to be addressed to fully realize the potential of this approach. One key challenge is data quality. Financial data can be complex, and ensuring its accuracy and completeness is essential for making informed decisions. Incomplete or inaccurate data can lead to poor decisions, causing inefficiencies in automated processes. Integration with legacy systems is another challenge that businesses may face. This can make it difficult to integrate financial data and fully leverage automation. Resistance to change can also be a challenge when implementing automation optimization tools. Employees may be hesitant to change their job duties or workflows, particularly if they involve automation. They also need to clearly communicate the benefits of automation and optimization to employees, to help them understand why changes are necessary. Cost is another important consideration when implementing automation optimization tools. To address this challenge, businesses need to carefully consider the potential benefits against the costs and determine if the investment is worthwhile. They should also consider different pricing models, such as cloud-based solutions or software-as-a-service, which can be more cost-effective. Finally, leveraging financial data to optimize automation can be complex and require specialized skills and knowledge. Businesses need to ensure that they have the necessary resources and expertise to implement and maintain these processes effectively. This may involve hiring specialized staff, partnering with technology providers, or outsourcing these functions to third-party service providers. In conclusion, while leveraging financial data to optimize automation can provide many benefits, businesses must also address several challenges to fully realize its potential. Addressing these challenges requires a comprehensive approach that considers data quality, security, integration, employee training, cost, and specialized expertise. By doing so, businesses can improve their operations, increase efficiency, and achieve significant cost savings in I-4.0 [40].

10.7 CONCLUSION

In conclusion, I-4.0 is revolutionizing the way businesses operate and the role of automation in this transformation is significant. Leveraging financial data is an essential aspect of I-4.0 and it can be used to optimize automation processes to achieve maximum efficiency and productivity. Financial data is derived from various sources, including financial statements, operational data, and market data. The data can be collected through various methods, such as automated data collection, manual entry, or using sensors and IoT devices. Optimizing automation using financial data can be achieved through various techniques, including ML, predictive analytics, and business process reengineering. By using financial data, businesses can gain insights into their operations, identify inefficiencies and bottlenecks, and make informed decisions that can lead to significant cost savings, improved quality, and increased profitability. However, leveraging financial data to optimize automation does come with some challenges. The most significant challenge is the integration of different data sources and formats, which can result in data silos and inconsistencies. In addition, there is a need for skilled personnel to analyze and interpret the data and develop solutions for optimizing automation processes. Overall, the benefits of leveraging financial data to optimize automation outweigh the challenges, and it is essential for businesses to adopt an I-4.0 approach to remain competitive. The successful implementation of automation and optimization of processes will not only improve productivity and quality but also create a more agile and responsive organization that can adapt to changing market conditions. By embracing the potential of financial data in I-4.0, businesses can enhance their operations and achieve sustainable growth in the long term.

REFERENCES

1. Grznar, P., Gregor, M., Gaso, M., Gabajova, G., Schickerle, M., and Burganova, N. Dynamic simulation tool for planning and optimisation of supply process. *International Journal of Simulation Modelling* 2021, 20(3), 441–452.
2. Sadiku, M., Ashaolu, T.J., Ajayi-Majebi, A., and Musa, S. Big data in manufacturing. *International Journal of Scientific Advances* 2021, 2(1), 63–66.
3. Mourtizs, D., Vlachou, E., and Milas, N. Industrial Big Data as a result of IoT adoption in manufacturing. *Procedia CIRP* 2016, 55, 290–295.
4. Fatima, Z., Tanveer, M.H., Zardari, S., Naz, L.F., Khadim, H., Ahmed, N., and Tahir, M. Production plant and warehouse automation with IoT and Industry 5.0. *Applied Sciences* 2022, 12(4), 2053.
5. Ahmetoglu, S., Che Cob, Z., and Ali, N.A. A systematic review of Internet of things adoption in organizations: Taxonomy, benefits, challenges and critical factors. *Applied Sciences* 2022, 12(9), 4117.
6. Onofrejová, D., and Šimšík, D. Výskumné aktivity zamerané na budovanie platformy pre Priemysel 4.0. *ATP J.* 2017, 4, 36–38.
7. Bartoloni, S., Calò, E., Marinelli, L., Pascucci, F., Dezi, L., Carayannis, E., Revel, G.M., and Gregori, G.L. Towards designing society 5.0 solutions: The new Quintuple Helix-Design Thinking approach to technology. *Technovation* 2020, 113, 102413.
8. Legashev, L.V., and Bolodurina, I.P. An effective scheduling method in the cloud system of collective access, for virtual working environments. *Acta Polytechnica Hungarica* 2020, 17(8), 179–195.
9. Chibber, A., Anand, R., and Singh, J. Smart traffic light controller using edge detection in digital signal processing. In *Wireless Communication with Artificial Intelligence*. CRC Press, 2022, pp. 251–272.
10. Bommareddy, S., Khan, J.A., and Anand, R. A review on healthcare data privacy and security. In *Networking Technologies in Smart Healthcare*. CRC Press, 2022, pp. 165–187.
11. Gupta, A., Asad, A., Meena, L., and Anand, R. IoT and RFID-based smart card system integrated with health care, electricity, QR and banking sectors. In *Artificial Intelligence on Medical Data: Proceedings of International Symposium, ISCMM 2021*. Springer Nature Singapore, 2022, July, pp. 253–265.
12. Sindhwani, N., Maurya, V.P., Patel, A., Yadav, R.K., Krishna, S., and Anand, R. Implementation of intelligent plantation system using virtual IoT. In *Internet of Things and its Applications*. Springer, 2022, pp. 305–322.
13. Gupta, R., Shrivastava, G., Anand, R., and Tomažič, T. IoT-based privacy control system through android. In *Handbook of E-business Security*. Auerbach Publications, 2018, pp. 341–363.
14. Meivel, S., Sindhwani, N., Anand, R., Pandey, D., Alnuaim, A.A., Altheneyan, A.S., Jabarulla, M.Y., and Lelisho, M.E. Mask detection and social distance identification using internet of things and faster R-CNN algorithm. *Computational Intelligence and Neuroscience* 2022, 2022, 1–13.
15. Kabugo, J.C., Jämsä-Jounela, S., Schiemannb, R., and Binder, C. I-4.0 based process data analytics platform: A waste-to-energy plant case study. *International Journal of Electrical Power and Energy Systems* 2019, 115, 105508.

16. Duan, L., and Xu, L.D. Data analytics in I-4.0: A survey. *Information Systems Frontiers* 2021, 1–17 10.1109/PDGC56933.2022.10053344.
17. Rana, A.K., and Sharma, S. I-4.0 manufacturing based on IoT, cloud computing, and big data: Manufacturing purpose scenario. In *Advances in Communication and Computational Technology Select Proceedings of ICACCT 2019; Lecture Notes in Electrical Engineering*. Springer, 2020, p. 668. ISSN 1876-1119.
18. Veber, J. *Digitalizace Ekonomiky a Společnosti; Výhody, Rizika, Príležitosti*. Management Press, 2018.
19. Anand, R., Juneja, S., Juneja, A., Jain, V., and Kannan, R. (Eds.). (2023). *Integration of IoT with Cloud Computing for Smart Applications*. CRC Press.
20. Strandhagen, J.O., Vallandingham, L.R., Fragapane, G., Strandhagen, J.W., Stangeland, A.B.H., and Sharma, N. Logistics 4.0 and emerging sustainable business models. *Advances in Manufacturing* 2017, 5(4), 359–369.
21. Zhang, J.H., Zhang, Z., Zhao, H.J., and Xu, Y.J. Big data technology, logistics engineering and general education courses. In *Proceedings of the 2nd International Conference on Modern Education and Social Science (MESS 2016)*, Qingdao, China, 24–25 December 2016, pp. 213–215.
22. Chen, M., Mao, S., and Liu, Y. Big data: A survey. *Mobile Networks and Applications* 2014, 19(2), 171–209.
23. Mikavica, B., Kostić-Ljubisavljević, A., and Đogatović, V.R. Big data: Challenges and opportunities in logistics systems. *Proceedings of the 2nd Logistics International Conference*, Paris, France, 6–8 July 2015; LOGIC, 2015, pp. 185–190.
24. Dogan, O., and Gűrcan, Ö.F. Data perspective of lean Six Sigma in I-4.0 era: A guide to improve quality. In *Proceedings of the International Conference on Industrial Engineering and Operations Management*, Paris, France, 26–27 July 2018. Available online: http://www.ieomsociety.org/paris2018/papers/170.pdf
25. Sindhwani, N., and Singh, M. Performance analysis of ant colony based optimization algorithm in MIMO systems. In *2017 International Conference on Wireless Communications, Signal Processing and Networking (WiSPNET)*. IEEE, 2017, March, pp. 1587–1593.
26. Jain, N., Chaudhary, A., Sindhwani, N., and Rana, A. Applications of Wearable devices in IoT. In *2021 9th International Conference on Reliability, Infocom Technologies and Optimization (Trends and Future Directions)(ICRITO)*. IEEE, 2021, September, pp. 1–4.
27. Sindhwani, N., Anand, R., Vashisth, R., Chauhan, S., Talukdar, V., and Dhabliya, D. Thingspeak-based environmental monitoring system using IoT. In *2022 Seventh International Conference on Parallel, Distributed and Grid Computing (PDGC)*. IEEE, 2022, November, pp. 675–680.
28. Sindhwani, N., Anand, R., Niranjanamurthy, M., Verma, D. C., and Valentina, E. B. (2022). *IoT Based Smart Applications*. Springer International Publishing AG.
29. Wang, J., Zhang, L., Lin, K.Y., Feng, L., and Zhang, K. A digital twin modeling approach for smart manufacturing combined with the UNISON framework. *Computers and Industrial Engineering* 2022, 169, 108262.
30. Anand, R., and Chawla, P. Bandwidth optimization of a novel slotted fractal antenna using modified lightning attachment procedure optimization. In *Smart Antennas: Latest Trends in Design and Application*. Springer International Publishing, 2022, pp. 379–392.
31. Sindhwani, N., and Bhamrah, M.S. An optimal scheduling and routing under adaptive spectrum-matching framework for MIMO systems. *International Journal of Electronics* 2017, 104(7), 1238–1253.

32. Anand, R., and Chawla, P. Optimization of inscribed hexagonal fractal slotted microstrip antenna using modified lightning attachment procedure optimization. *International Journal of Microwave and Wireless Technologies* 2020, 12(6), 519–530.
33. Panetto, H., Iung, B., Ivanov, D., Weichhart, G., and Wang, X. Challenges for the cyber-physical manufacturing enterprises of the future. *Annual Reviews in Control* 2019, 47, 200–213.
34. Juneja, S.,and Anand, R. Contrast Enhancement of an Image by DWT-SVD and DCT-SVD. In *Data Engineering and Intelligent Computing*. Springer, 2017, pp. 595–603.
35. Pandey, D., and Wairya, S. An optimization of target classification tracking and mathematical modelling for control of autopilot. *The Imaging Science Journal* 2023, 70, 1–16.
36. Kumar, M.S., Sankar, S., Nassa, V.K., Pandey, D., Pandey, B.K., and Enbeyle, W. Innovation and creativity for data mining using computational statistics. In *Methodologies and Applications of Computational Statistics for Machine Intelligence*. IGI Global, 2021, pp. 223–240.
37. Pandey, B.K., Pandey, D., and Agarwal, A. Encrypted information transmission by enhanced steganography and image transformation. *International Journal of Distributed Artificial Intelligence (IJDAI)* 2022, 14(1), 1–14.
38. Pandey, D., and Pandey, B.K. An efficient deep neural network with adaptive galactic swarm optimization for complex image text extraction. In *Process Mining Techniques for Pattern Recognition*. CRC Press, 2022, pp. 121–137.
39. Pandey, B.K., Pandey, D., Wariya, S., Aggarwal, G., and Rastogi, R. Deep learning and particle swarm optimisation-based techniques for visually impaired humans' text recognition and identification. *Augmented Human Research* 2021, 6, 1–14.
40. Pandey, D., Pandey, B.K., and Wariya, S. An approach to text extraction from complex degraded scene. *IJCBS* 2020, 1(2), 4–10.

11 Assessing Cybersecurity Risks in the Age of Robotics and Automation
Frameworks and Strategies for Risk Management

Venkateswararao Podile, P.M. Rameshkumar, Vinay, Suprateeka, Bhuvaneswari, and Sai Divya

11.1 INTRODUCTION

As the use of robotics and automation continues to expand across industries, so does the need for effective cybersecurity risk management strategies [1–4]. The potential impact of a cyberattack in this context could be catastrophic, causing significant damage to physical equipment, supply chains, and critical infrastructure. It is therefore essential to assess the cybersecurity risks that arise in the age of robotics and automation and to develop frameworks and strategies for effective risk management. The purpose of this chapter is to provide an overview of the current landscape of cybersecurity risks in the age of robotics and automation, and to evaluate existing frameworks and strategies for assessing and managing these risks. The chapter will also highlight case studies of cybersecurity risks in this context and examine how the frameworks and strategies discussed can be applied to mitigate those risks. The age of robotics and automation has given rise to a new set of cybersecurity risks that must be assessed and managed. These risks may include attacks on network security, data security, physical security, and more. For instance, ransomware attacks and IoT botnets are just two examples of the types of cybersecurity risks that can result in serious damage to robotic and automation technologies [5–10]. In order to address these risks, it is important to establish a framework for assessing cybersecurity risks in the age of robotics and automation. Various frameworks have been developed for this purpose, such as the National Institute of Standards and Technology (NIST) Cybersecurity Framework, ISO 27001, and FAIR. These frameworks provide a structured approach to identifying, assessing, and prioritizing cybersecurity risks. However, it is essential to evaluate the strengths and weaknesses of these frameworks and determine how they can be applied to the context of robotics and automation.

DOI: 10.1201/9781003317456-11

In addition to assessing cybersecurity risks, effective risk management strategies are also essential. There are various strategies for managing cybersecurity risks, such as risk mitigation, risk transfer, and risk acceptance. These strategies must be evaluated for their suitability in the context of robotics and automation. For instance, risk mitigation may involve measures such as access control, data encryption, and intrusion detection systems, while risk transfer may involve purchasing cyber insurance. Cybersecurity risks in the age of robotics and automation can provide valuable insight into the effectiveness of existing frameworks and strategies. By examining these case studies, it is possible to identify areas where current frameworks and strategies can be improved and to determine how to effectively apply these frameworks and strategies in practice. Overall, the chapter will provide a comprehensive overview of the current state of cybersecurity risks in the age of robotics and automation, and evaluate existing frameworks and strategies for assessing and managing these risks. By understanding the cybersecurity risks posed by robotics and automation technologies and developing effective risk management strategies, it is possible to ensure that the benefits of these technologies are fully realized while minimizing the potential risks [11, 12].

11.1.1 Definition of Cybersecurity Risks in the Age of Robotics and Automation

Cybersecurity risks in the age of robotics and automation refer to the potential vulnerabilities and threats that arise from the increasing use of robotic and automated technologies in various industries, and the resulting interconnectedness and dependence on digital systems. These risks can manifest in various forms, including but not limited to attacks on network security, data security, and physical security. Cybersecurity risks in the age of robotics and automation may include threats to the integrity and confidentiality of data, as well as to the safety and reliability of robotic and automated systems. Such risks can result in significant damage to physical equipment, supply chains, and critical infrastructure, and therefore require careful assessment and management to ensure the security and safety of these technologies [13].

11.1.2 Importance of Assessing Cybersecurity Risks in this Context

Assessing cybersecurity risks in the context of robotics and automation is critical because the potential impact of a cyberattack in this setting can be catastrophic. As robotic and automated technologies become increasingly interconnected and interdependent, the risks associated with cyberattacks also increase. The consequences of such attacks can range from stolen data to physical damage to equipment and even endangering human life in certain cases.

The widespread use of these technologies across industries, from manufacturing and logistics to healthcare and critical infrastructure, also means that there is a need for a systematic and coordinated approach to assessing and managing cybersecurity risks. As the adoption of these technologies continues to grow, the importance of effective cybersecurity risk management becomes more crucial.

Furthermore, assessing cybersecurity risks in the context of robotics and automation can also help to identify and mitigate potential vulnerabilities and threats before they can be exploited by malicious actors. By developing a clear understanding of the specific cybersecurity risks posed by these technologies, organizations can implement appropriate security measures and protocols to protect against potential threats. Finally, as new threats and vulnerabilities emerge, it is important to continue assessing and managing cybersecurity risks in the context of robotics and automation to stay ahead of potential attacks. This requires ongoing monitoring and evaluation of cybersecurity risks, as well as updating risk management strategies to ensure that they remain effective in the face of evolving threats.

Overall, the importance of assessing cybersecurity risks in the context of robotics and automation cannot be overstated. The potential consequences of a cyberattack in this setting are too great to ignore, and effective risk management strategies are critical to ensure the security and reliability of these technologies [14].

11.2 CYBERSECURITY RISKS IN THE AGE OF ROBOTICS AND AUTOMATION

Cybersecurity risks in the age of robotics and automation refer to the vulnerabilities and threats that arise from the increasing use of robotic and automated technologies in various industries, and the resulting interconnectedness and dependence on digital systems. The use of these technologies has created new types of cybersecurity risks that need to be identified and managed to ensure the safety, reliability, and security of these systems. These risks can manifest in various forms, including but not limited to attacks on network security, data security, and physical security. For instance, cybercriminals may attempt to exploit vulnerabilities in the software or hardware of robotic and automated systems to steal data, manipulate equipment, or cause physical damage. In addition, the use of the Internet of Things (IoT) devices in the context of robotics and automation introduces additional risks. These devices may lack appropriate security features, making them vulnerable to attack and compromise. Malicious actors may use these devices as entry points to launch attacks on the larger systems they are connected to, potentially causing significant damage. Other examples of cybersecurity risks in the age of robotics and automation include denial-of-service attacks, ransomware attacks, and insider threats. Denial-of-service attacks can overwhelm the systems, causing them to become non-operational. Ransomware attacks can encrypt data and demand payment to restore access. Insider threats can come from malicious employees or contractors with access to critical systems, who may cause damage intentionally or unintentionally. The potential impact of a cyberattack on robotic and automated systems can be catastrophic. A successful attack can cause physical damage to equipment and supply chains, endanger human lives, and cause significant financial losses. Therefore, assessing and managing cybersecurity risks in the age of robotics and automation is critical to ensure the safety and security of these systems. This requires the implementation of appropriate security measures and protocols to protect against potential threats, as well as the ongoing monitoring and evaluation of cybersecurity risks to stay ahead of evolving threats [15–17].

11.2.1 Overview of the Landscape of Cybersecurity Risks

The landscape of cybersecurity risks in the age of robotics and automation is complex and ever-changing. As robotics and automation technologies continue to advance, so do the potential threats and vulnerabilities that come with them. A comprehensive understanding of the risks is crucial to developing effective strategies for risk management. One major area of concern is network security. As digital systems become increasingly interconnected and automated, the risks associated with network security have grown. Malicious actors can exploit vulnerabilities in network infrastructure to gain unauthorized access to robotic and automated systems, potentially causing damage or theft of data.

Another important aspect of cybersecurity risks in the age of robotics and automation is data security. Physical security is also a key area of concern. As robotic and automated systems become more advanced, they also become more complex and susceptible to physical attacks. Sabotage, theft, environmental damage, and accidents can all pose a significant risk to the security of these systems and the data they contain. The Internet of Things (IoT) is another area where cybersecurity risks have become more prevalent. As the use of IoT devices continues to expand in the context of robotics and automation, so does the potential for vulnerabilities [18].

Finally, insider threats are another significant concern. Employees, contractors, or other individuals with access to robotic and automated systems can pose a threat to their security. Malicious actors with privileged access can cause significant damage to these systems, including data theft or sabotage.

In summary, the landscape of cybersecurity risks in the age of robotics and automation is complex and multifaceted. It is critical for organizations to understand the risks and vulnerabilities present in their systems and technologies to develop effective strategies for risk management. By doing so, they can better protect themselves from the potentially devastating consequences of a cybersecurity breach.

11.2.2 Types of Cybersecurity Risks

Cybersecurity risks come in many forms and can impact different aspects of an organization's systems and data. Here are some common types of cybersecurity risks [19, 20]:

- *Endpoint Security Risks*: These risks involve vulnerabilities in the devices used to access an organization's network or data. Endpoint security risks can include malware, phishing attacks, and unauthorized access to devices.
- *Cloud Security Risks*: These risks involve threats to the security and integrity of data stored in the cloud. Cloud security risks can include data breaches, unauthorized access to data, and insecure interfaces.
- *Physical Security Risks*: These risks involve threats to the physical infrastructure of an organization, such as theft or damage to servers, routers, or other equipment.

- *Human Error and Insider Threats*: These risks involve unintentional or intentional actions by employees or other authorized individuals that can result in data breaches or other security incidents. Human error and insider threats can include accidental data exposure, social engineering attacks, or malicious actions by employees.

Each of these types of cybersecurity risks requires a specific approach to risk management and mitigation. Organizations must be aware of the different types of risks they face and take steps to protect their systems and data from potential threats.

Some examples of types of cybersecurity risks in the age of robotics and automation are shown in Table 11.1.

Table 11.1 provides examples of network security risks that can pose a significant threat to robotic and automated systems. Man-in-the-middle attacks are a significant threat to network security. In this type of attack, a hacker intercepts communications between two systems, potentially gaining access to sensitive data or control of robotic and automated systems. Distributed denial-of-service attacks are used to overwhelm network systems, making them unavailable to users. Such attacks can cause significant disruption to operations or be used to steal data. Brute force attacks are another technique used by hackers to gain access to network systems or control of robotic and automated systems. In these attacks, hackers use automated trial and error methods to guess passwords or other login credentials.

TABLE 11.1
Examples of Network Security Risks

Network Security Risks	Description
Man-in-the-middle Attack	An attack where a hacker intercepts communications between two systems to gain access to sensitive data or to modify data. This type of attack can be particularly dangerous in the context of robotic and automated systems, as it can be used to hijack control of the systems.
Distributed Denial-of-Service (DDoS)	An attack where a hacker floods a network with traffic, overwhelming the system and causing it to become unavailable to users. This type of attack can be used to disrupt operations or steal data.
Brute Force Attack	An attack where a hacker attempts to gain access to a system or network by guessing passwords or other login credentials through automated trial and error methods. Brute force attacks can be used to gain access to sensitive data or control of robotic and automated systems.
Malware	Malicious software that can be used to gain access to a system or network, steal data, or cause physical damage to robotic and automated systems. Malware can be introduced to a network through email attachments, software downloads, or other means.

Malware is a type of malicious software that can be used to gain unauthorized access to network systems or cause physical damage to robotic and automated systems. Malware can be introduced to a network through email attachments, software downloads, or other means. Zero-day vulnerabilities are unknown vulnerabilities in network systems that can be exploited by attackers to gain access to network systems or to cause damage to robotic and automated systems. Organizations must remain vigilant to detect and respond to these vulnerabilities in order to mitigate the risks.

11.2.3 Specific Examples of Cybersecurity Risks in the Age of Robotics and Automation

There are numerous specific examples of cybersecurity risks in the age of robotics and automation, including [21]:

- *Ransomware*: Malicious actors can exploit vulnerabilities in robotic and automated systems to install ransomware, which encrypts data and demands payment in exchange for decryption. This can cause significant disruption to operations and result in financial losses.
- *Insider Threats*: Employees, contractors, or other individuals with access to robotic and automated systems can pose a threat to their security. Malicious insiders can steal sensitive data, cause physical damage to systems, or install malware.
- *Remote Exploits*: As many robotic and automated systems are designed for remote access and control, they can be vulnerable to remote exploits. Attackers can exploit vulnerabilities in remote access software or protocols to gain unauthorized access to systems.
- *Insecure Communications*: Robotic and automated systems often rely on wireless communication protocols to exchange data. These communication channels can be vulnerable to interception or tampering, which can result in data theft or physical damage to systems.
- *Physical Attacks*: Physical attacks on robotic and automated systems can cause significant damage, resulting in operational disruption, data theft, or financial losses. Such attacks can include vandalism, theft, or the use of physical force to gain access to systems.
- *Internet of Things (IoT) Exploits*: As the use of IoT devices continues to expand in the context of robotics and automation, so does the potential for vulnerabilities. Attackers can exploit vulnerabilities in IoT devices to gain unauthorized access to robotic and automated systems or steal sensitive data.

These are just a few examples of the many cybersecurity risks that organizations face in the age of robotics and automation. As technology continues to advance, it is critical to remain vigilant and proactive in identifying and mitigating these risks.

11.3 FRAMEWORKS FOR ASSESSING CYBERSECURITY RISKS IN THE AGE OF ROBOTICS AND AUTOMATION

As the use of robotics and automation grows, the need for effective cybersecurity measures becomes increasingly important. To help organizations identify, analyze, and manage cybersecurity risks, a variety of frameworks and models have been developed. Here are some popular frameworks used for assessing cybersecurity risks in the age of robotics and automation:

Organizations today face a growing number of cybersecurity risks in the age of robotics and automation.

By using FAIR, organizations can identify the factors that influence risk, such as threat frequency, vulnerability, and the potential impact of a breach. This allows organizations to make informed decisions about where to allocate their resources for risk mitigation. The CIS Controls, developed by the Center for Internet Security (CIS), provide a set of guidelines for securing an organization's IT systems and data. The framework consists of 20 controls that address key areas of cybersecurity risk, including identity and access management, software configuration, and incident response. By implementing the CIS Controls, organizations can identify potential vulnerabilities in their systems and implement best practices to reduce the risk of a cybersecurity incident.

In conclusion, organizations can use various frameworks and strategies to effectively manage cybersecurity risks in the age of robotics and automation. These frameworks provide a comprehensive approach to assessing and mitigating cybersecurity risks, allowing organizations to identify potential vulnerabilities and implement risk management strategies to improve their overall security posture.

Using a framework or model for assessing cybersecurity risks can help organizations effectively manage and reduce their risk exposure. These frameworks provide a structured approach to risk management, enabling organizations to identify and prioritize potential threats, assess the impact of those threats, and implement strategies for risk mitigation [22, 23].

11.3.1 OVERVIEW OF EXISTING FRAMEWORKS FOR ASSESSING CYBERSECURITY RISKS

One widely used framework is the Center for Internet Security Controls. The CIS Controls provide a set of guidelines for securing an organization's IT systems and data. The framework consists of 20 controls that address key areas of cybersecurity risk, including identity and access management, software configuration, and incident response. The CIS Controls are designed to be implementable by organizations of all sizes and are updated regularly to reflect emerging threats and best practices. Another framework specifically designed for organizations that handle payment card data is the Payment Card Industry Data Security Standard. This framework provides a set of requirements for protecting cardholder data, including physical security, access control, and encryption [24].

11.3.2 EVALUATION OF THE STRENGTHS AND WEAKNESSES OF EACH FRAMEWORK

The CIS Controls framework has the strength of providing a comprehensive set of guidelines for securing IT systems and data. The framework is updated regularly to reflect emerging threats and best practices, making it adaptable to the evolving cybersecurity landscape. However, the framework's focus on technical controls may limit its effectiveness in addressing human factors in cybersecurity risk management. Payment Card Industry Data Security Standard (PCI DSS) has the strength of being a widely recognized and mandatory framework for organizations that handle payment card data. Compliance is enforced through audits and assessments, making it an effective means of ensuring a baseline level of security for sensitive data. Its five core functions provide a holistic approach to cybersecurity risk management. The framework is regularly updated and incorporates input from a range of stakeholders. However, the framework's flexibility may also make it challenging for some organizations to implement effectively, and its high-level nature may require additional guidance for practical implementation. Overall, each framework has its own strengths and weaknesses, and organizations should carefully consider their specific needs and goals when selecting a framework for assessing cybersecurity risks. A comprehensive approach to cybersecurity risk management may involve integrating multiple frameworks to address different types of risks and to ensure a holistic approach to cybersecurity [25].

11.3.3 APPLICATION OF FRAMEWORKS IN ROBOTICS AND AUTOMATION

The growing dependence on robotics and automation across various industries has led to the emergence of new cybersecurity risks that were previously unknown. To mitigate these risks, a number of frameworks have been created to help organizations manage their cybersecurity risks. This chapter discusses how each framework can be implemented in the context of robotics and automation.

The NIST Cybersecurity Framework is a widely used framework in the United States that offers guidance on managing and minimizing cybersecurity risks. The framework is composed of five core functions: Identify, Protect, Detect, Respond, and Recover. In the case of robotics and automation, the Identify function would include identifying the vulnerable systems and assets that are prone to cybersecurity threats. The Protect function would require implementing safeguards to prevent cyberattacks. The Detect function would entail identifying cyber threats in real time, while the Respond function would demand an immediate response to cyber incidents. Lastly, the Recover function would require restoring systems to normal operation after a cyber incident.

CIS Controls is another framework that can be implemented to improve the cybersecurity posture of an organization. The controls are grouped into three categories: Basic, Foundational, and Organizational. In the case of robotics and automation, the Basic controls would include implementing basic cybersecurity measures such as firewalls and antivirus software. The Foundational controls would require implementing more advanced cybersecurity measures such as multi-factor authentication and regular security awareness training. The Organizational controls would entail establishing a culture of cybersecurity within the organization.

Lastly, the FAIR (Factor Analysis of Information Risk) framework provides a quantitative approach to assessing cybersecurity risks. In the context of robotics and automation, the framework would involve identifying critical assets to the operation of the system, assessing the likelihood and impact of a cyberattack, and calculating the associated risk with each asset. This would allow organizations to prioritize their cybersecurity efforts based on the level of risk associated with each asset.

To conclude, the use of robotics and automation has introduced new cybersecurity risks that organizations must manage. To manage these risks, various frameworks have been developed, each with its unique approach to managing cybersecurity risks. The NIST Cybersecurity Framework, ISO/IEC 27001:2013, CIS Controls, and FAIR framework are four frameworks that can be applied to the context of robotics and automation to manage cybersecurity risks. By applying these frameworks, organizations can better protect their systems and assets from cyber threats [26, 27].

11.4 STRATEGIES FOR MANAGING CYBERSECURITY RISKS IN THE AGE OF ROBOTICS AND AUTOMATION

Here are some strategies for managing cybersecurity risks in the age of robotics and automation [28–30]:

1. *Develop a comprehensive cybersecurity risk management plan*: This plan should involve a thorough assessment of the organization's technology infrastructure, identifying vulnerabilities and assessing the potential impact of cybersecurity risks. The plan should also include policies and procedures for risk mitigation and incident response.
2. *Implement a layered defense approach*: A layered defense approach involves implementing multiple layers of security controls to protect against cyber threats. This includes using firewalls, antivirus software, intrusion detection and prevention systems, and encryption.
3. *Conduct regular cybersecurity training*: Employees are often the weakest link in an organization's cybersecurity defenses. Therefore, it is essential to provide regular cybersecurity training to all employees, including training on identifying and reporting suspicious activity, password management, and safe internet browsing practices.
4. *Implement access control and monitoring*: Access control measures should be implemented to ensure that only authorized personnel have access to sensitive information and systems. Monitoring should also be conducted to detect any suspicious activity and potential security breaches.
5. *Regularly update and patch systems*: Cybersecurity threats are constantly evolving, and software vulnerabilities can be exploited by hackers to gain access to systems. Therefore, it is important to regularly update and patch systems to ensure that they are protected against the latest threats.

6. *Conduct regular cybersecurity risk assessments*: Cybersecurity risks are constantly changing, and new vulnerabilities may be introduced with the implementation of new technology. Therefore, it is essential to regularly conduct cybersecurity risk assessments to identify new risks and implement measures to mitigate them.
7. *Collaborate with industry and government partners*: Cybersecurity is a shared responsibility, and organizations can benefit from collaboration with industry and government partners. These partnerships can provide access to expertise, best practices, and threat intelligence.

By implementing these strategies, organizations can better manage cybersecurity risks in the age of robotics and automation. It is important to note that cybersecurity risk management is an ongoing process and requires a continuous effort to identify and address new risks as they emerge. Figure 11.1 shows strategies for managing cybersecurity risks in the age of robotics and automation.

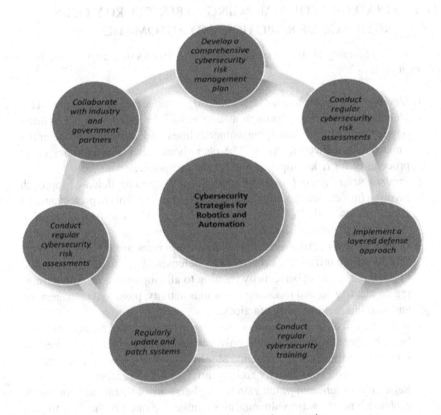

FIGURE 11.1 Cybersecurity strategies for robotics and automation.

11.5 TRENDS AND TECH AFFECTING CYBERSECURITY IN ROBOTICS AND AUTOMATION

Emerging technologies and trends have the potential to significantly impact cybersecurity risks in the age of robotics and automation. Here is a discussion of some of the emerging trends and technologies that are likely to have a significant impact on cybersecurity risks. Emerging trends and technologies are significantly impacting cybersecurity risks in the age of robotics and automation. Artificial Intelligence and Machine Learning have great potential to improve the efficiency and productivity of robotics and automation systems. However, they also introduce new cybersecurity risks. Attackers may target AI and ML models or use these technologies to automate cyberattacks. The Internet of Things is rapidly expanding, with the number of connected devices expected to reach 50 billion by 2025. The proliferation of connected devices introduces new cybersecurity risks, including distributed denial-of-service attacks, data breaches, and attacks on critical infrastructure. As more devices become connected, the potential for cyberattacks also increases. Therefore, it is important to ensure that these devices are secured from attacks and that data privacy is maintained. The introduction of 5G networks is expected to significantly increase the speed and bandwidth of internet connections. However, this increased connectivity also creates new cybersecurity risks. Hackers may gain access to more devices and data, which could lead to serious data breaches. Therefore, it is important to ensure that 5G networks are secured and that adequate cybersecurity measures are in place to protect against potential threats [31, 32]. Quantum computing is an emerging technology that has the potential to solve complex problems at a much faster rate than traditional computing methods. However, it also has the potential to undermine the security of traditional encryption methods, which could make it easier for hackers to access sensitive data. This technology has the potential to significantly impact the field of cybersecurity, and it is important to stay updated on its potential risks and vulnerabilities. Blockchain is a decentralized ledger technology that has the potential to improve the security of transactions and data storage [33].

As these technologies continue to develop and become more prevalent in robotics and automation, it is likely that cybersecurity risks will continue to evolve and become more complex. Organizations that are using these technologies will need to be vigilant in identifying and addressing these risks.

To address these emerging cybersecurity risks, organizations should consider implementing a proactive and holistic cybersecurity strategy. This strategy should include regular risk assessments to identify potential threats, as well as the development of policies and procedures for responding to cybersecurity incidents. Additionally, organizations should consider implementing technologies such as artificial intelligence and machine learning to help identify and respond to threats in real time [34–37]. In conclusion, emerging technologies and trends have the potential to significantly impact cybersecurity risks in the age of robotics and automation. As these technologies continue to develop and become more prevalent, organizations must be prepared to identify and address new and complex cybersecurity risks. By implementing a proactive and holistic cybersecurity strategy, organizations can help mitigate the risks associated with these emerging technologies and trends.

11.6 CONCLUSION

In conclusion, as robotics and automation technologies become more prevalent in various industries, the importance of assessing and managing cybersecurity risks cannot be overstated. The emergence of new technologies and trends, such as AI and machine learning, IoT, 5G networks, quantum computing, and blockchain, introduce new challenges for risk management. To effectively manage these risks, several frameworks and strategies have been proposed, each with its own strengths and weaknesses. The NIST Cybersecurity Framework provides a comprehensive approach to risk management, while the FAIR framework provides a quantitative method for risk assessment. The Zero Trust model and Defence in Depth strategy offer different approaches to security architecture. However, it is important to evaluate each framework and strategy's suitability for the context of robotics and automation [38]. Effective risk management requires a multifaceted approach that combines preventive, detective, and responsive measures. Preventive measures such as access controls, network segmentation, and secure coding practices can reduce the likelihood of cyberattacks. Detective measures such as intrusion detection systems and security analytics can detect threats early, while responsive measures such as incident response plans and disaster recovery plans can minimize the impact of attacks [39]. Finally, it is important to continuously monitor and update risk management strategies to address new threats and vulnerabilities that arise. Cybersecurity risks in the age of robotics and automation are an ever-evolving landscape that requires constant vigilance and adaptation. By implementing appropriate frameworks and strategies and staying informed of emerging trends and technologies, organizations can effectively manage cybersecurity risks and ensure the safety and security of their operations.

REFERENCES

1. Sindhwani, N., Anand, R., Niranjanamurthy, M., Verma, D. C., and Valentina, E. B. (Eds.). *IoT Based Smart Applications*. Springer Nature, 2022.
2. Chibber, A., Anand, R., and Singh, J. Smart traffic light controller using edge detection in digital signal processing. In *Wireless Communication with Artificial Intelligence*. CRC Press, 2022, pp. 251–272.
3. Dahiya, A., Anand, R., Sindhwani, N., and Kumar, D. A novel multi-band high-gain slotted fractal antenna using various substrates for X-band and Ku-band applications. *MAPAN* 2022, 37(1), 175–183.
4. Raghavan, R., Verma, D. C., Pandey, D., Anand, R., Pandey, B. K., and Singh, H. Optimized building extraction from high-resolution satellite imagery using deep learning. *Multimedia Tool. Appl.* 2022, 81(29), 42309–42323.
5. Kumar Pandey, B., Pandey, D., Nassa, V. K., Ahmad, T., Singh, C., George, A. S., and Wakchaure, M. A. Encryption and steganography-based text extraction in IoT using the EWCTS optimizer. *Imaging Sci. J.* 2022, 69, 1–19.
6. Pandey, B. K., Pandey, D., Wairya, S., Agarwal, G., Dadeech, P., Dogiwal, S. R., and Pramanik, S. Application of integrated steganography and image compressing

techniques for confidential information transmission. *Cyber Sec. Netw. Sec.* 2022, 169–191. ieeexplore.ieee.org/abstract/document/10112282

7. Kumar, M. S., Sankar, S., Nassa, V. K., Pandey, D., Pandey, B. K., and Enbeyle, W. Innovation and creativity for data mining using computational statistics. In *Methodologies and Applications of Computational Statistics for Machine Intelligence.* IGI Global, 2021, pp. 223–240.

8. Pandey, D., and Pandey, B. K. An efficient deep neural network with adaptive galactic swarm optimization for complex image text extraction. In *Process Mining Techniques for Pattern Recognition.* CRC Press, 2022, pp. 121–137.

9. Jain, N., Chaudhary, A., Sindhwani, N., and Rana, A. Applications of Wearable devices in IoT. In *2021 9th International Conference on Reliability, Infocom Technologies and Optimization (Trends and Future Directions)(ICRITO) 2021.* IEEE, 2021, September, pp. 1–4.

10. Gupta, B., Chaudhary, A., Sindhwani, N., and Rana, A. Smart shoe for detection of electrocution using Internet of things (IoT). In *2021 9th International Conference on Reliability, Infocom Technologies and Optimization (Trends and Future Directions) (ICRITO) 2021.* IEEE, 2021, September, pp. 1–3.

11. Abouzakhar, N. Critical infrastructure cybersecurity: A review of recent threats and violations. In *Proceedings of the European Conference on Information Warfare and Security*, Jyväskylä, Finland, 11–12 July 2013.

12. Marvell, S. *The Real and Present Threat of a Cyber Breach Demands Real-Time Risk Management; Acuity Risk Management.* London, 2015.

13. Adar, E., and Wuchner, A. Risk management for critical infrastructure protection (CIP) challenges, best practices and tools. In *Proceedings of the First IEEE International Workshop on Critical Infrastructure Protection (IWCIP'05)*, Darmstadt, Germany, 3–4 November 2005.

14. Marvell, S. Real-time cyber security risk management. *ITNOW* 2015, 57(4), 26–27.

15. Harvey, J., and Service, T. I. Introduction to managing risk. Available online: http://www.cimaglobal.com/Documents/ImportedDocuments/cid_tg_intro_to_managing_rist.apr07.pdf.

16. Georgieva, K., Farooq, A., and Dumke, R. R. Analysis of the risk assessment methods–A survey. In *International Workshop on Software Measurement.* Springer, 2009.

17. Cherdantseva, Y., Burnap, P., Blyth, A., Eden, P., Jones, K., Soulsby, H., and Stoddart, K. A review of cyber security risk assessment methods for SCADA systems. *Comput. Secur.* 2016, 56, 1–27.

18. Patel, S. C., Graham, J. H., and Ralston, P. A. Quantitatively assessing the vulnerability of critical information systems: A new method for evaluating security enhancements. *Int. J. Inf. Manag.* 2008, 28(6), 483–491.

19. Hahn, A., Ashok, A., Sridhar, S., and Govindarasu, M. Cyber-physical security testbeds: Architecture, application, and evaluation for smart grid. *IEEE Trans. Smart Grid* 2013, 4(2), 847–855.

20. Cárdenas, A. A., Amin, S., Lin, Z. S., Huang, Y. L., Huang, C. Y., and Sastry, S. Attacks against process control systems: Risk assessment, detection, and response. In *Proceedings of the 6th ACM Symposium on Information, Computer and Communications Security*, Hong Kong, China, 22–24 March 2011.

21. Peng, Y., Lu, T., Liu, J., Gao, Y., Guo, X., and Xie, F. Cyber-physical system risk assessment. In *Proceedings of the Ninth International Conference on Intelligent Information Hiding and Multimedia Signal Processing*, Beijing, China, 6–18 October 2013.

22. Cardenas, A., Amin, S., Sinopoli, B., Giani, A., Perrig, A., and Sastry, S. Challenges for securing cyber physical systems. In *Proceedings of the Workshop on Future Directions in Cyber-Physical Systems Security*, Newark, NJ, 23–24 July 2009.

23. Sridhar, S., Hahn, A., and Govindarasu, M. Cyber–physical system security for the electric power grid. *Proceedings IEEE* 2012, 100(1), 210–224.
24. Yoneda, S., Tanimoto, S., Konosu, T., Sato, H., and Kanai, A. Risk assessment in cyber-physical system in office environment. In *Proceedings of the 2015 18th International Conference on Network-Based Information Systems (NBiS)*, Taipei, Taiwan, 2–4 September 2015.
25. Ten, C.-W., Manimaran, G., and Liu, C.-C. Cybersecurity for critical infrastructures: Attack and defense modeling. *IEEE Trans. Syst. Man Cybern. Part A Syst. Hum.* 2010, 40(4), 853–865.
26. Gai, K., Qiu, M., Ming, Z., Zhao, H., and Qiu, L. Spoofing-jamming attack strategy using optimal power distributions in wireless smart grid networks. *IEEE Trans. Smart Grid* 2017, 8(5), 2431–2439.
27. Gai, K., Qiu, M., Zhao, H., Tao, L., and Zong, Z. Dynamic energy-aware cloudlet-based mobile cloud computing model for green computing. *J. Netw. Comput. Appl.* 2016, 59, 46–54.
28. Gai, K., and Qiu, M. Blend arithmetic operations on tensor-based fully homomorphic encryption over real numbers. *IEEE Trans. Ind. Inform.* 2017, 14(8), 3590–3598.
29. Ray, P. D., Harnoor, R., and Hentea, M. Smart power grid security: A unified risk management approach. In *Proceedings of the 2010 IEEE International Carnahan Conference on Security Technology (ICCST)*, San Jose, CA, 5–8 October 2010.
30. Yadav, D., and Mahajan, A. R. Smart grid cyber security and risk assessment: An overview. *Int. J. Sci. Eng. Technol. Res.* 2015, 4, 3078–3085.
31. Anand, R., Ahamad, S., Veeraiah, V., Janardan, S. K., Dhabliya, D., Sindhwani, N., and Gupta, A. Optimizing 6G wireless network security for effective communication. In *Innovative Smart Materials Used in Wireless Communication Technology*. IGI Global, 2023, pp. 1–20.
32. Sindhwani, N., and Singh, M. A joint optimization based sub-band expediency scheduling technique for MIMO communication system. *Wirel. Personal Commun.* 2020, 115(3), 2437–2455.
33. Rice, E. B., and AlMajali, A. Mitigating the risk of cyber attack on smart grid systems. *Procedia Comput. Sci.* 2014, 28, 575–582.
34. Pandey, B. K., Pandey, D., Wariya, S., Aggarwal, G., and Rastogi, R. Deep learning and particle swarm optimisation-based techniques for visually impaired humans' text recognition and identification. *Augmented Hum. Res.* 2021, 6, 1–14.
35. Tripathi, R. C., Gupta, P., Anand, R., Jayashankar, R. J., Mohanty, A., Michael, G., & Dhabliya, D. (2023). Application of Information Technology Law in India on IoT/IoE With Image Processing. In *Handbook of Research on Thrust Technologies' Effect on Image Processing*. IGI Global. pp. 135–150.
36. Pandey, D., Pandey, B. K., Sindhwani, N., Anand, R., Nassa, V. K., and Dadheech, P. An interdisciplinary approach in the post-COVID-19 pandemic era. In *An Interdisciplinary Approach in the Post-COVID-19 Pandemic Era*, 2022, NOVA SCIENCE, pp. 1–290.
37. Kumar, M. S., Sankar, S., Nassa, V. K., Pandey, D., Pandey, B. K., and Enbeyle, W. Innovation and creativity for data mining using computational statistics. In *Methodologies and Applications of Computational Statistics for Machine Intelligence*. IGI Global, 2021, pp. 223–240.
38. Saini, M. K., Nagal, R., Tripathi, S., Sindhwani, N., and Rudra, A. PC interfaced wireless robotic moving arm. In *AICTE Sponsored National Seminar on Emerging Trends in Software Engineering*, 2008, Vol. 50, pp. 787–788.
39. Anand, R., Shrivastava, G., Gupta, S., Peng, S. L., and Sindhwani, N. Audio watermarking with reduced number of random samples. In *Handbook of Research on Network Forensics and Analysis Techniques*. IGI Global, 2018, pp. 372–394.

12 The Role of Education in Addressing the Workforce Challenges of Automation
A Global Perspective

Deepti Sharma, P. Krishna Priya,
Sachin Tripathi, G. Bhuvaneswari,
M. Kavitha, and V. Lakshmi Prasanna

12.1 INTRODUCTION

Automation has the potential to create new jobs and increase productivity, but it also poses significant challenges for workers who may be displaced or require new skills to remain employable [1–4]. The workforce must be equipped with the necessary knowledge and skills to adapt to the changing demands of the labor market, and education has a critical role in providing such preparation. Education has always been essential to developing the workforce's capacity to adapt to changing economic, social, and technological trends. The advent of automation has placed even more significant emphasis on the importance of education as a tool for addressing workforce challenges. Education institutions must take a leading role in preparing the workforce for the changing demands of the labor market, and they must do so in a way that is globally relevant. This chapter explores the role of education in addressing the workforce challenges of automation from a global perspective. It examines the extent of automation in the workforce, the skills and knowledge required for the new jobs created by automation, and the education systems' response to automation. The chapter will analyze the role of vocational education in addressing the workforce challenges of automation and provide a global perspective on the education system's response to automation. The analysis will evaluate the effectiveness of education systems in preparing the workforce for automation and identify potential areas for future research. In conclusion, this chapter provides a critical analysis of the role of education in addressing the workforce challenges of automation from a global perspective. It is essential to develop and implement comprehensive education policies and practices that are responsive to the changing demands of the labor market. Education institutions must be innovative and proactive in equipping the workforce with the necessary skills and knowledge to remain relevant in the era of automation. The findings of this chapter are expected to contribute to the ongoing policy debates and discussions on the future of work and education in the context of automation [5].

12.1.1 Automation in Education for the Workforce

The topic of the role of education in addressing the workforce challenges of automation is critical in today's rapidly changing world. As automation continues to advance, it has the potential to create new jobs and increase productivity, but it also poses significant challenges for workers who may be displaced or require new skills to remain employable. Education plays a critical role in equipping the workforce with the necessary knowledge and skills to adapt to the changing demands of the labor market. The importance of this topic is further heightened by the global nature of the workforce challenges of automation. The effects of automation are not limited to specific regions or countries but have global implications, making it essential to explore the role of education in addressing these challenges from a global perspective. This chapter seeks to provide a comprehensive analysis of the current state of education in relation to automation challenges, identify good practices and gaps, and explore possible future directions in a global context. Moreover, the topic of education's role in addressing workforce challenges posed by automation is timely, given the ongoing discussions on the future of work and education. As automation technologies continue to advance, they will continue to transform the labor market, requiring education systems to be innovative and proactive in responding to these challenges. Therefore, the findings of this chapter are expected to contribute to ongoing policy debates and discussions on the future of work and education in the context of automation [6].

In summary, the importance of exploring the role of education in addressing the workforce challenges of automation from a global perspective cannot be overstated. As the world continues to adapt to the changing nature of work brought about by automation, it is essential to develop and implement comprehensive education policies and practices that are responsive to the changing demands of the labor market. This chapter aims to provide valuable insights and recommendations for policymakers, education leaders, and other stakeholders to enhance the effectiveness of education in addressing the challenges posed by automation [7–9].

12.1.2 Overview

Automation is transforming the way we work, with significant implications for the global workforce. As automation technologies continue to advance, they are creating new jobs and increasing productivity, but they are also posing significant challenges for workers who may be displaced or require new skills to remain employable. Job displacement is one of the most significant workforce challenges caused by automation. The rise of automation has led to the displacement of many traditional jobs in industries such as manufacturing, transportation, and retail [10–13]. This displacement has resulted in unemployment and job insecurity for many workers, and it poses a significant challenge for the workforce. To address this challenge, educational institutions must take a leading role in preparing workers for the changing demands of the labor market. Another challenge posed by automation is the skills mismatch. As new jobs are created by automation, workers may not possess the necessary skills and knowledge to perform these new roles effectively. This skills mismatch can lead to a shortage of qualified workers in certain industries, making it difficult for companies to fill critical roles. Education institutions must take a proactive approach to address this challenge by equipping workers with the necessary skills and knowledge to remain employable. Automation has the potential to exacerbate income inequality. While it may create high-skilled jobs that pay well, it can also create low-skilled jobs that do not pay enough to support a family. This income inequality can lead to social and economic instability, and it poses a significant challenge for the workforce. To address this challenge, education institutions must focus on providing workers with the skills and knowledge they need to succeed in high-skilled jobs, while also providing support for those who may be displaced by automation. Worker well-being is another challenge posed by automation. The increase in the pace of work and the reduction in worker autonomy can lead to burnout, stress, and dissatisfaction. This challenge can be addressed by providing workers with the necessary skills and knowledge to manage their workloads effectively and by promoting work-life balance. Education institutions can play a critical role in addressing this challenge by incorporating well-being and stress management into their training programs. Finally, automation can impact the social fabric of communities. As jobs disappear and economic and social inequality increases, the social and economic stability of communities can be disrupted. To address this challenge, education institutions must focus on providing workers with the skills and knowledge they need to succeed in the changing labor market while also promoting social and economic stability. This can be achieved through training programs that focus on skills that are in high demand and that provide workers with the knowledge they need to succeed in the modern workforce [14].

12.2 THE IMPACT OF AUTOMATION ON THE WORKFORCE

Automation has had a significant impact on the workforce in recent years, and this impact is expected to continue as automation technologies continue to advance. The impact of automation on the workforce has resulted in several challenges. One of the most significant challenges is job displacement, with automation threatening to replace many traditional jobs with machines and software. This trend is most pronounced in industries and sectors where jobs are repetitive, routine, and predictable, and it has led to significant job losses in these areas. Another challenge is the skills mismatch that automation creates. As new jobs are created by automation, workers may not possess the necessary skills and knowledge to perform these new roles effectively, leading to a skills mismatch. This can result in a mismatch between the skills that workers possess and the skills that employers require, which can make it difficult for workers to find employment and can result in labor market polarization.

Automation also has the potential to exacerbate income inequality, as it may create high-skilled jobs that pay well, but also low-skilled jobs that do not pay enough to support a family. This can result in a widening income gap between high-skilled and low-skilled workers, as well as a reduction in middle-skill jobs. This trend has been observed in several developed and developing countries and poses significant challenges for policymakers seeking to promote more equitable and inclusive economic growth. Worker well-being is also an important consideration in the context of automation. Automation can increase the pace of work and reduce worker autonomy, which can lead to worker burnout, stress, and dissatisfaction. The loss of jobs and the disruption of traditional work practices can also create uncertainty and anxiety among workers, leading to negative impacts on their physical and mental health. The social impact of automation is also a key concern. As jobs disappear and economic and social inequality increases, automation can impact the social fabric of communities. This can result in the erosion of social cohesion and the rise of social unrest, as well as the decline of local economies and the loss of community identity. These challenges are particularly acute in rural areas and smaller communities that are heavily dependent on specific industries or sectors that are vulnerable to automation [15]. The impact of automation on the workforce is complex and multifaceted, with several challenges emerging as a result of this transformative technology. Addressing these challenges will require a coordinated response from policymakers, educators, and employers to ensure that workers are equipped with the skills and knowledge they need to thrive in the rapidly changing world of work.

12.2.1 Overview of the Extent of Automation in the Workforce

Automation has been rapidly expanding in the workforce over the past few decades, and it is expected to continue to do so in the future. In manufacturing, for example, automation has led to the widespread use of robots and other machines that can perform tasks more efficiently and effectively than human workers. Automation has also been adopted in many other sectors, including transportation, retail, and healthcare, to name a few. The extent of automation in the workforce varies widely across different countries and industries. Some countries, such as Japan and South Korea, have been early adopters of automation and have invested heavily in robotics and other technologies to improve productivity and competitiveness. Other countries, such as India and many African nations, have lower levels of automation due to lower labor costs and a lack of infrastructure to support automation. In terms of industries, manufacturing has been at the forefront of automation, with robots and other machines performing many tasks that were once done by human workers. Transportation is another sector where automation is rapidly expanding, with the development of self-driving cars and trucks. Retail is also seeing a shift towards automation, with the use of self-checkout kiosks and automated warehouses.

While automation has led to increased efficiency and productivity, it has also led to significant job displacement in some industries. In the future, it is expected that automation will continue to expand in the workforce, but it will also create new opportunities for workers with the necessary skills and knowledge to succeed in a more automated workplace.

Education: Addressing the Workforce Challenges of Automation

12.2.2 Explanation of the Effects of Automation on Employment

Automation has had a significant impact on employment, with both positive and negative effects. To better understand these effects, both positive and negative effects have been considered [16]:

12.2.2.1 Positive Effects of Automation on Employment:
- *Increased productivity*: automation can perform tasks more quickly and efficiently than human workers, leading to increased productivity and output.
- *New job opportunities*: while automation may displace some jobs, it can also create new opportunities for workers with the necessary skills to design, build, and maintain automated systems.
- *Safer working conditions*: automation can perform dangerous and hazardous tasks that may be too risky for human workers, leading to safer working conditions.

12.2.2.2 Negative Effects of Automation on Employment:
- *Job displacement*: automation has the potential to replace many traditional jobs with machines and software, leading to job displacement in certain industries and sectors.
- *Skills mismatch*: as new jobs are created by automation, workers may not possess the necessary skills and knowledge to perform these new roles effectively, leading to a skills mismatch.
- *Income inequality*: automation has the potential to exacerbate income inequality as it may create high-skilled jobs that pay well, but also low-skilled jobs that do not pay enough to support a family.
- *Social impact*: automation can impact the social fabric of communities as jobs disappear and economic and social inequality increases.

Overall, automation has had a complex impact on employment, with both positive and negative effects. While automation has led to increased productivity, safer working conditions, and new job opportunities, it has also led to job displacement, a skills mismatch, and income inequality. It is important for policymakers, educators, and business leaders to address these challenges and develop strategies to help workers adapt to a more automated workplace [17, 18]. Figure 12.1 shows the effects of automation on employment.

12.2.3 Identification of the Skills and Knowledge Required for the New Jobs Created by Automation

As automation continues to transform the workforce, new jobs are being created that require a different set of skills and knowledge than traditional roles. Some of the key skills and knowledge required for these new jobs include [19, 20]:

1. *Technical skills*: with the increased use of automation and technology, workers need to possess technical skills such as coding, data analysis, and software development.

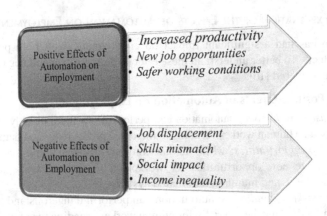

FIGURE 12.1 Effects of automation on employment.

2. *Critical thinking*: as automation takes over routine tasks, workers need to be able to analyze complex information, solve problems, and think creatively to complete more complex tasks.
3. *Adaptability*: Workers need to be able to adapt to new technologies and work environments as the pace of technological change increases.
4. *Emotional intelligence*: as automation increases, workers in certain roles may need to interact more with customers or clients. As a result, emotional intelligence is becoming an increasingly important skill for workers to possess.
5. *Communication skills*: as new jobs require more collaboration and teamwork, workers need to possess strong communication skills to effectively work with others.
6. *Cybersecurity skills*: with the increased use of automation and technology, cybersecurity skills are becoming increasingly important to ensure the protection of sensitive information and data.
7. *Creativity*: while automation can handle routine tasks, creativity remains a uniquely human trait. Workers in fields such as design, content creation, and marketing need to possess creativity to excel in their roles.

These are just a few of the skills and knowledge areas that are becoming increasingly important for workers in the age of automation. As the pace of technological change accelerates, it is important for workers to continually upskill and reskill to remain relevant in the workforce.

12.3 THE ROLE OF EDUCATION IN ADDRESSING AUTOMATION CHALLENGES

The challenges posed by automation require a comprehensive response, and education is a critical component in addressing these challenges. Education can help prepare workers with the skills and knowledge necessary to succeed in new jobs created by automation, as well as equip workers with the adaptability and resilience necessary to navigate a changing work environment [21].

- *Upskilling and reskilling*: education can provide workers with the opportunity to upskill and reskill to remain relevant in the workforce. Through programs such as vocational training, apprenticeships, and online learning, workers can gain the technical and soft skills necessary to succeed in new roles created by automation.
- *Building a diverse skill set*: education can also help workers build a diverse skill set that includes technical, interpersonal, and problem-solving skills. By developing a range of skills, workers can adapt to changing work environments and take on new roles as the workforce continues to evolve.
- *Collaboration and teamwork*: education can promote collaboration and teamwork skills that are necessary in many new roles created by automation. By fostering a collaborative learning environment, education can prepare workers to work effectively with others, solve complex problems, and communicate effectively.
- *Encouraging creativity and innovation*: education can also encourage creativity and innovation by providing workers with the opportunity to develop new ideas and approaches. This can be done through design thinking, project-based learning, and other forms of experiential education that allow workers to explore new ideas and ways of working.
- *Addressing social impact*: education can also play a role in addressing the social impact of automation. By promoting values such as social responsibility, empathy, and equity, education can help workers understand the social impact of automation and develop solutions that minimize negative impacts on communities.

12.3.1 Analysis of the Education System's Response to Automation

The education system has an important role to play in responding to the challenges posed by automation. One way that the education system can respond is by updating curricula to include new skills that are in demand in the workforce. This could include skills related to data analysis, computer programming, and artificial intelligence (AI), among others. By providing students with the skills and knowledge needed to perform new jobs that are created by automation, the education system can help to address the skills mismatch and reduce job displacement [22].

In addition to updating curricula, the education system can also provide training and education programs for workers who are displaced by automation. These programs could be focused on helping workers develop new skills that are in demand in the workforce, or on providing workers with the knowledge and resources needed to start their own businesses. By providing these types of programs, the education system can help to reduce the impact of automation on workers and communities. The education system can also play a role in promoting lifelong learning and upskilling.

As the pace of technological change continues to accelerate, workers will need to be able to adapt and learn new skills throughout their careers. By updating curricula, providing training and education programs, promoting lifelong learning, and ensuring equal access to education and training opportunities, the education system can help address the workforce challenges of automation and promote social and economic mobility [23].

12.3.2 The Role of Vocational Education in Automation Workforce Challenges

Vocational education can play a crucial role in addressing the workforce challenges of automation by equipping workers with the skills and knowledge they need to perform the new jobs created by automation. Vocational education programs focus on developing practical skills and knowledge that are directly relevant to specific industries and occupations, making them well-suited to addressing the skills mismatch caused by automation. By providing training in areas such as coding, robotics, and AI, vocational education can help workers gain the skills they need to work with and operate the new technologies that are increasingly being used in the workplace. Vocational education can also provide workers with the soft skills that are essential for success in the modern workplace, such as communication, teamwork, and problem-solving. In addition to providing training and skills development, vocational education can also play a role in supporting workers who have been displaced by automation. This can include providing career guidance and counseling, helping workers identify new career paths and training opportunities, and supporting workers through the process of reskilling and upskilling. However, to be effective in addressing the workforce challenges of automation, vocational education programs must be responsive to the changing needs of the labor market. This requires close collaboration between vocational education providers, industry stakeholders, and policymakers, as well as a commitment to ongoing innovation and development. Overall, the role of vocational education in addressing the workforce challenges of automation is a critical one. By providing workers with the skills and knowledge they need to thrive in the new economy, vocational education can help mitigate the negative impacts of automation on the workforce and ensure that workers are equipped to succeed in the jobs of the future [24].

12.4 GLOBAL PERSPECTIVE ON EDUCATION AND AUTOMATION

Automation is a global phenomenon, and the impact of automation on the workforce is a challenge faced by countries around the world. As such, there is a need for a global perspective on education and automation, and the role of education in addressing the challenges of automation. Many countries have recognized the need to address the challenges of automation through education and have implemented various policies and programs aimed at equipping the workforce with the skills and knowledge necessary to adapt to the changing nature of work. For example, in Germany, the dual vocational training system combines classroom education with on-the-job training to prepare students for specific occupations, including those in the technology and automation sectors. Similarly, Singapore has implemented a comprehensive skills development framework that aims to ensure that workers have the skills and knowledge necessary to succeed in a rapidly changing economy. In the United States, community colleges and vocational training programs have been identified as key components in addressing the skills mismatch caused by automation. These programs offer training and education in a range of industries, including those related to automation and technology. The ILO has called for a "human-centered approach" to automation that prioritizes the needs and well-being of workers, while the OECD has emphasized the need for policies that promote lifelong learning and skills development [25].

The challenges posed by automation are a global phenomenon, and the role of education in addressing these challenges is a pressing issue facing countries around the world. Vocational education and training programs, as well as broader policies aimed at promoting lifelong learning and skills development, are key components in addressing the workforce challenges of automation in a global context.

12.4.1 Approaches to Automation in Global Education Systems

Various countries have different approaches to automation in global education systems [26, 27].

- Different countries have varying approaches to addressing the workforce challenges posed by automation. Some countries are more proactive in investing in education and vocational training, while others may be slower to adapt.
- In some countries, such as Germany and Switzerland, vocational education plays a central role in preparing students for careers in the manufacturing industry. These countries have developed strong apprenticeship programs and vocational schools to provide students with the practical skills they need to succeed in the workforce.
- Some countries, such as the United States and the United Kingdom, have a more decentralized approach to education and may rely more heavily on private-sector partnerships to provide training and education for new jobs created by automation.
- In developing countries, there is often a lack of access to quality education and training programs, which can make it difficult for workers to adapt to the changing demands of the workforce.

12.4.2 Comparison of Education Systems for Automation Workforce Preparation

A comparison of the effectiveness of education systems in preparing the workforce for automation is shown in Table 12.1.

TABLE 12.1
Comparison of Education Systems for Automation Workforce Preparation

Country	Education System	Vocational Training	Technological Integration	Industry-Academia Collaboration	Overall Effectiveness
United States	Emphasizes general education over vocational training	Limited opportunities for vocational training	High levels of technological integration	Limited collaboration between industry and academia	Moderately effective
Germany	Strong emphasis on vocational education and apprenticeships	Extensive vocational training programs	High levels of technological integration	Strong collaboration between industry and academia	Highly effective
Japan	Balanced emphasis on general and vocational education	Comprehensive vocational training programs	High levels of technological integration	Strong collaboration between industry and academia	Highly effective
China	Emphasizes general education over vocational training	Increasing opportunities for vocational training	Moderate levels of technological integration	Increasing collaboration between industry and academia	Moderately effective
South Korea	Balanced emphasis on general and vocational education	Comprehensive vocational training programs	High levels of technological integration	Strong collaboration between industry and academia	Highly effective

In Table 12.1, we have included additional columns to compare the education systems of different countries in terms of their emphasis on vocational training, technological integration, and industry-academia collaboration. We have also included an overall effectiveness rating based on the combined factors. This table highlights the differences between countries in how they approach preparing their workforce for automation. For example, Germany's strong emphasis on vocational education and extensive training programs has proven to be highly effective, while the United States' emphasis on general education over vocational training has resulted in a moderately effective approach. Japan and South Korea have a more balanced approach to education that combines both general and vocational training, with a strong emphasis on collaboration between industry and academia, resulting in highly effective approaches to automation preparedness. Meanwhile, China is currently in the process of increasing opportunities for vocational training and collaboration between industry and academia, which has resulted in a moderately effective approach.

12.4.3 Evaluation of the Role of International Cooperation in Addressing Automation Challenges

As automation continues to impact the workforce globally, the role of international cooperation in addressing its challenges has become increasingly important. International cooperation can help countries learn from one another's experiences, collaborate on best practices, and address common challenges. One way that international cooperation can be effective is through the sharing of information and knowledge. By sharing information about the effects of automation on the workforce, countries can learn from one another's experiences and better prepare for the challenges that lie ahead.

Another way that international cooperation can be effective is through the development of joint initiatives and programs. By collaborating on initiatives that address the challenges of automation, countries can leverage their collective strengths and resources to develop more effective solutions. For example, countries could collaborate on the development of training programs for workers in high-growth sectors, or on the creation of job-matching platforms that connect workers with available positions. International cooperation can also help to address the challenges of automation by promoting greater investment in education and training. By working together, countries can share resources and best practices for developing effective training programs that can help workers acquire the skills and knowledge needed for the new jobs that are being created by automation. In addition, international cooperation can help to ensure that the benefits of automation are shared fairly across countries and populations. By working together to address income inequality and other social and economic challenges, countries can help mitigate the negative impacts of automation and promote greater equality and social cohesion. Overall, international cooperation has the potential to play a critical role in addressing the challenges of automation. By sharing information and best practices, developing joint initiatives and programs, investing in education and training, and promoting greater equality, countries can work together to ensure that the benefits of automation are realized by all [28–30].

12.4.4 Potential Future Developments in Automation

Automation has already significantly impacted the way we work, and its effects will only become more pronounced in the coming years. The field of AI is rapidly evolving, and we can expect to see continued advancements in machine learning and natural language processing. These developments could lead to the creation of more complex and sophisticated AI systems that can perform a wider range of tasks, from customer service to medical diagnosis. As AI technology becomes more advanced, we may also see an increase in the use of chatbots and virtual assistants to handle customer inquiries and perform other tasks that were previously performed by humans [31, 32]. The use of robotics technology is also becoming more affordable and accessible, which means we are likely to see an increase in the use of robots in a wide range of industries. This could include everything from manufacturing and construction to healthcare and hospitality. In the healthcare industry, robots can be used to assist with surgeries, while in the hospitality industry, they can be used to deliver room service and perform other tasks. As automation continues to replace traditional jobs, we may see an increase in the number of workers who turn to freelance and contract work. This could create new opportunities for individuals to work in flexible roles, but it could also result in greater income insecurity and a lack of employment benefits. We may also see an increase in the number of workers who turn to gig work platforms, such as Uber and Lyft, to earn a living. As the skills required in the workforce continue to shift, there will be a growing need for workers to continually update their skills and knowledge. This could lead to a greater emphasis on vocational education and training programs, as well as a shift towards lifelong learning. In order to stay competitive in the workforce, workers will need to continually update their skills and knowledge to keep up with technological advancements. As automation becomes more pervasive, there will be a growing need to address the ethical implications of its use. This could include everything from ensuring that AI systems are not biased or discriminatory to addressing the impact of automation on worker well-being. As AI systems become more advanced, there is a risk that they could be used to make decisions that could have a negative impact on individuals or groups [33, 34]. While automation has the potential to replace many jobs, there is also an opportunity to create new roles that combine automation with human labor. For example, automation could be used to perform routine or dangerous tasks, while humans focus on more complex and creative work. This could create new opportunities for workers to use their skills and knowledge in innovative ways.

As automation continues to transform the workforce, there will be ongoing debates about its impact on employment. Some argue that automation will create new jobs and drive economic growth, while others are concerned about the potential for widespread job displacement and income inequality. As automation technology continues to advance, it is likely that these debates will become more complex and nuanced, and will require ongoing discussion and analysis.

12.5 CONCLUSION

In conclusion, the rise of automation presents a significant challenge for the global workforce, and education has a critical role to play in addressing these challenges. The education systems around the world must adapt to the changing demands of the workforce and prepare students for the new types of jobs that are emerging. This chapter has highlighted the various ways in which education can address these challenges, including vocational education, lifelong learning, and the development of critical thinking and problem-solving skills. However, no single solution can address the diverse challenges posed by automation, and collaboration between education institutions, industry, and government is essential. This collaboration can help to bridge the gap between education and the needs of the workforce, leading to the development of effective training programs and the creation of new job opportunities. Additionally, it is important to address the ethical implications of automation and to ensure that the benefits of new technologies are distributed fairly across society. As automation continues to transform the global workforce, ongoing dialogue and debate will be necessary to ensure that the workforce is prepared for the changes to come. In summary, education is an essential tool for addressing the challenges of automation, and a global perspective is necessary to ensure that education systems are adequately prepared for the changes to come [35]. By embracing new technologies and collaboration between different stakeholders, we can build a workforce that is prepared for the challenges of the future.

REFERENCES

1. Sindhwani, N., Maurya, V. P., Patel, A., Yadav, R. K., Krishna, S., and Anand, R. Implementation of intelligent plantation system using virtual IoT. In *Internet of Things and its Applications*, 2022, Springer, pp. 305–322.
2. Chibber, A., Anand, R., and Singh, J. Smart traffic light controller using edge detection in digital signal processing. In *Wireless Communication with Artificial Intelligence*. CRC Press, 2022, pp. 251–272.
3. Shukla, R., Dubey, G., Malik, P., Sindhwani, N., Anand, R., Dahiya, A., and Yadav, V. Detecting crop health using machine learning techniques in smart agriculture system. *Journal of Scientific and Industrial Research* 2021, 80(8), 699–706.
4. Anand, R., Arora, S., and Sindhwani, N. A miniaturized UWB antenna for high speed applications. In *2022 International Conference on Computing, Communication and Power Technology (IC3P)*, 2022, IEEE: Visakhapatnam, India, pp. 264–267.
5. Manyika, J., Chui, M., Bughin, J., Dobbs, R., Bisson, P., and Marrs, A. *Disruptive Technologies: Advances That Will Transform Life, Business, and the Global Economy*. McKinsey Global Institute, San Francisco, CA, 2013.
6. Wolfgang, M. *The Robotics Market—Figures and Forecasts*. RoboBusiness; Boston Consulting Group, Boston, MA, 2016.
7. Brynjolfsson, E., and McAfee, A. *Race Against the Machine: How the Digital Revolution Is Accelerating Innovation, Driving Productivity, and Irreversibly Transforming Employment and the Economy*. Digital Frontier Press, Lexington, MA, 2011.
8. Ford, M. *The Rise of the Robots: Technology and the Threat of Mass Unemployment*. Oneworld Publications, Oxford, 2015.

9. Frey, C. B., and Osborne, M. The future of employment: How susceptible are jobs to computerisation? *Technological Forecasting and Social Change* 2017, *114*, 254–280.
10. Pandey, D., Pandey, B. K., Sindhwani, N., Anand, R., Nassa, V. K., and Dadheech, P. An interdisciplinary approach in the post-COVID-19 pandemic era. *An Interdisciplinary Approach in the Post-COVID-19 Pandemic Era* 2022, NOVA SCIENCE, 1–290.
11. Singh, P., Kaiwartya, O., Sindhwani, N., Jain, V., and Anand, R. (Eds.). *Networking Technologies in Smart Healthcare: Innovations and Analytical Approaches*. CRC Press, 2022.
12. Pandey, B. K., Pandey, D., Wariya, S., Aggarwal, G., and Rastogi, R. Deep learning and particle swarm optimisation-based techniques for visually impaired humans' text recognition and identification. *Augmented Human Research* 2021, *6*, 1–14.
13. Kumar Pandey, B., Pandey, D., Nassa, V. K., Ahmad, T., Singh, C., George, A. S., and Wakchaure, M. A. Encryption and steganography-based text extraction in IoT using the EWCTS optimizer. *The Imaging Science Journal* 2022, 1–19.
14. Acemoglu, D., and Restrepo, P. *The Race Between Machine and Man: Implications of Technology for Growth, Factor Shares and Employment*. NBER Working Paper; Social Science Electronic Publishing, Inc., Rochester, NY, 2016.
15. Acemoglu, D., and Restrepo, P. *Robots and Jobs: Evidence from US Labor Markets*. NBER Working Paper; Social Science Electronic Publishing, Inc., Rochester, NY, 2017.
16. Acemoglu, D., and Restrepo, P. *Artificial Intelligence, Automation and Work*. NBER Working Paper; Social Science Electronic Publishing, Inc., Rochester, NY, 2018.
17. Autor, D. H. Why are there still so many jobs? The history and future of workplace automation. *Journal of Economic Perspectives* 2015, *29*, 3–30.
18. Autor, D., Salomons, A.M. Is automation labor-displacing? Productivity growth, employment, and the labor share. In *Proceedings of the BPEA Conference Drafts*, Detroit, MI, 8–9 March 2018.
19. Graetz, G., and Michaels, G. *Robots at Work*. CEP Discussion Papers; Centre for Economic Performance, London School of Economics and Political Science, London, 2015.
20. Gregory, T., Salomons, A., and Zierahn, U. *Racing With or Against the Machine? Evidence from Europe*. Discussion Paper No. 16-053; ZEW, Mannheim, Germany, 2016.
21. Stewart, I., De, D., and Cole, A. *Technology and People: The Great Job-Creating Machine*. Deloitte, London, 2015.
22. Vivarelli, M. *Innovation and Employment Technological Unemployment Is Not Inevitable—Some Innovation Creates Jobs, and Some Job Destruction Can Be Avoided*. IZA Technical Report Forschungsinstitut zur Zukunft der Arbeit (IZA); Bonn, Germany, 2007.
23. Vivarelli, M. Innovation, employment, and skills in advanced and developing countries: A survey of the economic literature. *Journal of Economic Issues* 2014, *48*(1), 123–154.
24. Mokyr, J., Vickers, C., and Ziebarth, N. L. The history of technological anxiety and the future of economic growth: Is this time different? *Journal of Economic Perspectives* 2015, *29*(3), 31–50.
25. Pasinetti, L. *Structural Change and Economic Growth: A Theoretical Essay on the Dynamics of the Wealth of Nations*. Cambridge University Press, Cambridge, 1981.
26. Saviotti, P. P., and Pyka, A. Economic development, qualitative change and employment creation. *Structural Change and Economic Dynamics* 2004, *15*(3), 265–287.
27. Saviotti, P. P., and Pyka, A. Product variety, competition and economic growth. *Journal of Evolutionary Economics* 2008, *18*(3–4), 323–347.

28. Autor, D. H.; Levy, F., and Murnane, R. J. The skill content of recent technological change: An empirical exploration. *Quarterly Journal of Economics* 2003, *118*(4), 1279–1333.
29. Levy, F., and Murnane, R. J. How computerized work and globalization shape human skill demands. In *Learning in the Global Era: International Perspectives on Globalization and Education*, Suarez-Orozco, M., Ed.; University of California Press, Berkeley, CA, 2007, pp. 158–174.
30. Autor, D. H., Katz, L. F., and Kearney, M. S. *The Polarization of the US Labor Market*. NBER Working Paper; Social Science Electronic Publishing, Inc., Rochester, NY, 2006.
31. Pandey, B. K., Pandey, D., and Agarwal, A. Encrypted information transmission by enhanced steganography and image transformation. *International Journal of Distributed Artificial Intelligence (IJDAI)* 2022, *14*(1), 1–14.
32. Chaudhary, A., Bodala, D., Sindhwani, N., and Kumar, A. Analysis of customer loyalty using artificial neural networks. In *2022 International Mobile and Embedded Technology Conference (MECON)*, 2022, IEEE, pp. 181–183.
33. Janani, S., Sivarathinabala, M., Anand, R., Ahamad, S., Ahmer Usmani, M. and Mahabub Basha, S. in Machine Learning Analysis on Predicting Credit Card Forgery. In *2023* International Conference On Innovative Computing And Communication, 2023, Springer, pp. 137–48.
34. Chawla, P., Sindhwani, N., and Anand, R. Smart coal payload delivery system using pic microcontroller. *International Journal of Advanced Science and Technology* 2020, *29*(10s), 1485–1490.
35. Pandey, D., Gul, R., Canete, J. J. O., Rocha, I. C. N., Gowwrii, G., Pandey, B. K., and Peter, S. N. Mental stress in online learning during the pandemic: An assessment of learners' perception. *Asian Journal of Advances in Research* 2021, 37–49. https://doi.org/10.1007/978-981-19-0151-5_15

13 The Intersection of Human Resource Management and Automation
Opportunities and Challenges for HR Professionals

Ajay Sidana, Juliet Gladies Jayasuria,
Luigi Pio Leonardo Cavaliere, S. Ramesh
Babu, M. Kavitha, and P. Balaji

13.1 INTRODUCTION

In the present-day business landscape, where technology is progressing at an unprecedented pace, the integration of automation and human resource management has become a vital subject of discussion. The amalgamation of these two fields has gained significant importance due to its potential to streamline HR operations and increase efficiency. As companies strive to keep up with the fast-paced business environment, automating routine HR tasks can provide a competitive edge by freeing up resources that can be utilized for strategic decision-making.

This intersection of automation and human resource management (HRM) has thus become a crucial consideration for organizations seeking to improve their operational effectiveness in the modern era. HR professionals are facing significant challenges and opportunities as automation continues to disrupt traditional HR practices. As such, it is imperative to understand the implications of automation for HR management and the workforce as a whole. The purpose of this chapter is to examine the intersection of human resource management and automation and to highlight the opportunities and challenges that HR professionals face in this context. On one hand, it can bring numerous benefits such as increased efficiency, cost reduction, and enhanced accuracy in HR operations.

On the other hand, it can pose a significant threat to jobs that were traditionally performed by HR professionals. As such, HR professionals must be equipped with the knowledge and skills to adapt to these changes and remain relevant in the workforce We will also discuss how automation is transforming the nature of HR jobs and the skills required of HR professionals. This includes the need for HR professionals to develop expertise in areas such as data analytics, artificial intelligence, and digital technology. Additionally, we will examine the ethical implications of automation in HR management, such as bias and discrimination in recruitment and decision-making processes [1–4].

Moreover, this chapter will provide practical recommendations for HR professionals on how to navigate the intersection of HR management and automation. We will discuss how HR professionals can leverage automation to enhance their skills, remain competitive in the job market, and add value to their organizations. We will also offer insights into how HR professionals can balance the use of automation with the need for human interaction and empathy in HR management. In conclusion, this chapter will shed light on the intersection of human resource management and automation, its impact on the workforce, and the opportunities and challenges that HR professionals face in this context. It is crucial for HR professionals to stay informed and updated on the latest trends and developments in automation and to equip themselves with the necessary skills and knowledge to remain competitive and relevant in the workforce. By doing so, HR professionals can enhance their value to their organizations and play a critical role in shaping the future of work [5, 6]. Figure 13.1 shows the functions of HRM.

13.1.1 Definition of Human Resource Management (HRM)

The management and development of an organization's workforce are the primary responsibilities of the function known as Human Resource Management (HRM). HRM encompasses a broad spectrum of tasks, which includes identifying and attracting suitable candidates, providing training and development opportunities, evaluating employee performance, determining compensation and benefits, managing employee relationships, and ensuring workplace safety and well-being. The primary aim of HRM is to ensure that the organization has the necessary talent pool to achieve its objectives in an efficient and effective manner. Moreover, HRM also has a vital role to play in ensuring the organization's compliance with pertinent employment laws and regulations while fostering a positive and productive workplace culture [7].

13.1.2 Importance of HRM and Automation in the Modern Workplace

Human resource management (HRM) and automation are two critical aspects of the modern workplace that play a significant role in enhancing organizational efficiency, productivity, and competitiveness. Both HRM and automation are important in the modern

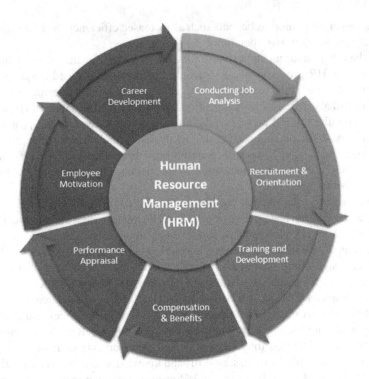

FIGURE 13.1 Functions of Human Resource Management (HRM).

workplace, and they complement each other in many ways. The importance of HRM in the modern workplace lies in the fact that the workforce is the most critical asset of any organization. A well-designed compensation and benefits system can help the organization attract and retain the best talent in the industry and motivate employees to perform at their best. Automation, on the other hand, plays a crucial role in enhancing the efficiency and effectiveness of HRM processes. Automation can also enhance the accuracy and consistency of HRM processes and reduce the likelihood of errors and mistakes. For example, automation can be used to streamline the recruitment process by using software to screen resumés, conduct initial interviews, and schedule appointments. This can help HR professionals save time and reduce the workload associated with manual screening and scheduling. Similarly, automation can be used to monitor and analyze employee performance data, enabling HR professionals to identify patterns, trends, and opportunities for improvement. This can help HR professionals develop targeted training and development programs that address employees' specific needs and improve their performance. Moreover, automation can be used to enhance employee engagement and communication by using digital tools such as social media, mobile apps, and chatbots. These tools can help HR professionals communicate more effectively with employees, provide them with relevant information, and facilitate collaboration and teamwork. In conclusion, the importance of HRM and automation in the modern workplace cannot be overstated. HRM plays a critical role in managing the organization's most important asset, its workforce, while automation can help enhance the efficiency and effectiveness of HRM processes. By combining HRM and automation, organizations can achieve greater efficiency, productivity, and competitiveness, while providing employees with a better work experience [8, 9].

13.2 OPPORTUNITIES FOR HR PROFESSIONALS IN THE INTERSECTION OF HRM AND AUTOMATION

Here are some of the opportunities for HR professionals in the intersection of HRM and automation:

- *Strategic planning*: HR professionals can use automation to streamline HR processes and free up time for strategic planning. This can involve analyzing workforce data to identify talent gaps, develop succession plans, and design training and development programs that address employees' specific needs. By doing so, HR professionals can contribute to the organization's strategic objectives and improve its competitiveness.
- *Talent acquisition*: automation can help HR professionals attract and select the best talent in the industry. This can involve using software to screen resumés, conduct initial interviews, and schedule appointments. Automation can also help HR professionals create more targeted and effective recruitment campaigns that reach a broader pool of qualified candidates.
- *Performance management*: HR professionals can use automation to monitor and analyze employee performance data and provide feedback and recognition in real-time. This can involve using software to collect and analyze performance data, identify patterns and trends, and develop targeted training and development programs that improve employees' skills and competencies. By doing so, HR professionals can help employees perform at their best and contribute to the organization's performance.
- *Employee engagement*: automation can help HR professionals communicate more effectively with employees, provide them with relevant information, and facilitate collaboration and teamwork. This can involve using digital tools such as social media, mobile apps, and chatbots to engage with employees and create a more positive and productive work environment.
- *Compliance*: HR professionals can use automation to ensure that the organization complies with relevant employment laws and regulations. This can involve using software to track and manage employee records, monitor compliance with labor laws, and generate reports for regulatory agencies.
- *Data analytics*: automation can help HR professionals collect, analyze, and interpret data to identify trends and opportunities for improvement. This can involve using software to analyze employee performance data, turnover rates, compensation and benefits data, and other workforce metrics. By doing so, HR professionals can develop targeted interventions that address specific workforce issues and improve organizational performance.

The intersection of HRM and automation presents numerous opportunities for HR professionals to add value to their organizations and advance their careers. By leveraging automation, HR professionals can streamline HR processes, improve employee performance and engagement, and contribute to the organization's strategic objectives. To succeed in this new environment, HR professionals will need to acquire new skills, such as data analysis, software implementation, and digital communication, and adapt to the changing needs of the modern workplace [10].

13.2.1 STREAMLINED RECRUITMENT PROCESSES

Streamlined recruitment processes are one of the key benefits of automation in the intersection of human resource management (HRM) and technology. Recruitment is a critical aspect of HRM that involves sourcing, selecting, and hiring the best candidates for the job. However, traditional recruitment processes can be time-consuming, inefficient, and costly, especially for organizations that receive a large volume of resumés and job applications. Automation can help streamline recruitment processes by using software to automate repetitive, manual, and time-consuming tasks. This can include resumé screening, candidate assessments, interview scheduling, and onboarding. By doing so, automation can help HR professionals save time, reduce workload, and enhance the quality and efficiency of the recruitment process.

Streamlined recruitment processes can help organizations reduce the time-to-hire by enabling HR professionals to quickly screen and assess candidates. By automating manual and repetitive tasks, such as resumé screening and candidate assessments, organizations can expedite the recruitment process and fill vacant positions faster. This can help reduce the cost and disruption associated with prolonged recruitment processes, and provide the organization with the talent it needs to achieve its goals. Additionally, faster time-to-hire can enhance the organization's competitiveness in the job market by enabling it to attract and retain top talent. Automation can create a more positive candidate experience by providing timely and personalized communication and feedback. With automated software, candidates can receive instant notifications and updates on their job application status. This can help improve the employer brand of the organization and create a more positive candidate experience. In turn, this can help attract and retain top talent, who will want to work for an organization that values their time and communicates with them efficiently. By automating manual and repetitive tasks, organizations can save costs associated with recruitment, such as advertising, travel, and administrative expenses. This can help reduce the workload of HR professionals and free up their time for more strategic activities. Organizations can invest in automation software, which can help save costs in the long run by reducing the time-to-hire and improving the quality of recruitment. This, in turn, can help the organization achieve its goals more efficiently and cost-effectively. Streamlined recruitment processes can improve the accuracy and consistency of recruitment decisions. By using software to screen resumés and assess candidate skills and competencies, organizations can make more informed and data-driven recruitment decisions. This can help avoid biases and human error in recruitment decisions and ensure that the best candidate is selected for the job.

Additionally, the use of automation can provide a level of consistency in the recruitment process that can help organizations make more objective decisions and avoid potential legal liabilities. Automation can also provide HR professionals with better data analytics by collecting and analyzing recruitment data, such as the number of applicants, time-to-hire, and cost-per-hire. By analyzing this data, HR professionals can identify trends and opportunities for improvement and make more informed and data-driven recruitment decisions. This can help organizations continuously improve their recruitment processes and achieve better results over time. In turn, this can help the organization achieve its strategic goals and improve its overall performance. In conclusion, streamlined recruitment processes are one of the key benefits of automation in the intersection of HRM and technology. By automating manual and repetitive tasks, organizations can reduce the time-to-hire, improve the candidate experience, save costs, and enhance the accuracy and consistency of recruitment processes. To succeed in this new environment, HR professionals will need to acquire new skills, such as data analytics, software implementation, and digital communication, and adapt to the changing needs of the modern workplace [11–13].

13.2.2 Automated Employee Data Management

The utilization of technology to simplify the management of employee data is known as automated employee data management. This can encompass various activities, including but not limited to onboarding, performance management, payroll, and benefits administration. Through the automation of these processes, companies can enhance their efficiency, decrease errors, and guarantee adherence to legal and regulatory obligations. One of the primary benefits of automated employee data management is the ability to store and access employee information in a central location. This can help HR professionals to easily access and analyze employee data, such as performance reviews, compensation, and benefits. It can also help to ensure that employee information is accurate and up-to-date, reducing the risk of errors and compliance issues. Automation can also help to streamline the onboarding process, making it more efficient for both HR professionals and new employees. For example, an online onboarding portal can automate tasks such as sending offer letters, collecting employee information, and conducting background checks. This can help to reduce the time and resources required for onboarding, while also ensuring a consistent and standardized process for all new employees. Automated performance management systems can also help to simplify the process of setting goals, and tracking progress. This can help to improve employee engagement and retention, while also ensuring that the organization is able to meet its performance goals. Payroll and benefits administration can also be automated, reducing the time and resources required for these functions. For example, online payroll systems can automate tasks such as calculating employee pay and taxes, while also providing self-service options for employees to access their pay and benefits information. This can help to reduce errors and ensure compliance with legal and regulatory requirements. Overall, automated employee data management can provide a range of benefits for HR professionals and organizations. The use of automation can lead to a reduction in expenses, enhancement in efficiency, and assurance of legal and regulatory compliance by simplifying procedures and enhancing precision. Furthermore, it can promote a more favorable employee experience by providing a more efficient and seamless process for onboarding, performance management, and benefits administration. The increasing accessibility and affordability of automated HR software will likely result in more organizations adopting this approach to employee data management in the future [14, 15].

13.2.3 Improved Employee Engagement

Improved employee engagement is another opportunity for HR professionals in the intersection of HRM and automation. Employee engagement refers to the level of commitment and emotional investment that employees have in their work and their organization. High levels of employee engagement are associated with a range of positive outcomes, such as increased productivity, lower turnover, and improved customer satisfaction. Automation can help to improve employee engagement by providing tools and resources that support employee development and well-being. For example, learning management systems can provide employees with access to training and development opportunities that are customized to their individual needs and interests. This can help to increase employee skills and knowledge, while also providing opportunities for career growth and advancement.

In addition, wellness programs can be automated to promote employee well-being and improve engagement. For example, software can be used to track employee participation in wellness activities, such as fitness classes or meditation sessions. This can help to promote healthy habits and reduce stress, which can lead to increased engagement and productivity. Another way that automation can improve employee engagement is by providing real-time feedback and recognition. For example, software can be used to collect and analyze employee feedback, allowing managers to identify areas for improvement and recognize employees for their achievements. This can help to improve employee motivation and satisfaction, while also providing data that can be used to improve overall performance. All in all, automation can help to improve employee engagement by providing tools and resources that support employee development, well-being, and recognition [16].

13.2.4 Enhanced Training and Development Programs

Enhanced training and development programs are another opportunity for HR professionals in the intersection of HRM and automation. By leveraging technology, HR professionals can create training and development programs that are more effective, efficient, and engaging. One way that automation can enhance training and development programs is by providing personalized learning experiences. Learning management systems can be used to deliver customized content that is tailored to the needs and preferences of individual employees. This can help to increase employee engagement and retention, as well as improve the effectiveness of training programs. Another way that automation can enhance training and development programs is by providing real-time feedback and assessment. Software can be used to track employee progress and provide feedback on their performance, allowing employees to identify areas for improvement and receive support from their managers or trainers. This can help to improve employee confidence and competence, while also providing data that can be used to improve training programs. Automation can also be used to gamify training and development programs, making them more engaging and interactive. For example, software can be used to create simulations or games that allow employees to practice new skills in a safe and controlled environment. This can help to increase employee motivation and retention, as well as provide valuable data on employee performance and engagement. Finally, automation can be used to provide just-in-time training and support. Mobile apps or chatbots can be used to deliver training content or answer employee questions on demand, allowing employees to access information and support when and where they need it. This can help to improve employee performance and reduce the time and cost associated with traditional training programs. Overall, automation can enhance training and development programs by providing personalized learning experiences, real-time feedback and assessment, gamification, and just-in-time training and support. By leveraging technology to create more effective, efficient, and engaging training programs, HR professionals can help to improve employee skills, performance, and engagement, leading to a more productive and successful workforce [17].

13.2.5 Potential Cost Savings

Cost savings is another opportunity for HR professionals in the intersection of HRM and automation. By automating manual and repetitive tasks, organizations can reduce the workload of HR professionals and free up their time for more strategic activities, leading to potential cost savings in various areas. For example, streamlined recruitment processes can help organizations save costs associated with recruitment, such as advertising, travel, and administrative expenses. By automating candidate screening and assessment, organizations can reduce the need for expensive recruitment agencies or external consultants. In addition, by reducing the time-to-hire, organizations can save costs associated with prolonged recruitment processes, such as lost productivity or revenue. Automation can also help to reduce costs associated with employee data management. By automating the collection, storage, and analysis of employee data, organizations can reduce the need for manual data entry and processing, which can be time-consuming and error-prone. This can help to reduce administrative expenses, as well as improve data accuracy and consistency. Another way that automation can lead to cost savings is by improving employee engagement and retention. By providing personalized learning experiences, real-time feedback and assessment, and just-in-time training and support, organizations can improve employee skills and performance, leading to reduced turnover and associated costs, such as recruitment and training expenses. Finally, automation can help organizations to reduce costs associated with compliance and legal issues. By automating the tracking and reporting of employee data, organizations can ensure compliance with legal and regulatory requirements, such as data privacy laws or labor regulations.

This can help to reduce the risk of costly fines or legal disputes, as well as improve the overall compliance posture of the organization. In sum, automation can lead to potential cost savings in various areas, such as recruitment, employee data management, employee engagement and retention, and compliance and legal issues. By leveraging technology to automate manual and repetitive tasks, HR professionals can free up their time for more strategic activities and help their organizations save costs and improve performance [18].

13.3 CHALLENGES FOR HR PROFESSIONALS IN THE INTERSECTION OF HRM AND AUTOMATION

While there are significant opportunities for HR professionals in the intersection of HRM and automation, there are also several challenges that they must navigate. Some of the main challenges include [19]:

1. *Resistance to change*: one of the biggest challenges that HR professionals may face when introducing automation into HRM processes is resistance to change from employees and stakeholders. People may be resistant to automation because they fear job loss or feel that the technology is too complex or impersonal. HR professionals must be prepared to address these concerns and communicate the benefits of automation to gain buy-in from employees and stakeholders.
2. *Data privacy and security*: the process of automation includes gathering, retaining, and scrutinizing confidential employee data, which can result in apprehension regarding data security and privacy. HR professionals must ensure that the automation technology they use is compliant with data privacy regulations and that proper security measures are in place to protect the confidentiality of employee data.
3. *Skills gap*: the introduction of automation technology may require HR professionals to have new or enhanced technical skills, such as data analysis, programming, or system administration. HR professionals may need to undergo training to acquire these skills, or organizations may need to hire new HR staff with specialized technical expertise.
4. *Integration with legacy systems*: many organizations have legacy HR systems in place, which can be difficult to integrate with new automation technology. HR professionals may need to work closely with IT departments or software vendors to ensure seamless integration of the new automation technology with existing systems.
5. *Ethical considerations*: automation can introduce ethical considerations in HRM processes, such as the use of algorithms to screen and select candidates or the use of AI to make decisions related to employee performance or compensation. HR professionals must ensure that these processes are fair and unbiased and do not perpetuate or amplify existing biases or discrimination.

The Intersection of Human Resource Management and Automation

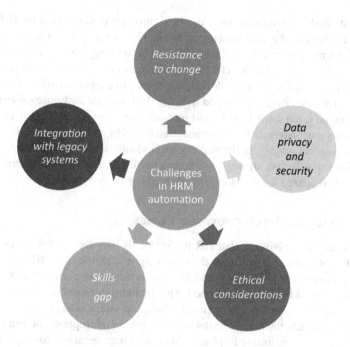

FIGURE 13.2 Challenges in HRM automation.

HR professionals must be prepared to navigate challenges such as resistance to change, data privacy and security concerns, the skills gap, integration with legacy systems, and ethical considerations when introducing automation into HRM processes. By addressing these challenges, HR professionals can ensure that automation enhances HRM practices and benefits both employees and the organization as a whole. Challenges in HRM automation are as shown in Figure 13.2.

13.3.1 Job Displacement and Retraining

Another important challenge that HR professionals may face in the intersection of HRM and automation is job displacement and the need for employee retraining. As automation technology becomes more prevalent, certain jobs may become redundant or may be automated, resulting in job displacement for some employees. HR professionals must be prepared to manage this process with sensitivity and compassion, while also ensuring that the organization continues to operate effectively. Retraining employees for new roles that are less likely to be automated can be an effective way to mitigate the impact of job displacement. HR professionals can work with managers and employees to identify new roles that align with their skills and interests and provide training opportunities to develop new skills.

However, retraining programs can be costly and time-consuming, and not all employees may be interested or able to transition to new roles. HR professionals must balance the need to support employees with the need to maintain the organization's productivity and financial viability. Another important consideration for HR professionals is the impact of automation on the broader job market. As automation technology becomes more prevalent, certain industries or job categories may be particularly vulnerable to job displacement. HR professionals may need to work with policymakers and industry stakeholders to develop strategies to support affected workers and ensure that the workforce is equipped with the skills needed for future job opportunities. In summary, job displacement and retraining are significant challenges that HR professionals must navigate at the intersection of HRM and automation. By working with employees, managers, policymakers, and industry stakeholders, HR professionals can develop effective strategies to support affected workers and ensure that the workforce is prepared for the changing demands of the job market [20].

13.3.2 Ensuring Data Security and Privacy

Another important challenge that HR professionals may face in the intersection of HRM and automation is ensuring data security and privacy. As HR departments rely more heavily on technology and automation, there is a greater risk of sensitive employee data being compromised. This includes personal information such as Social Security numbers, performance reviews, and salary information.

HR professionals must ensure that they are taking appropriate measures to protect employee data, including implementing strong data security protocols and complying with relevant data privacy regulations. This may include using secure data storage and transmission methods, limiting access to sensitive information, and regularly reviewing and updating security policies and procedures. HR professionals must also be aware of the potential for human error or insider threats, which can pose a significant risk to data security. This may include providing regular training to employees on data security best practices, conducting background checks on new employees, and implementing appropriate access controls to limit access to sensitive data. In addition to these internal measures, HR professionals may need to work with external partners, such as technology vendors or data processing firms, to ensure that their data security practices are up to par. This may involve conducting audits or assessments of third-party vendors to ensure that they are meeting relevant data security standards.

In summary, ensuring data security and privacy is a significant challenge for HR professionals in the intersection of HRM and automation. By implementing strong data security protocols, providing regular training to employees, and working with external partners, HR professionals can help protect sensitive employee data and minimize the risk of data breaches or other security threats.

13.3.3 Maintaining Personal Connections with Employees

Another challenge that HR professionals may face in the intersection of HRM and automation is maintaining personal connections with employees. As automation technology becomes more prevalent in HR departments, there is a risk that employees may feel disconnected or disengaged from the organization. Automated processes may lack the personal touch and empathy that are essential for building strong relationships between employees and their employers. HR professionals must be aware of this risk and take steps to maintain personal connections with employees. This may include incorporating more opportunities for face-to-face interactions, such as in-person meetings or team-building events, and using communication tools that allow for real-time feedback and collaboration. Additionally, HR professionals can use automation technology in a way that complements personal connections rather than replacing them. For example, automated performance management tools can be used to collect data on employee performance, but HR professionals should still make time for personal conversations and feedback sessions with employees. Another way to maintain personal connections with employees is by promoting a culture of open communication and transparency. HR professionals can encourage employees to share their thoughts and feedback on company policies, procedures, and other issues, and demonstrate that their opinions are valued and taken into consideration. In summary, maintaining personal connections with employees is a challenge that HR professionals must navigate at the intersection of HRM and automation. By incorporating more opportunities for face-to-face interactions, using communication tools that allow for real-time feedback, and promoting a culture of open communication and transparency, HR professionals can help ensure that employees feel valued and engaged with the organization [21].

13.3.4 BALANCING AUTOMATION WITH THE HUMAN TOUCH

Balancing automation with the human touch is another challenge that HR professionals may face at the intersection of HRM and automation. While automation can improve efficiency, reduce costs, and enhance accuracy in HR processes, it can also lead to a loss of the human touch. This can result in decreased employee satisfaction, engagement, and trust in the organization. To address this challenge, HR professionals must find a way to strike a balance between automation and the human touch. They can do this by identifying areas of HR where automation can be most effective, while still maintaining a human element in other areas. For example, HR professionals can use automation to manage administrative tasks, such as scheduling and time-tracking, and use the time saved to focus on building relationships with employees, providing career development support, and creating a positive workplace culture. Additionally, HR professionals can use automation to provide employees with more self-service options, such as online learning and development programs or benefits enrolment portals, while still providing opportunities for personal interaction and support. HR professionals can also leverage technology to enhance the human touch. For example, they can use video conferencing and virtual collaboration tools to facilitate remote communication and connection or use chatbots to provide personalized support and assistance to employees. In summary, balancing automation with the human touch is a challenge that HR professionals must navigate at the intersection of HRM and automation. By identifying areas where automation can be most effective and maintaining a human element in other areas, providing self-service options while still providing opportunities for personal interaction and support, and leveraging technology to enhance the human touch, HR professionals can help ensure that automation is used in a way that benefits both the organization and its employees [22].

13.3.5 Overcoming Resistance to Change

Overcoming resistance to change is another challenge that HR professionals may face at the intersection of HRM and automation. Resistance to change can be a common response to automation initiatives, as employees may feel that their job security, autonomy, and skillsets are threatened by the introduction of new technology. To overcome this challenge, HR professionals must communicate the benefits of automation to employees and involve them in the implementation process. This can help build support for the initiative and increase employee buy-in and adoption. HR professionals can also provide training and development opportunities to help employees develop the skills and knowledge needed to work effectively with automation technology. By investing in employee development, organizations can help alleviate concerns around job displacement and demonstrate a commitment to employee growth and career development. Additionally, HR professionals can work to create a culture of innovation and continuous improvement, where employees are encouraged to embrace change and experiment with new technologies and processes. By fostering a culture of innovation, organizations can help employees feel more comfortable with change and encourage them to explore new ways of working. Finally, HR professionals can leverage change management frameworks and best practices to guide the implementation of automation initiatives. These frameworks can help HR professionals anticipate and address employee concerns and facilitate a smoother transition to new ways of working. In summary, overcoming resistance to change is a challenge that HR professionals must address when implementing automation initiatives. By communicating the benefits of automation, involving employees in the implementation process, providing training and development opportunities, creating a culture of innovation, and leveraging change management frameworks, HR professionals can help ensure that automation initiatives are successful and well-received by employees [23, 24]

13.4 ETHICAL CONSIDERATIONS IN HRM AUTOMATION

Table 13.1 outlines some of the ethical considerations that HR professionals should be aware of when implementing HRM automation systems. These considerations include potential bias and discrimination, data privacy and security, transparency and accountability, employee autonomy and privacy, and the ethical use of data. By addressing these considerations, HR professionals can ensure that automation systems are used ethically and responsibly [25, 26].

These ethical considerations are important for HR professionals to consider when implementing HRM automation systems. By addressing these considerations, HR professionals can ensure that automation systems are designed and used in an ethical and responsible manner that promotes fairness, transparency, and respect for employee rights.

TABLE 13.1
Ethical Considerations in HRM Automation

Ethical Consideration	Description
Bias and Discrimination	HRM automation systems may perpetuate existing biases and discrimination in recruitment, selection, and performance management processes. This can have negative impacts on marginalized groups and perpetuate inequality in the workplace. HR professionals must ensure that automated systems are designed to reduce bias and discrimination and incorporate human oversight and intervention when needed.
Data Privacy and Security	HRM automation systems may collect and process sensitive employee data, such as performance evaluations and health information. HR professionals must ensure that such data is protected and secured from unauthorized access and breaches. Additionally, HR professionals must be transparent about how data is collected, used, and shared with employees and obtain their consent when necessary.
Transparency and Accountability	HRM automation systems may operate with little transparency or oversight, making it difficult to understand how decisions are made and to challenge decisions if they are unfair or unjust. HR professionals must ensure that automated systems are transparent, accountable, and subject to human review and oversight.
Employee Autonomy and Privacy	HRM automation systems may infringe on employee autonomy and privacy, as they may monitor and collect data on employees without their knowledge or consent. HR professionals must balance the benefits of automation with the need to respect employee autonomy and privacy and ensure that automated systems are designed with employee rights in mind.
Ethical Use of Data	HRM automation systems may be used to make decisions that have significant impacts on employees' lives, such as hiring, promotion, and termination. HR professionals must ensure that automated systems are used ethically and that decisions made by automated systems are subject to human review and oversight. Additionally, HR professionals must ensure that employees are informed of how data is used to make decisions and have the opportunity to challenge decisions if they feel they are unfair or unjust.

13.5 CONCLUSION

In conclusion, the intersection of human resource management and automation presents numerous opportunities and challenges for HR professionals. By leveraging automation technologies, HR professionals can streamline recruitment processes, improve employee engagement, enhance training and development programs, and achieve potential cost savings. However, this intersection also poses several challenges, including the potential for job displacement, data privacy and security concerns, maintaining personal connections with employees, balancing automation with the human touch, and overcoming resistance to change. To successfully navigate these challenges, HR professionals must prioritize ethics and be mindful of the potential impact of automation on employees. Ethical considerations, such as potential bias and discrimination, data privacy and security, transparency and accountability, employee autonomy and privacy, and ethical use of data, must be taken into account when implementing HRM automation systems [27]. Overall, the intersection of HRM and automation is a complex and rapidly evolving field that requires HR professionals to adapt to technological advances and changing workforce needs. By embracing automation technologies while remaining mindful of ethical considerations, HR professionals can harness the power of automation to improve HRM practices and enhance the overall employee experience.

REFERENCES

1. Pandey, D., Pandey, B. K., Sindhwani, N., Anand, R., Nassa, V. K., Dadheech, P. An interdisciplinary approach in the post-COVID-19 pandemic era. *An Interdisciplinary Approach in the Post-COVID-19 Pandemic Era* 2022, 1–290.
2. Pandey, B. K., Pandey, D., Anand, R., Singh, H., Sindhwani, N., Sharma, Y. The impact of digital change on student learning and mental anguish in the COVID era. *An Interdisciplinary Approach in the Post-COVID-19 Pandemic Era* 2022, 197–206.
3. Gupta, A., Anand, R., Pandey, D., Sindhwani, N., Wairya, S., Pandey, B. K., Sharma, M. Prediction of breast cancer using extremely randomized clustering forests (ERCF) technique: Prediction of breast cancer. *International Journal of Distributed Systems and Technologies (IJDST)* 2021, *12*(4), 1–15.
4. Meivel, S., Sindhwani, N., Anand, R., Pandey, D., Alnuaim, A. A., Altheneyan, A. S., Jabarulla, M. Y., Lelisho, M. E. Mask detection and social distance identification using internet of things and faster R-CNN algorithm. *Computational Intelligence and Neuroscience* 2022, *2022*, 1–13.
5. Fan, D., Zhu, C. J., Huang, X., Kumar, V. Mapping the terrain of international human resource management research over the past fifty years: A bibliographic analysis. *Journal of World Business* 2021, *56*(2), 101185.
6. Van Lancker, E., Knockaert, M., Audenaert, M., Cardon, M. HRM in entrepreneurial firms: A systematic review and research agenda. *Human Resource Management Review* 2021, *32*(3), 100850.
7. Armstrong, M. *Ř ízení Lidských Zdrojů* . Grada: Praha, Czech Republic, 2015; ISBN 978-80-247-1407-3.
8. Vetráková, M. S. *Ľudské Zdroje a ich Riadenie*. Ekonomická Fakulta UMB, Banská Bystrica, Slovakia, 2011; ISBN 80-8055-581-8.
9. Andrejcak, M. Personnel audit. In *Proceedings of the 13th International Scientific Conference on Hradec Economic Days, Hradec Králové, Czeck Republic*, 3–4 February 2015, *5*, 11–16.
10. Potkány, M. Personnel controlling. In *Human Resources and Ergonomics*. Available online: http://frcatel.fri.uniza.sk/hrme/files/2007/2007_1_04.pdf
11. Šikýř, M. *Personalistika Pro Manažéry a Personalisty*, 2nd ed. Grada Publishing, Praha, Czech Republic, 2016, p. 208. ISBN 978-80-271-9527-5.
12. Kubalák, M. *Efektívne Riadenie Ľudských Zdrojov*. Eurokódex, Žilina, Slovakia, 2013; ISBN 978-80-8155-0164.
13. Beránek, M. *Audit, Kontrola a VKS, Ř ízení Rizik*. VŠH, Praha, Czech Republic, 2016.
14. Grenčíková, A. *Riadenie Ľudských Zdrojov*. TnUAD, Trenčín, Slovakia, 2008; ISBN 978-80-8075-319-1.

15. Urban, J. Personální a Organizační Audity Podniku—Cíle, Metody a Výsledky. In *Práce a Mzda*. Available online: https://is.vstecb.cz/do/vste/archiv_starych_dokumentu /akreditacni_zadosti/EPM_NMgr/studijni_opory_ekonomika_podniku_a_management_-_nmgr/Rizeni_lidskych_zdroju.pdf?lang=cs.
16. Arens, A., Best, P., Shailer, G., Fielder, B., Elder, R., Beasley, M. Auditing and assurance services. In *Australia—An Integrated Approach*, 6th ed. Pearson Education, Frenchs Forest, 2005.
17. Chocholouš, I. Personálny Audit: Kto Pôjde z Kola von? Available online: https://www.vsemba.sk/portals/0/Subory/casopis_stud%2032%20-%20web.pdf.
18. Cantera, J. *Del Control Externo a la Auditoria de Recursos Humanos*. La Nueva Gestión De Los Recursos Humanos, Madrid, Spain, 2000, pp. 369–397.
19. Dobiasová, P. Význam personálneho a organizačného auditu pre. In *Personálny a Mzdový Poradca pre Podnikateľa*. Poradca Podnikateľa, Zilina, Slovakia, 2004.
20. Ficek, J. Personálny audit. Available online: http://labartthospitality.com/media/messages/pdf/4e9808582de3707_2011_Personalny_audit__Top_Hotelir.pdf.
21. Zagorsek, B., Szarkova, M. Personnel audit as a function of personnel marketing and personnel management. In *Ekonomicky Casopis*. Slovak Academic Press Ltd., Bratislava, Slovakia, 2015, *63*, 551–552.
22. Coffee, J. C. Why do auditors fail? What might work? What won't? *Account Business Research* 2019, *49*, 540–561.
23. Loughran, M. *Auditing for Dummies*. Wiley Publishing Inc., Hoboken, NJ, 2010; ISBN 978-0-470-53071-9.
24. Vojtovič, S. *Personálny Manažment v Organizácii*. Aleš Čeněk, Plzeň, Czech Republic, 2013; ISBN 978-80-7380-483-1.
25. Ragsdell, G., Probets, S., Ahmed, G., Murray, I. Knowledge audit: Findings from the energy sector. *Knowledge and Process Management* 2014, *21*(4), 270–279.
26. Volkova, M., Solomatina, E., Shabutskaya, N., Sabetova, T., Shubina, E. History of views on audit of commercial structures management quality. In *Integration and Clustering for Sustainable Economic Growth*, Springer International Publishing, Volgograd, Russia, 2017, pp. 81–90.
27. Pandey, D., Pandey, B. K., Wariya, S. An approach to text extraction from complex degraded scene. *IJCBS* 2020, *1*(2), 4–10.

14 Data-Driven Insights for Agricultural Management
Leveraging Industry 4.0 Technologies for Improved Crop Yields and Resource Optimization

Ashok Kumar Koshariya, P.M. Rameshkumar,
P. Balaji, Luigi Pio Leonardo Cavaliere,
Venkata Harshavardhan Reddy
Dornadula, and Barinderjit Singh

14.1 INTRODUCTION

The purpose of this chapter is to investigate the utilization of data-driven insights in agricultural management, utilizing Industry 4.0 technologies to enhance crop yields and optimize resource allocation [1–4]. Data-driven insights involve the utilization of data analytics and artificial intelligence to extract insights and inform decision-making. Such insights empower farmers to make informed decisions, such as the ideal planting time, optimal quantity of fertilizer to use, and when to irrigate, founded on data from different sources, such as soil sensors, weather stations, and satellite imagery. The first section of the chapter provides an overview of data-driven insights for agricultural management, highlighting their benefits and applications in agriculture. The second section discusses Industry 4.0 technologies and their role in agriculture. The third and fourth sections examine how data-driven insights can be leveraged to improve crop yields and optimize resources, respectively. The fifth section presents case studies that illustrate the application, analyzing the lessons learned and the potential for future applications. There are also challenges and limitations that must be addressed. The chapter's penultimate section discusses these challenges, including the ethical considerations that must be taken into account. Finally, the chapter concludes by summarizing the key points, highlighting the implications for agricultural management, and suggesting future directions for research. By providing a comprehensive overview of the topic, this chapter contributes to the growing body of literature on precision agriculture, data analytics, and AI in agriculture [5].

14.1.1 Background and Significance of the Topic

The agricultural sector is facing significant challenges that threaten food security, environmental sustainability, and economic growth. These challenges include climate change, water scarcity, soil degradation, population growth, and the need to produce more food with fewer resources. The agricultural sector must therefore adopt innovative approaches to optimize crop yields, manage resources efficiently, and reduce waste. The use of data analytics and AI can help farmers optimize inputs such as water, fertilizer, and pesticides, thereby reducing waste and improving efficiency. This information can be used by farmers to make informed decisions based on data-driven insights. By utilizing these devices, farmers can make more precise judgments about when to plant, irrigate, and apply fertilizers, resulting in better crop yields and resource optimization. The potential benefits of data-driven insights for agricultural management are significant. The significance of the topic of data-driven insights for agricultural management is driven by the urgent need to address the challenges facing the agricultural sector, including climate change, resource depletion, and food insecurity. By leveraging Industry 4.0 technologies, farmers can optimize crop yields, manage resources efficiently, and improve sustainability, thereby contributing to global food security and environmental sustainability [6, 7].

14.2 DATA-DRIVEN INSIGHTS FOR AGRICULTURAL MANAGEMENT

Data-driven insights can be employed to predict crop yields, allowing farmers to plan accordingly and optimize their production processes. Overall, data-driven insights have significant potential to transform the agricultural sector and drive greater efficiency, productivity, and sustainability. Data collection is a critical component of data-driven insights for agricultural management. Collecting data on weather patterns, soil moisture, crop health, and other factors can provide valuable insights for farmers. This data can be collected through sensors, drones, satellite imagery, and other technologies. Similarly, weather data can be used to forecast weather patterns, helping farmers prepare for weather events such as droughts or floods. Once data has been collected, it can be analyzed to identify patterns and trends. This is known as data analysis. In agriculture, data analysis can help farmers to optimize inputs such as water, fertilizer, and pesticides, and to identify areas for improvement. For example, data analysis can be used to identify areas of a farm that are particularly fertile, enabling farmers to focus their efforts on those areas. Predictive modeling involves using data to predict future outcomes. In agriculture, predictive modeling can be used to forecast crop yields, identify areas at risk of pests and diseases, and manage risks associated with weather events. For example, predictive modeling can be used to forecast the yield of a particular crop, enabling farmers to make informed decisions about planting and harvesting. Similarly, predictive modeling can be used to identify areas at risk of pests and diseases, enabling farmers to take preventative measures such as crop rotation or pesticide application. This is known as decision support. For example, data-driven insights can be used to identify the optimal time to plant a particular crop, based on weather patterns and soil conditions. Finally, data-driven insights can be used to optimize the supply chain, enabling farmers to reduce waste and improve efficiency. This is known as supply chain management. For example, data-driven insights can be used to optimize transportation routes, ensuring that crops are delivered to market in a timely and cost-effective manner. Similarly, data-driven insights can be used to optimize storage and distribution, reducing waste and ensuring that crops reach consumers in good condition [8].

14.2.1 Definition of Data-driven Insights

Data-driven insights refer to the knowledge or understanding gained from the analysis of data. These insights are derived from the interpretation of patterns, trends, and correlations within large datasets, which can provide valuable information to inform decision-making and strategic planning. This involves collecting, analyzing, and interpreting data from various sources, such as sensors, drones, satellite imagery, and other technologies [9]. By leveraging data-driven insights, farmers can make informed decisions to improve their operations, reduce waste, and ultimately, increase profitability [10].

14.2.2 Benefits of Data-driven Agricultural Management

There are many benefits of data-driven agricultural management, which include [11, 12]:

- *Improved crop yields*: by optimizing inputs such as water, fertilizer, and pesticides, data-driven agricultural management can help to increase crop yields. This can ultimately lead to higher profitability for farmers.
- *Better risk management*: data-driven insights can help farmers manage risks associated with weather events, pests, and diseases. By identifying areas at risk of pest infestations or weather events, farmers can take preventative measures to mitigate these risks.
- *Increased efficiency*: one key benefit of data-driven agricultural management is increased profitability for farmers. By leveraging data and technology to make informed decisions, farmers can improve their crop yields and reduce waste, which can ultimately lead to higher profits. For example, by analyzing data on soil composition and nutrient levels, farmers can optimize their fertilizer use, resulting in healthier crops and higher yields. Similarly, by tracking weather patterns and soil moisture levels, farmers can make informed decisions about irrigation, reducing the risk of over-watering or under-watering crops and improving overall efficiency.
- *Better decision-making*: Overall, data-driven agricultural management can provide many benefits to farmers, including increased profitability, reduced

waste, improved resource management, better risk management, increased efficiency, and better decision-making.

14.2.3 Applications of Data-driven Insights in Agriculture

The various applications of the data-driven insights in agriculture are [13–15]:

- *Precision agriculture*: data-driven insights can optimize input utilization, including water, fertilizer, and pesticides, to reduce waste and increase crop yields, based on factors such as soil moisture, crop health, and weather patterns.
- *Crop forecasting*: predictive modeling is a technique that enables farmers to anticipate future crop yields by using data analysis and statistical algorithms. With the help of predictive modeling, farmers can make informed decisions about planting and harvesting, as they are able to forecast the potential crop yields. This information is vital as it allows farmers to optimize their crop production, minimize waste, and increase profitability. By knowing the expected crop yields in advance, farmers can adjust their input usage such as fertilizer and water to optimize the crop yield for specific areas of the farm. This approach can also help farmers to plan and adjust their harvest time, enabling them to maximize the potential yield and minimize losses due to delayed harvesting.
- *Supply chain optimization*: data-driven insights can be used to optimize the supply chain, enabling farmers to reduce waste and improve efficiency. This can help to ensure that crops are delivered to market in a timely and cost-effective manner.
- *Sustainable agriculture*: data-driven insights can help farmers adopt more sustainable farming practices by optimizing inputs, reducing waste, and increasing efficiency. This can help to reduce the environmental impact of agriculture and improve long-term sustainability.

The applications of data-driven insights in agriculture are vast and can help farmers make informed decisions that optimize their operations, reduce waste, and increase profitability. The benefits and applications of data-driven insights in agriculture are shown in Figure 14.1.

14.3 INDUSTRY 4.0 TECHNOLOGIES IN AGRICULTURE

By analyzing data on crop yields and fertilizer inputs, farmers can optimize their use of fertilizers, reducing waste and increasing efficiency. AI involves developing intelligent algorithms that can learn from data and make predictions. In agriculture, AI can be used to develop predictive models for crop forecasting and risk management. For example, by using AI to analyze weather patterns, farmers can make informed decisions about when to plant and harvest crops, reducing the risk of crop failure. This can help to increase efficiency and reduce labor costs. For example, by using automated systems to plant and harvest crops, farmers can reduce the need for manual labor, increasing efficiency and reducing costs. The

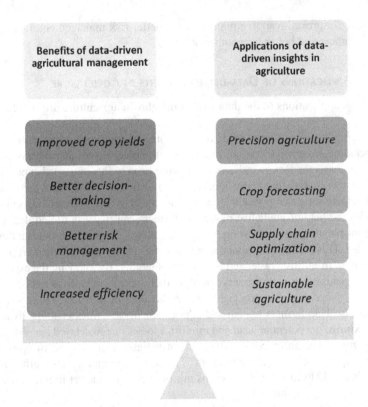

FIGURE 14.1 Benefits and applications of data-driven insights in agriculture.

utilization of drones in agriculture is an effective method, as well as conducting tasks like crop spraying. This is helpful in optimizing inputs and boosting crop yields. By monitoring crop health using drones, farmers can pinpoint areas of the field that need more attention, such as extra fertilization or irrigation, leading to improved crop health and yield. Blockchain technology, on the other hand, involves creating a secure, decentralized database that can be utilized to record transactions and track goods as they move throughout the supply chain. In the agricultural industry, blockchain can help optimize the supply chain, providing farmers with increased transparency and traceability. This can lead to reduced waste and increased efficiency in the system. For instance, through the use of blockchain technology, farmers can track their crops from the farm to the market, ensuring that the produce is delivered promptly and efficiently, which minimizes waste and enhances profitability [16, 17].

14.3.1 INDUSTRY 4.0 TECHNOLOGIES

Industry 4.0 technologies refers to the integration of advanced technologies such as the IoT, big data analytics, AI, robotics and automation, drones, and blockchain into

Data-Driven Insights for Agricultural Management

industrial processes [18–21]. These technologies are characterized by their ability to collect and analyze vast amounts of data, enabling organizations to optimize their operations, increase efficiency, and reduce waste. Industry 4.0 technologies are often seen as a new era in industrial production, with the potential to revolutionize manufacturing, supply chain management, and other aspects of industry. In agriculture, Industry 4.0 technologies can be used to optimize inputs, increase crop yields, and improve resource management.

14.3.2 Benefits of Industry 4.0 Technologies in Agriculture

Industry 4.0 technologies offer many benefits for agriculture [22, 23] including:

- *Increased efficiency*: by using technology to automate tasks and optimize inputs, farmers can increase efficiency and reduce labor costs.
- *Improved crop yields*: by using data to optimize inputs such as fertilizer, water, and pesticides, farmers can improve crop yields and reduce waste.
- *Better risk management*: predictive modeling and AI can be used to manage risks associated with weather events, pests, and other factors that can affect crop growth.
- *Greater transparency and traceability*: blockchain technology can be used to provide farmers with greater transparency and traceability throughout the supply chain, helping to reduce waste and increase efficiency.
- *Reduced environmental impact*: by optimizing inputs and reducing waste, Industry 4.0 technologies can help to reduce the environmental impact of agriculture.

14.4 IMPROVED CROP YIELDS THROUGH DATA-DRIVEN INSIGHTS

Improving crop yields is a critical goal in agriculture, as it helps to ensure food security, increase profitability, and promote sustainability. Data-driven insights can play an important role in achieving this goal by providing farmers with valuable information on soil health, weather patterns, and crop growth. By collecting and analyzing this data, farmers can optimize their inputs, reduce waste, and improve overall crop yields. By collecting data on soil moisture, temperature, and other factors, farmers can develop a more nuanced understanding of their fields and adjust their inputs accordingly. This can lead to more efficient use of resources such as water, fertilizer, and pesticides, ultimately resulting in higher crop yields. Data-driven insights can also be used to predict future crop yields and identify areas at risk of pests and diseases. By analyzing historical data and using predictive modeling, farmers can better anticipate potential problems and take proactive steps to mitigate them. This can lead to a more resilient agricultural system, capable of withstanding a range of challenges and producing more food with fewer resources. Overall, data-driven insights are an important tool for improving crop yields in agriculture [24].

14.4.1 Importance of Crop Yields in Agriculture

Crop yields are a critical factor in agriculture, as they determine the amount of food and other products that can be produced from a given area of land. Higher crop yields mean more food and other agricultural products can be produced per unit of land, which is essential to meet the needs of a growing population. Crop yields are a critical aspect of agriculture as they determine how much food can be produced from a given area of land. With the world population expected to reach 9.7 billion by 2050, it is imperative to increase crop yields to meet the growing demand for food. Higher crop yields can help to ensure food security and reduce the risk of hunger, malnutrition, and food shortages.

In addition to providing food security, higher crop yields can also lead to increased profitability for farmers. By producing more food with fewer resources, farmers can reduce their costs and increase their revenue. This can help to support rural livelihoods and contribute to economic growth in the agriculture sector. Increasing crop yields is also essential for sustainability. By producing more food without expanding the amount of land under cultivation, farmers can help to reduce pressure to clear forests and other natural habitats for agriculture. This can help to conserve biodiversity, reduce greenhouse gas emissions, and mitigate the impacts of climate change.

Finally, higher crop yields can also increase the resilience of agricultural systems. By producing more food per unit of land, farmers can better withstand weather events, pests, and other challenges that can impact crop production. This can help to ensure the long-term viability of agriculture and support the livelihoods of farmers and rural communities [25, 26].

14.4.2 Factors Affecting Crop Yields

It is important for farmers to monitor and manage these factors to optimize crop yields and maintain sustainable agricultural practices. By using data-driven insights, farmers can gain a better understanding of these factors and make informed decisions to improve their yields and profitability. Table 14.1 shows the various factors affecting crop yields.

There are several factors that can affect crop yields, and soil quality is one of the most important. The quality of the soil, including its nutrient content, structure, pH level, and texture, can all affect crop yields. Soil that is rich in nutrients and has good structure and texture can help crops grow strong and healthy. Soil that is too acidic or too alkaline can negatively affect crop growth and yield. Water availability is another important factor that can affect crop yields. The amount and timing of water available to crops can greatly affect yields. Water-stressed crops may produce lower yields, while crops with adequate water may produce higher yields. Proper irrigation management is essential to ensure crops have enough water at critical growth stages. Climate is also an important factor that can affect crop yields. Weather patterns such as temperature, rainfall, and humidity can all impact crop yields. Crops require specific conditions to grow optimally, and variations from ideal conditions can affect yields. For example, crops grown in areas with a

TABLE 14.1
Factors Affecting Crop Yields

Factor	Description
Soil quality	The physical and chemical properties of the soil can impact crop yields. Factors such as soil texture, pH, nutrient levels, and organic matter content can all affect plant growth and productivity.
Water availability	Adequate water is essential for plant growth, and water stress can lead to reduced yields. Factors such as rainfall patterns, irrigation systems, and soil moisture levels can all impact water availability for crops.
Climate	Temperature, sunlight, and other climate factors can impact plant growth and productivity. Frost, heat waves, and other extreme weather events can also damage crops and reduce yields.
Pests and diseases	Pests and diseases can damage crops and reduce yields. Effective pest and disease management is essential for maintaining crop health and productivity.
Genetics	The genetic makeup of plants can impact their growth and productivity. Plant breeding and genetic modification can be used to develop crops with improved yields and other desirable traits.
Management practices	Farming practices such as crop rotation, fertilizer application, and pest management can all impact crop yields. Effective management practices can help to optimize inputs and increase productivity.

shorter growing season may have lower yields compared to those grown in areas with longer growing seasons. Pest and disease pressure can greatly reduce crop yields, often by causing damage to plants or reducing their ability to photosynthesize. Farmers must manage pests and diseases through practices such as crop rotation, the use of resistant crop varieties, and proper application of pesticides and fungicides. Genetics is another factor that can affect crop yields. The genetic makeup of a crop can affect its ability to produce high yields. Modern plant breeding techniques have allowed for the development of crop varieties with higher yields and better resistance to pests and diseases. Farmers can select the right varieties to grow based on their local growing conditions and yield potential. Finally, input management is an essential factor that can affect crop yields. Proper management of inputs such as fertilizers, pesticides, and herbicides can greatly affect crop yields. Too little or too much of these inputs can impact crop health and yield. Proper management of inputs can help farmers optimize crop growth and yield potential while reducing negative impacts on the environment [27, 28].

14.4.3 ROLE OF DATA-DRIVEN INSIGHTS IN IMPROVING CROP YIELDS

By collecting and analyzing data on various factors that affect crop growth, such as soil quality, water availability, climate, and pest and disease pressure, farmers can make informed decisions about how to manage their crops. For example, data on soil quality can help farmers determine what types of fertilizers and soil amendments to use to ensure that their crops are receiving the nutrients they need to grow optimally. Data on water availability can inform decisions about irrigation scheduling and the use of water-efficient technologies. Data on climate can help farmers predict weather patterns and adjust their planting and harvesting schedules accordingly. Data-driven insights can also help farmers manage pest and disease pressure by providing information on which pests and diseases are most prevalent in their area and which treatments are most effective in controlling them. Furthermore, data-driven insights can help farmers optimize their input management by providing information on the optimal levels of fertilizers, pesticides, and herbicides to use for a given crop and soil type. This can help farmers reduce waste and costs associated with excess inputs, while still ensuring that their crops are receiving the necessary nutrients and protection from pests and diseases. Overall, data-driven insights have the potential to revolutionize the way farmers manage their crops.

14.5 RESOURCE OPTIMIZATION THROUGH DATA-DRIVEN INSIGHTS

Data-driven insights are crucial for optimizing various aspects of business operations. By analyzing data, organizations can gain valuable insights that can help them identify inefficiencies, bottlenecks, and opportunities for improvement. These insights can be used to optimize supply chain management, manufacturing processes, or reduce energy consumption. To obtain data-driven insights, organizations must collect, analyze, and interpret data. This involves using statistical analysis, machine learning, and data mining techniques to extract valuable information from large datasets. By analyzing this data, organizations can gain a better understanding of their operations and make informed decisions that improve their performance. Data-driven insights can also be used to predict future outcomes and optimize operations accordingly. Predictive analytics can be used to forecast demand for a product, allowing organizations to adjust their production schedules to meet demand and avoid stockouts. Data-driven insights are an essential tool for optimizing business operations. By leveraging data analysis techniques and tools, organizations can gain valuable insights that help them reduce costs, increase efficiency, and improve performance [29].

14.5.1 Resource Management in Agriculture

Resource management in agriculture involves the efficient use of various resources to achieve maximum crop yields while minimizing waste and environmental impact. The key resources managed in agriculture include land, water, energy, and inputs such as fertilizers, pesticides, and herbicides. Effective resource management in agriculture requires the use of various strategies, including precision agriculture, conservation agriculture, and integrated pest management. This data is then used to optimize inputs, improve crop yields, and reduce waste. Conservation agriculture is another approach to resource management in agriculture. It involves practices such as minimal tillage, cover cropping, and crop rotation to improve soil health, reduce

Data-Driven Insights for Agricultural Management 269

erosion, and conserve water. By improving soil health, conservation agriculture can help to increase crop yields and reduce the need for inputs such as fertilizers and pesticides. Integrated pest management (IPM) is a third strategy for resource management in agriculture. IPM involves the use of a variety of methods to control pests, including biological control, cultural practices, and the judicious use of chemical pesticides. By using multiple approaches to pest control, IPM can reduce the number of pesticides needed and minimize their impact on the environment.

Overall, effective resource management in agriculture is essential for sustainable crop production. By optimizing the use of land, water, energy, and inputs, farmers can achieve maximum crop yields while minimizing waste and environmental impact.

14.5.2 CHALLENGES IN RESOURCE OPTIMIZATION

1. *Limited access to technology*: many farmers may not have access to the latest technologies and tools needed to optimize resources, such as precision agriculture equipment and sensors.
2. *Lack of data*: without access to accurate and timely data on crop health, weather patterns, and other important factors, farmers may struggle to make informed decisions about resource management.
3. *Cost*: implementing resource optimization strategies can be expensive, particularly for small-scale farmers who may not have the financial resources to invest in new technologies or equipment.
4. *Complexity*: resource optimization strategies can be complex and require specialized knowledge and skills to implement effectively. This can be a barrier for some farmers, particularly those with limited education or training.
5. *Environmental concerns*: some resource optimization strategies, such as the use of fertilizers and pesticides, can have negative environmental impacts if not managed carefully. Farmers must balance the need to increase yields with the need to protect the environment and promote sustainability.

14.5.3 ROLE OF DATA-DRIVEN INSIGHTS IN RESOURCE OPTIMIZATION

Data-driven insights play a critical role in resource optimization in agriculture by providing farmers with accurate and timely information that can inform decision-making. By collecting and analyzing data on factors such as soil moisture, weather patterns, and crop health, farmers can make more informed decisions about how to allocate resources such as water, fertilizers, and pesticides. For example, data on soil moisture levels can help farmers determine the optimal time to irrigate crops, reducing water waste and ensuring that crops have adequate water for optimal growth. Similarly, data on weather patterns can inform decisions about when to apply pesticides and herbicides, reducing the amount of chemicals needed and minimizing environmental impacts, ultimately leading to improved yields and profitability. In addition, data-driven insights can inform decisions about crop selection and rotation, helping farmers optimize resource use over the long term. By understanding which crops are best suited to their specific growing conditions, farmers can maximize yields and reduce inputs, ultimately leading to more sustainable and profitable agricultural systems [30, 31].

14.5.4 Challenges in Data Collection and Analysis

Data collection and analysis are essential components of decision-making in almost every field, including agriculture. In the agricultural industry, data is used to make informed decisions about resource management, crop production, and more. However, collecting and analyzing data can be a complex and challenging process, and there are several challenges that farmers and other stakeholders must overcome to make the most of their data. One of the most significant challenges in data collection and analysis is the sheer volume of data that is available. With the advent of new technologies such as IoT devices, drones, and other sensors, there is an almost endless stream of data that can be collected about agricultural operations. However, not all of this data is useful or relevant to decision-making, and it can be difficult to sift through the noise to find meaningful insights. Another challenge is data quality. Inaccurate or incomplete data can lead to incorrect conclusions and poor decision-making. Data privacy and security is another major concern in data collection and analysis. Agricultural data is often sensitive and can contain valuable information about crop yields, soil quality, and more. Farmers and other stakeholders must take steps to protect this data from unauthorized access and theft, which can be a challenge in an era of increasing cyber threats. Another challenge in data collection and analysis is the need for specialized skills and expertise. Collecting and analyzing data requires a range of skills, including data science, statistics, and machine learning. Many farmers and other stakeholders may not have these skills, which can make it difficult to effectively collect and analyze data. In addition to these challenges, there are also logistical challenges associated with data collection and analysis. For example, data may be spread across multiple platforms and systems, making it difficult to consolidate and analyze. There may also be issues with data standardization and interoperability, which can make it difficult to compare data from different sources. Despite these challenges, there are many benefits to collecting and analyzing agricultural data, and new technologies and tools are emerging to help farmers and other stakeholders overcome these challenges. For example, data visualization tools can help to make sense of large volumes of data, while cloud-based data storage and analysis platforms can help to improve data accessibility and collaboration [32].

In conclusion, data collection and analysis are critical for decision-making in agriculture, but there are several challenges that must be overcome to make the most of this data. These challenges include the volume and quality of data, data privacy and security, specialized skills and expertise, and logistical challenges. However, by leveraging new technologies and tools, farmers and other stakeholders can overcome these challenges and unlock the many benefits of data-driven decision-making in agriculture.

14.5.5 LIMITATIONS OF INDUSTRY 4.0 TECHNOLOGIES IN AGRICULTURE

While Industry 4.0 technologies offer numerous benefits to agriculture, there are also several limitations and challenges associated with their implementation. Some of the limitations of Industry 4.0 technologies in agriculture include:

- *Infrastructure*: the most important challenge is the lack of adequate infrastructure, particularly in rural areas where connectivity may be limited. Without access to reliable internet and other necessary technologies, data collection and analysis may not be possible.
- *Data quality*: one of the limitations of Industry 4.0 technologies in agriculture is related to data quality. While data can be collected from various sources including sensors, the accuracy and reliability of such data may be influenced by several factors. These factors can include environmental conditions that may affect the performance of the sensors, malfunctioning of the sensors themselves, or human error during data collection or analysis. Thus, it is essential to ensure that data is accurate and reliable before it is used to make any critical decisions in the agricultural sector
- *Data privacy and security*: farmers and other stakeholders may be hesitant to share data due to concerns about confidentiality and the potential for data breaches.
- *Limited applicability*: not all Industry 4.0 technologies may be applicable to all agricultural operations. Some technologies may be better suited for certain crops or regions, and others may require specific expertise to implement effectively.
- *Lack of training and education*: many farmers may not have the necessary training or education to effectively use and benefit from Industry 4.0 technologies. The complexity of these technologies may require specialized knowledge and training, which may not be readily available or accessible to all farmers.

Since Industry 4.0 technologies offer significant potential for improving agricultural productivity, profitability, and sustainability, their successful implementation requires addressing these limitations and challenges. Strategies such as providing financial support, improving infrastructure, ensuring data quality and privacy, and providing training and education can help overcome these limitations and ensure that farmers can effectively benefit from these technologies [33, 34].

14.6 CONCLUSION

In conclusion, the utilization of data analytics and artificial intelligence in agriculture can provide farmers with valuable insights, especially in decision-making regarding planting, irrigation, fertilization, and pest management. By analyzing and interpreting data from various sources such as sensors and satellite imagery, farmers can make informed decisions to optimize the use of inputs and resources while reducing waste [35, 36]. For instance, the appropriate amounts of irrigation and fertilization needed in different areas of the farm. Similarly, data on pest populations and disease prevalence can be used to develop targeted and efficient pest management strategies. Such data-driven approaches can improve the overall efficiency of agricultural operations, leading to increased profitability for farmers. However, there are also limitations and challenges to the adoption of these technologies. Data collection and analysis can be difficult in remote areas, and the cost of implementing these technologies can be prohibitive for some farmers. Furthermore, there are concerns about data privacy and security [37, 38]. Nonetheless, with ongoing advancements in technology and increased adoption, it is likely that the benefits of Industry 4.0 technologies in agriculture will continue to outweigh the challenges. The potential for increased food production, profitability, and sustainability makes it imperative for the agricultural industry to embrace these technologies and continue to explore innovative solutions for improved agricultural management.

REFERENCES

1. Shukla, R.; Dubey, G.; Malik, P.; Sindhwani, N.; Anand, R.; Dahiya, A.; Yadav, V. Detecting crop health using machine learning techniques in smart agriculture system. *J. Sci. Ind. Res.* 2021, 80(8), 699–706.
2. Singh, P.; Kaiwartya, O.; Sindhwani, N.; Jain, V.; Anand, R. (Eds.). *Networking Technologies in Smart Healthcare: Innovations and Analytical Approaches.* CRC Press, 2022.
3. Sindhwani, N.; Anand, R.; Niranjanamurthy, M.; Verma, D.C.; Valentina, E.B. (Eds.). *IoT Based Smart Applications.* Springer Nature, 2022.
4. Pandey, D.; Pandey, B.K.; Sindhwani, N.; Anand, R.; Nassa, V.K.; Dadheech, P. *An Interdisciplinary Approach in the Post-COVID-19 Pandemic Era*, 2022, pp. 1–290.
5. Martín-Martín, A.; Orduna-Malea, E.; Thelwall, M.; López-Cózar, E.D. Google Scholar, Web of Science, and Scopus: A systematic comparison of citations in 252 subject categories. *J. Inf.* 2018, 12, 1160–1177.
6. Mongeon, P.; Paul-Hus, A. The journal coverage of Web of Science and Scopus: A comparative analysis. *Scientometrics* 2016, 106, 213–228.
7. Bergman, E.M.L. Finding citations to social work literature: The relative benefits of using Web of Science, Scopus, or Google Scholar. *J. Acad. Libr.* 2012, 38, 370–379.
8. Aramyan, L.H.; Lansink, A.O.; Van Der Vorst, J.G.A.J.; Van Kooten, O. Performance measurement in agri-food supply chains: A case study. *Supply Chain Manag. Int. J.* 2007, 12, 304–315.
9. Goyal, B.; Dogra, A.; Khoond, R.; Gupta, A.; Anand, R. Infrared and visible image fusion for concealed weapon detection using transform and spatial domain filters. In *2021 9th International Conference on Reliability, Infocom Technologies and Optimization (Trends and Future Directions)(ICRITO)*. IEEE, 2021, September, pp. 1–4.
10. Talavera, J.M.; Tobon, L.; Gómez, J.A.; Culman, M.; Aranda, J.; Parra, D.T.; Quiroz, L.A.; Hoyos, A.; Garreta, L.E. Review of IoT applications in agro-industrial and environmental fields. *Comput. Electron. Agric.* 2017, 142, 283–297.
11. Patrício, D.I.; Rieder, R. Computer vision and artificial intelligence in precision agriculture for grain crops: A systematic review. *Comput. Electron. Agric.* 2018, 153, 69–81.
12. Kamilaris, A.; Fonts, A.; Prenafeta-Boldú, F. X. The rise of blockchain technology in agriculture and food supply chains. *Trends Food Sci. Technol.* 2019, 91, 640–652.

13. Bibi, F.; Guillaume, C.; Gontard, N.; Sorli, B. A review: RFID technology having sensing aptitudes for food industry and their contribution to tracking and monitoring of food products. *Trends Food Sci. Technol.* 2017, 62, 91–103.
14. Kumar, I.; Rawat, J.; Mohd, N.; Husain, S. Opportunities of artificial intelligence and machine learning in the food industry. *J. Food Qual.* 2021, 2021, 1–10.
15. Bécue, A.; Praça, I.; Gama, J. Artificial intelligence, cyber-threats and Industry 4.0: Challenges and opportunities. *Artif. Intell. Rev.* 2021, 54(5), 3849–3886.
16. Mavani, N.R.; Ali, J.M.; Othman, S.; Hussain, M.A.; Hashim, H.; Rahman, N.A. Application of artificial intelligence in food industry—A guideline. *Food Eng. Rev.* 2021 14, 1–42.
17. Massaro, A.; Galiano, A. Re-engineering process in a food factory: An overview of technologies and approaches for the design of pasta production processes. *Prod. Manuf. Res.* 2020, 8(1), 80–100.
18. Pandey, B.K.; Pandey, D.; Nassa, V.K.; Ahmad, T.; Singh, C.; George, A.S.; Wakchaure, M.A. Encryption and steganography-based text extraction in IoT using the EWCTS optimizer. *Imaging Sci. J.* 2022, 69, 1–19.
19. Pandey, B.K.; Pandey, D.; Agarwal, A. Encrypted information transmission by enhanced steganography and image transformation. *Int. J. Distrib. Artif. Intell. (IJDAI)* 2022, 14(1), 1–14.
20. Pandey, B.K.; Pandey, D.; Gupta, A.; Nassa, V.K.; Dadheech, P.; George, A.S. Secret data transmission using advanced morphological component analysis and steganography. In *Role of Data-Intensive Distributed Computing Systems in Designing Data Solutions*. Springer International Publishing, 2023, pp. 21–44.
21. Pandey, B.K.; Pandey, D.; Wariya, S.; Aggarwal, G.; Rastogi, R. Deep learning and particle swarm optimisation-based techniques for visually impaired humans' text recognition and identification. *Augmented Hum. Res.* 2021, 6, 1–14.
22. Di Vaio, A.; Boccia, F.; Landriani, L.; Palladino, R. Artificial intelligence in the agri-food system: Rethinking sustainable business models in the COVID-19 scenario. *Sustainability* 2020, 12(12), 4851.
23. Su, Y.; Wang, X. Innovation of agricultural economic management in the process of constructing smart agriculture by big data. *Sustain. Comput. Inform. Syst.* 2021, 31, 100579.
24. Arora, D. Demand prognosis of Industry 4.0 to agriculture sector in India. *Int. J. Knowl. Based Intell. Eng. Syst.* 2021, 25(1), 129–138.
25. Sindhwani, N., Anand, R., Meivel, S., Shukla, R., Yadav, M.P., Yadav, V. Performance Analysis of Deep Neural Networks Using Computer Vision. *inis* 2013, 21(29), e3
26. Islam, S.; Manning, L.; Cullen, J.M. A hybrid traceability technology selection approach for sustainable food supply chains. *Sustainability* 2021, 13(16), 9385.
27. Gallo, A.; Accorsi, R.; Goh, A.; Hsiao, H.; Manzini, R. A traceability-support system to control safety and sustainability indicators in food distribution. *Food Control* 2021, 124, 107866.
28. Schmetz, A.; Lee, T.H.; Hoeren, M.; Berger, M.; Ehret, S.; Zontar, D.; Min, S.-H.; Ahn, S.-H.; Brecher, C. Evaluation of Industry 4.0 data formats for digital twin of optical components. *Int. J. Precis. Eng. Manuf. Technol.* 2020, 7(3), 573–584.
29. Niknejad, N.; Ismail, W.; Bahari, M.; Hendradi, R.; Salleh, A.Z. Mapping the research trends on blockchain technology in food and agriculture industry: A bibliometric analysis. *Environ. Technol. Innov.* 2020, 21, 101272.
30. Kayikci, Y.; Subramanian, N.; Dora, M.; Bhatia, M.S. Food supply chain in the era of Industry 4.0: Blockchain technology implementation opportunities and impediments from the perspective of people, process, performance, and technology. *Prod. Plan. Control* 2020, 33, 1–10.

31. Lin, C.-F. Blockchainizing food law: Promises and perils of incorporating distributed ledger technologies to food safety, traceability, and sustainability governance. *Food Drug Law J.* 2019, 74, 586–612.
32. Li, X.; Huang, D. Research on value integration mode of agricultural e-commerce industry chain based on Internet of things and blockchain technology. *Wirel. Commun. Mob. Comput.* 2020, 2020, 1–11.
33. Defraeye, T.; Tagliavini, G.; Wu, W.; Prawiranto, K.; Schudel, S.; Kerisima, M.A.; Verboven, P.; Bühlmann, A. Digital twins probe into food cooling and biochemical quality changes for reducing losses in refrigerated supply chains. *Resour. Conserv. Recycl.* 2019, 149, 778–794.
34. Demestichas, K.; Peppes, N.; Alexakis, T.; Adamopoulou, E. Blockchain in agriculture traceability systems: A review. *Appl. Sci.* 2020, 10(12), 4113.
35. Meivel, S.; Sindhwani, N.; Anand, R.; Pandey, D.; Alnuaim, A.A.; Altheneyan, A.S.; Jabarulla, M.Y.; Lelisho, M.E. Mask detection and social distance identification using internet of things and faster R-CNN algorithm. *Comp. Intell. Neurosci.* 2022, 2022, 1–13.
36. Srivastava, A.; Gupta, A.; Anand, R. Optimized smart system for transportation using RFID technology. *Math. Eng. Sci. Aerosp. (MESA)* 2021, 12(4), 953–965.
37. Sindhwani, N.; Rana, A.; Chaudhary, A. Breast cancer detection using machine learning algorithms. In *2021 9th International Conference on Reliability, Infocom Technologies and Optimization (Trends and Future Directions)(ICRITO)*. IEEE, 2021, September, pp. 1–5.
38. Verma, S.; Bajaj, T.; Sindhwani, N.; Kumar, A. Design and development of a driving assistance and safety system using deep learning. In *Advances in Data Science and Computing Technology*. Apple Academic Press, 2022, pp. 35–45.

15 Data Analytics and ML for Optimized Performance in Industry 4.0

K.V. Daya Sagar, K.K. Ramachandran,
Purnendu Bikash Acharjee, Pratibha Singh,
Harish Satyala, and Barinderjit Singh

15.1 INTRODUCTION

The fourth industrial revolution, commonly referred to as Industry 4.0, is marked by the incorporation of cutting-edge technologies like artificial intelligence (AI), big data, and the Internet of Things (IoT) into industrial operations. This integration of advanced technologies has led to a significant transformation in the manufacturing sector, enabling businesses to enhance operational efficiency, improve product quality, and reduce costs [1–5]. A thorough review of existing literature indicates that Industry 4.0 represents a major shift in the way organizations approach industrial production and is likely to have far-reaching implications for businesses across various industries. As a result, the manufacturing industry is becoming increasingly data-driven, with vast amounts of data being generated every day. DA and ML are two essential technologies that can be utilized in Industry 4.0 to process and analyze this data to optimize performance. DA and ML are complementary technologies that can be used together to achieve even greater benefits in Industry 4.0. By combining DA with ML, manufacturers can develop more sophisticated predictive models that can anticipate equipment failures, optimize production processes, and improve supply chain performance. For example, manufacturers can use ML algorithms to identify patterns in data from production processes and use these patterns to develop predictive models that can anticipate equipment failures. For example, the sheer volume of data generated by industrial processes can be overwhelming, making it difficult to manage and analyze. Additionally, there may be issues related to data privacy and security, as well as a shortage of skilled professionals with expertise in DA and ML. We will explore the use cases of these technologies in optimizing performance in manufacturing and supply chain processes, as well as the challenges and limitations associated with their implementation. Furthermore, we will examine the benefits of integrating DA and ML and provide examples of how these technologies can be combined to achieve even greater performance optimization [6, 7].

15.1.1 Overview

The combination of DA and ML can enable manufacturers to develop more sophisticated predictive models that can anticipate equipment failures, optimize production processes, and improve supply chain performance. In addition to these applications, DA and ML can also be used in Industry 4.0 to improve product design, enhance quality control, and reduce waste. For example, manufacturers can use DA to identify patterns in customer feedback and use this information to design products that better meet customer needs and ML algorithms can be used to identify defects in products during the manufacturing process, enabling manufacturers to reduce waste and improve product quality. Additionally, by meeting customer demands effectively, manufacturers can enhance customer satisfaction, which can lead to increased revenue and loyalty. Furthermore, these benefits can give manufacturers a competitive advantage in the market by enabling them to produce customized products on a mass scale and meet the unique needs of individual customers while still achieving economies of scale. However, implementing these models presents several challenges, including infrastructure and talent requirements. To use these technologies effectively, organizations need to invest in the necessary hardware and software and develop the talent needed to operate them. Moreover, ethical and privacy concerns arise as companies collect and analyze vast amounts of data, which must be handled responsibly and transparently. Despite these challenges, the use of predictive models in Industry 4.0 has the potential to transform the manufacturing industry by enhancing efficiency and reducing costs, ultimately leading to increased customer satisfaction and competitiveness [8, 9].

15.2 DATA ANALYTICS FOR OPTIMIZED PERFORMANCE IN INDUSTRY 4.0

DA plays a critical role in optimizing performance in I-4.0. The integration of advanced technologies and the collection of vast amounts of data from various stages of the production process provide an opportunity to optimize manufacturing processes. The use of DA enables manufacturers to gain insights into their production processes and identify areas where improvements can be made. This includes analyzing data to identify patterns and trends, detecting anomalies, predicting machine failures, and identifying opportunities for process optimization. DA also enables manufacturers to optimize production processes. By analyzing data on production metrics such as throughput, cycle time, and yield, manufacturers can identify areas where improvements can be made and implement changes to increase efficiency and reduce waste. Furthermore, DA can be used to improve product quality by identifying potential defects early in the production process. Overall, the use of DA in I-4.0 is essential for optimizing performance and improving efficiency, identifying areas for improvement, and implementing changes that increase efficiency, reduce downtime, and improve product quality [10].

15.2.1 Data Analytics Techniques and Tools for Industry 4.0

Table 15.1 shows DA techniques and tools. Each of these techniques and tools plays a unique role in optimizing performance in I-4.0. Predictive maintenance can help reduce downtime by identifying potential equipment failures before they occur. Data visualization tools can be used to identify inefficiencies in production processes and monitor performance in real-time. Overall, these techniques and tools allow for data to be turned into actionable insights, helping to improve efficiency, reduce costs, and increase productivity in I-4.0.

DA techniques and tools used in I-4.0 include ML, artificial intelligence, predictive maintenance, data visualization, digital twins, natural language processing, edge computing, and cloud computing. ML and AI are used for predictive analysis and automated decision-making. Predictive maintenance helps in proactive maintenance scheduling. Data visualization provides clear and intuitive insights from complex datasets. Digital twins are used to optimize processes and identify inefficiencies. Natural language processing analyzes unstructured data like customer reviews to provide insights into customer preferences. Edge computing reduces latency and enables real-time decision-making, while cloud computing allows for centralized storage and processing of large amounts of data [11–13].

TABLE 15.1
DA Techniques and Tools

DA Techniques and Tools	Explanation
ML	ML algorithms are used to identify patterns and make predictions based on large datasets. This is useful for predicting equipment failures and optimizing production processes.
Artificial intelligence	AI systems are used to make automated decisions based on data. This can be useful for real-time decision-making and monitoring systems for anomalies.
Data visualization	Data visualization tools allow for complex datasets to be presented in a clear and intuitive way, making it easier to identify patterns and trends.
Digital twins	Digital twins are virtual replicas of physical systems, used to simulate and optimize processes. They can be used to identify inefficiencies and predict future performance.
Natural language processing	Natural language processing is used to analyze unstructured data such as text data from customer reviews or social media. This can provide insights into customer preferences and sentiment.
Edge computing	Edge computing involves processing data at the edge of the network, close to the devices generating the data. This reduces latency and allows for real-time decision-making.
Cloud computing	Cloud computing allows for large amounts of data to be stored and processed in a centralized location, making it easier to scale up DA operations.

15.2.2 Applications of Data Analytics in Industry 4.0 for Optimized Performance

DA plays a crucial role in I-4.0 by helping to optimize production processes, reduce downtime, improve product quality, and lower costs. Here are some use cases of DA in Industry 4.0 [14–16] as shown in Figure 15.1:

- *Quality control*: DA is used to identify patterns and trends in production data, enabling manufacturers to identify quality issues early and take corrective action before they become major problems.
- *Supply chain optimization*: DA is used to monitor and optimize supply chain processes, improving efficiency and reducing costs.
- *Energy management*: DA is used to monitor energy consumption and identify opportunities to reduce energy usage, resulting in lower costs and improved sustainability.
- *Process optimization*: DA is used to optimize production processes, reducing waste and improving product quality.
- *Predictive analytics*: DA is used to predict future demand and sales, enabling manufacturers to adjust production levels and avoid overproduction.

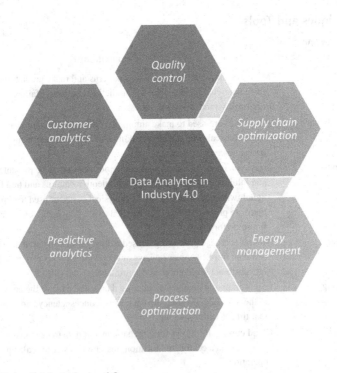

FIGURE 15.1 DA in Industry 4.0.

- *Customer analytics*: DA is used to analyze customer data, enabling manufacturers to better understand customer needs and preferences and develop products and services that better meet those needs.

15.2.3 Challenges for Data Analytics in Industry 4.0

One of the biggest challenges in DA is maintaining the quality and reliability of the data being used. Ensuring that data is accurate, complete, and consistent across different sources is essential for generating meaningful insights. Additionally, as more data is collected and analyzed in I-4.0, data privacy and security have become major concerns. Protecting sensitive data and implementing proper security measures are critical to preventing data breaches [17–20]. Figure 15.2 shows the various challenges for DA in Industry 4.0.

Another issue is the lack of standardization in data collection and analysis methods. The absence of a common framework makes it difficult to compare data across different sources and industries. Furthermore, DA can be a complex and technical process that requires specialized knowledge and expertise, making it difficult for smaller companies to implement DA solutions. Integrating DA solutions with existing systems can also be challenging, particularly when legacy systems are in use.

Moreover, algorithms used in DA can be biased, resulting in inaccurate results or discriminatory outcomes. It is particularly crucial to address this concern in

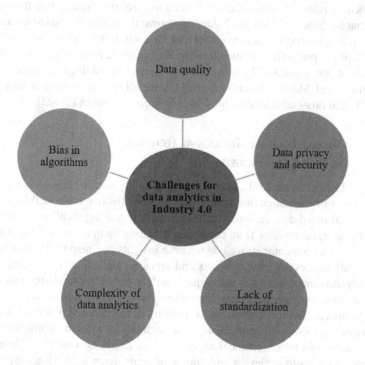

FIGURE 15.2 Challenges for DA in Industry 4.0.

industries such as manufacturing, where decisions made based on DA can have significant impacts on production processes and outcomes.

15.3 MACHINE LEARNING FOR OPTIMIZED PERFORMANCE IN INDUSTRY 4.0

ML algorithms can be trained on data from past production runs to identify patterns that indicate quality issues. By using ML to identify these patterns, manufacturers can quickly identify and address quality issues, reducing waste and improving customer satisfaction. ML can also be used for process optimization. This can lead to cost savings and improved productivity. Another important application of ML in I-4.0 is supply chain optimization. ML algorithms can be used to analyze data from suppliers, production processes, and customers to optimize the supply chain. This can help manufacturers to better predict demand, manage inventory, and improve delivery times. In some cases, data may be incomplete, inaccurate, or biased, which can lead to poor results. Another challenge is the need for specialized skills and expertise. ML requires expertise in areas such as data science, statistics, and computer science. Many manufacturers may not have the necessary expertise in-house, which can make it difficult to implement ML solutions. Finally, there are also concerns about the potential impact of ML on jobs. As machines become more intelligent, there is a risk that they may replace human workers, particularly in repetitive or low-skilled roles. Manufacturers will need to carefully manage this transition to ensure that the benefits of ML are balanced against the potential costs to workers and society. In conclusion, ML is a powerful tool for optimizing performance in I-4.0. It can be used for predictive maintenance, quality control, process optimization, and supply chain optimization. There is a need for specialized skills, and concerns about job displacement. Manufacturers will need to carefully consider these factors as they adopt ML and other advanced technologies in their operations [21–23].

15.3.1 Machine Learning Tools and Techniques and Tools for Industry 4.0

In today's fast-paced world, businesses are constantly seeking innovative solutions to stay ahead of the competition. One of the most effective tools in this pursuit is the use of advanced data analytics techniques. These techniques allow businesses to extract meaningful insights from their data, enabling them to make more informed decisions and gain a competitive edge. These insights can help businesses improve their operations, develop new products and services, and enhance customer experiences [24]. Advanced analytics techniques, such as predictive modeling and natural language processing, are becoming increasingly accessible to businesses of all sizes. This is thanks in part to the growing availability of cloud-based platforms and tools that enable businesses to implement these techniques without significant investment in infrastructure or technical expertise. As a result, businesses are leveraging advanced analytics to achieve a wide range of goals, from improving supply chain efficiency to enhancing marketing campaigns. The adoption of advanced analytics techniques is transforming the way businesses operate and compete. By harnessing the power of data, businesses can gain new insights into their operations, customers, and markets, and use these insights to drive growth and innovation. As data becomes an ever more valuable asset, the use of advanced analytics will only continue to increase, making it an essential tool for businesses looking to stay ahead in the next generation of industry [25–28].

15.3.2 Use Cases of Machine Learning in Industry 4.0 for Optimized Performance

ML algorithms are increasingly being used in manufacturing and production systems to enable predictive maintenance. These algorithms analyze sensor data from industrial equipment to identify patterns that indicate potential failures. By using predictive maintenance, manufacturers can schedule maintenance before a breakdown occurs, reducing downtime and extending the lifespan of equipment. This approach not only improves efficiency but also reduces costs associated with unplanned maintenance. In addition, it provides insights into areas that require further improvements or upgrades.

Another application of ML in the manufacturing industry is in quality control. By analyzing images or sensor data, ML algorithms can detect defects in products. This allows manufacturers to quickly identify and address quality issues, improving overall product quality and reducing waste. ML can also be used to detect anomalies in production processes that can lead to defects [29, 30].

ML algorithms can also be used to optimize supply chain operations. By analyzing data from suppliers, logistics providers, and customers, manufacturers can predict demand and identify inefficiencies. This helps in reducing lead times and inventory costs while improving customer satisfaction. The ability to predict demand and manage inventory more effectively enables manufacturers to improve the efficiency of their supply chain and reduce the risk of overproduction or stockouts. Another application of ML is in energy management. By analyzing data from sensors and meters, ML can optimize energy consumption in factories. The algorithms can predict energy demand and identify areas of waste, allowing manufacturers to reduce energy costs and improve sustainability. This approach not only saves costs but also enables manufacturers to become more environmentally responsible. Finally, ML can be used for predictive analytics. By analyzing large datasets from various sources, including sensors, social media, and customer feedback, ML algorithms can identify patterns and trends. This information enables manufacturers to make data-driven decisions that improve operational efficiency and customer satisfaction. Predictive analytics can be applied in a wide range of areas, including production planning, sales forecasting, and maintenance scheduling. In conclusion, ML algorithms have numerous applications in the manufacturing industry, including predictive maintenance, quality control, supply chain optimization, energy management, and predictive analytics. By using ML, manufacturers can improve efficiency, reduce costs, improve sustainability, and enhance customer satisfaction. The continued advancement of ML and DA is expected to drive further improvements in the performance and competitiveness of the manufacturing industry [31].

15.3.3 Challenges and Limitations of Machine Learning in Industry 4.0

One of the main challenges is the need for high-quality data to train ML models. I-4.0 generates vast amounts of data, but this data is often noisy, incomplete, or of poor quality. ML models are only as good as the data they are trained on, so ensuring that the data is clean and accurate is crucial.

Another challenge is the complexity of ML models. Some models can be difficult to interpret, making it hard to understand how they arrived at a particular decision. This can be a problem in safety-critical applications, where it is important to be able to explain why a particular decision was made.

For example, some models may require significant computing resources to run, which can be a problem in resource-constrained environments. Additionally, deploying ML models on edge devices can be challenging, as these devices may have limited processing power or connectivity.

Another challenge is the need to ensure that ML models are robust and resilient to attacks or adversarial examples. In some cases, it may be possible to fool an ML model by feeding it data that has been specifically crafted to cause it to make a mistake. Ensuring that models are robust and secure is important for safety-critical applications. Finally, there is the challenge of balancing the benefits of ML with the potential risks. ML models can be incredibly powerful, but they can also be used in ways that are unethical or harmful. Ensuring that ML is used ethically and responsibly is an important consideration in I-4.0 [32].

15.4 INTEGRATION OF DATA ANALYTICS AND MACHINE LEARNING FOR OPTIMIZED PERFORMANCE IN INDUSTRY 4.0

The integration of DA and ML techniques can lead to even greater optimization of performance in I-4.0. By combining the power of advanced data analysis with the ability of ML algorithms to make predictions and identify patterns, manufacturers can achieve even greater efficiencies and cost savings. One way in which DA and ML can be integrated is through the use of predictive maintenance. By combining DA with ML, manufacturers can use digital twins to identify inefficiencies and predict future performance, allowing for continuous optimization of production processes. Furthermore, DA can provide the input data necessary for ML algorithms to make accurate predictions. For example, data on customer behavior and preferences can be analyzed using DA tools, and the resulting insights can be used to train ML algorithms to make personalized product recommendations or optimize pricing strategies. One key challenge is ensuring the quality and reliability of the data being used. Poor quality or biased data can lead to inaccurate predictions and flawed decision-making. Another challenge is the need for skilled data scientists and engineers to develop and deploy effective DA and ML solutions. Despite these challenges, the integration of DA and ML is a promising area for optimizing performance in I-4.0. By combining these powerful technologies, manufacturers can achieve even greater efficiencies and cost savings, while also improving product quality and customer satisfaction [33, 34].

15.4.1 Explanation of the Benefits of Integrating Data Analytics and Machine Learning

By combining DA with ML, manufacturers can create more accurate predictive models, allowing them to predict equipment failures, maintenance needs, and production output with greater accuracy. This, in turn, can help manufacturers to plan and optimize production processes, reducing downtime and increasing productivity. The integration of DA and ML can also lead to better decision-making. ML algorithms can analyze vast amounts of data quickly and accurately, allowing decision-makers to identify patterns and insights that might otherwise go unnoticed. By using these insights to inform decisions, manufacturers can improve their operations, streamline processes, and identify opportunities for growth and innovation. By optimizing production processes and predicting equipment failures, manufacturers can reduce downtime and increase productivity. This, in turn, can lead to cost savings, as manufacturers are able to produce more output with the same resources. In addition to increased efficiency, integrating DA and ML can also improve product quality. By using these technologies to identify patterns and insights, manufacturers can gain a better understanding of their production processes and identify opportunities for improvement. This can lead to higher-quality products and fewer defects, ultimately improving customer satisfaction. Finally, the integration of DA and ML enables manufacturers to produce customized products on a mass scale. By using DA to gain insights into customer preferences and ML to optimize production processes, manufacturers can meet the unique needs of individual customers while still achieving economies of scale. This can help manufacturers to differentiate themselves from competitors and build stronger relationships with their customers [35–39].

15.4.2 Challenges and Limitations

While integrating DA and ML can bring many benefits to I-4.0, there are also several challenges to consider [40]:

1. *Data quality*: the accuracy and reliability of ML models depend heavily on the quality of data used to train them. If the data is incomplete, inconsistent, or biased, the resulting models may be inaccurate or unreliable.
2. *Data integration*: in I-4.0, data is generated from many different sources and in different formats. Integrating this data into a unified format can be a challenging task, requiring significant time and effort.

3. *Human expertise*: while ML algorithms can analyze large amounts of data quickly, they still require human expertise to interpret the results and make informed decisions based on them.
4. *Cost*: implementing and maintaining DA and ML systems can be costly, requiring investment in hardware, software, and skilled personnel.
5. *Ethical considerations*: as with any technology, there are ethical considerations to take into account. For example, the use of ML to automate decision-making may raise questions about accountability and transparency.

Integrated DA and ML offer many benefits to I-4.0; it is important to address these challenges and limitations to ensure that these technologies are implemented effectively and responsibly.

15.5 CONCLUSION

In conclusion, the integration of DA and ML has become an essential aspect of I-4.0 for optimized performance. Industries have been able to harness the power of real-time data collection and analysis to make informed decisions, optimize their operational efficiency, and improve the quality of their products. With the advent of advanced technologies such as the IoT, big DA, and AI, industries have been able to collect and analyze large amounts of data from their operations. By leveraging the insights obtained from this data, industries have been able to fine-tune their operations, reduce downtime, and enhance their overall efficiency [41–43]. One key area where the adoption of ML algorithms has been particularly impactful is predictive maintenance. By analyzing historical data and using predictive models, industries have been able to detect potential equipment failures before they occur and schedule maintenance proactively. This has resulted in a significant reduction in downtime, allowing industries to operate more efficiently and ultimately improve their bottom line. Another important application of ML algorithms in I-4.0 is demand forecasting. By leveraging historical data and using ML algorithms, industries have been able to forecast future demand for their products with a high degree of accuracy. This has enabled them to optimize their inventory management, reduce waste, and ensure that they are able to meet customer demands in a timely manner.

In addition to predictive maintenance and demand forecasting, ML algorithms have also been instrumental in detecting anomalies in industrial processes. By analyzing real-time data, industries have been able to quickly detect anomalies and take corrective action before they result in more significant issues. This has led to increased productivity, improved product quality, and cost savings for industrial manufacturers. Overall, the integration of real-time DA and ML algorithms has transformed the way industries approach their manufacturing processes. By embracing a data-driven approach, industries have been able to optimize their operations, reduce costs, and ultimately increase their competitiveness in the market. However, the implementation of these technologies requires significant investments in infrastructure, talent acquisition, and training, and there are also ethical and privacy concerns associated with the use of data. The continued advancement of these technologies is expected to drive further improvements in the performance and competitiveness of the industries.

REFERENCES

1. Pandey, D., Wairya, S., Sharma, M., Gupta, A. K., Kakkar, R.; Pandey, B. K. (2022). An approach for object tracking, categorization, and autopilot guidance for passive homing missiles. *Aerospace Systems*, 5(4), 553–566.
2. Pandey, D.; Wairya, S. (2022). An optimization of target classification tracking and mathematical modelling for control of autopilot. *The Imaging Science Journal*, 70(6), 371–386.
3. Kumar, M.S.; Sankar, S.; Nassa, V.K.; Pandey, D.; Pandey, B.K.; Enbeyle, W. Innovation and creativity for data mining using computational statistics. In *Methodologies and Applications of Computational Statistics for Machine Intelligence*. IGI Global, 2021, pp. 223–240.
4. Pandey, D.; Pandey, B.K.; Sindhwani, N.; Anand, R.; Nassa, V.K.; Dadheech, P. *An Interdisciplinary Approach in the Post-COVID-19 Pandemic Era*, 2022, pp. 1–290.
5. Singh, P.; Kaiwartya, O.; Sindhwani, N.; Jain, V.; Anand, R. (Eds.). *Networking Technologies in Smart Healthcare: Innovations and Analytical Approaches*. CRC Press, 2022.
6. Mourtzis, D.; Fotia, S.; Boli, N.; Vlachou, E. Modelling and quantification of Industry 4.0 manufacturing complexity based on information theory: A robotics case study. *Int. J. Prod. Res.* 2019, 57(22), 6908–6921.
7. Galin, R.; Meshcheryakov, R.; Kamesheva, S.; Samoshina, A. Cobots and the benefits of their implementation in intelligent manufacturing. *IOP Conf. Ser. Mater. Sci. Eng.* 2020, 862(3), 032075.
8. May, M.C.; Schmidt, S.; Kuhnle, A.; Stricker, N.; Lanza, G. Product generation module: Automated Production Planning for optimized workload and increased efficiency in Matrix Production Systems. *Procedia CIRP* 2020, 96, 45–50.
9. Lu, Y. Industry 4.0: A survey on technologies, applications and open research issues. *J. Ind. Inf. Integr.* 2017, 6, 1–10.
10. Miqueo, A.; Torralba, M.; Yagüe-Fabra, J.A. Lean manual assembly 4.0: A systematic review. *Appl. Sci.* 2020, 10(23), 8555.
11. Wuest, T.; Weimer, D.; Irgens, C.; Thoben, K.D. Machine learning in manufacturing: Advantages, challenges, and applications. *Prod. Manuf. Res.* 2016, 4(1), 23–45.
12. Rai, R.; Tiwari, M.K.; Ivanov, D.; Dolgui, A. Machine learning in manufacturing and Industry 4.0 applications. *Int. J. Prod. Res.* 2021, 59(16), 4773–4778.
13. Bertolini, M.; Mezzogori, D.; Neroni, M.; Zammori, F. Machine Learning for industrial applications: A comprehensive literature review. *Expert Syst. Appl.* 2021, 175, 114820.
14. Wang, J.; Ma, Y.; Zhang, L.; Gao, R.X.; Wu, D. Deep learning for smart manufacturing: Methods and applications. *J. Manuf. Syst.* 2018, 48, 144–156.
15. Dogan, A.; Birant, D. Machine learning and data mining in manufacturing. *Expert Syst. Appl.* 2021, 166, 114060.
16. Alshangiti, M.; Sapkota, H.; Murukannaiah, P.K.; Liu, X.; Yu, Q. Why is developing machine learning applications challenging? A study on stack overflow posts. In *Proceedings of the 2019 ACM/IEEE International Symposium on Empirical Software Engineering and Measurement (ESEM)*, Porto de Galinhas, Brazil, 19–20 September 2019, pp. 1–11.

17. Zeller, V.; Hocken, C.; Stich, V.; Acatech Industrie 4.0 maturity index—A multidimensional maturity model. In *Proceedings of the IFIP International Conference on Advances in Production Management Systems*, Seoul, Republic of Korea, 26–30 August 2018; Springer, 2018, pp. 105–113.
18. Yang, L.; Fan, J.; Huo, B.; Li, E.; Liu, Y. A nondestructive automatic defect detection method with pixelwise segmentation. *Knowl. Based Syst.* 2022, 242, 108338.
19. Wang, Y.; Gao, L.; Gao, Y.; Li, X. A new graph-based semi-supervised method for surface defect classification. *Robot. Comput. Integr. Manuf.* 2021, 68, 102083.
20. Kim, S.H.; Kim, C.Y.; Seol, D.H.; Choi, J.E.; Hong, S.J. Machine learning-based process-level fault detection and part-level fault classification in semiconductor etch equipment. *IEEE Trans. Semicond. Manuf.* 2022, 35(2), 174–185.
21. Peng, S.; Feng, Q.M. Reinforcement learning with Gaussian processes for condition-based maintenance. *Comput. Ind. Eng.* 2021, 158, 107321.
22. Zheng, W.; Liu, Y.; Gao, Z.; Yang, J. Just-in-time semi-supervised soft sensor for quality prediction in industrial rubber mixers. *Chemom. Intell. Lab. Syst.* 2018, 180, 36–41.
23. Kang, P.; Kim, D.; Cho, S. Semi-supervised support vector regression based on self-training with label uncertainty: An application to virtual metrology in semiconductor manufacturing. *Expert Syst. Appl.* 2016, 51, 85–106.
24. Srivastava, A.K.; Patra, P.K.; Jha, R. AHSS applications in Industry 4.0: Determination of optimum processing parameters during coiling process through unsupervised machine learning approach. *Mater. Today Commun.* 2022, 31, 103625.
25. Antomarioni, S.; Ciarapica, F.E.; Bevilacqua, M. Association rules and social network analysis for supporting failure mode effects and criticality analysis: Framework development and insights from an onshore platform. *Saf. Sci.* 2022, 150, 105711.
26. Pan, R.; Li, X.; Chakrabarty, K. Semi-supervised root-cause analysis with co-training for integrated systems. In *Proceedings of the 2022 IEEE 40th VLSI Test Symposium (VTS)*, San Diego, CA, 25–27 April 2022.
27. Chen, R.; Lu, Y.; Witherell, P.; Simpson, T.W.; Kumara, S.; Yang, H. Ontology-driven learning of bayesian network for causal inference and quality assurance in additive manufacturing. *IEEE Robot. Autom. Lett.* 2021, 6(3), 6032–6038.
28. Sikder, S.; Mukherjee, I.; Panja, S.C. A synergistic Mahalanobis–Taguchi system and support vector regression based predictive multivariate manufacturing process quality control approach. *J. Manuf. Syst.* 2020, 57, 323–337.
29. Cerquitelli, T.; Ventura, F.; Apiletti, D.; Baralis, E.; Macii, E.; Poncino, M. Enhancing manufacturing intelligence through an unsupervised data-driven methodology for cyclic industrial processes. *Expert Syst. Appl.* 2021, 182, 115269.
30. Kolokas, N.; Vafeiadis, T.; Ioannidis, D.; Tzovaras, D. A generic fault prognostics algorithm for manufacturing industries using unsupervised machine learning classifiers. *Simul. Modell. Pract. Theor.* 2020, 103, 102109.
31. Verstraete, D.; Droguett, E.; Modarres, M. A deep adversarial approach based on multisensor fusion for remaining useful life prognostics. In *Proceedings of the 29th European Safety and Reliability Conference (ESREL 2019)*, Hannover, Germany, 22–26 September 2020, pp. 1072–1077.
32. Wu, D.; Jennings, C.; Terpenny, J.; Gao, R.X.; Kumara, S. A comparative study on machine learning algorithms for smart manufacturing: Tool wear prediction using random forests. *J. Manuf. Sci. Eng. Trans. ASME* 2017, 139(7), 071018.
33. Viharos, Z.J.; Jakab, R. Reinforcement learning for statistical process control in manufacturing. *Meas. J. Int. Meas. Confed.* 2021, 182, 109616.
34. Luis, M.; Rodríguez, R.; Kubler, S.; Giorgio, A.D.; Cordy, M.; Robert, J.; Le, Y. Multi-agent deep reinforcement learning based predictive maintenance on parallel machines. *Robot. Comput. Integr. Manuf.* 2022, 78, 102406.

35. Paraschos, P.D.; Koulinas, G.K.; Koulouriotis, D.E. Reinforcement learning for combined production-maintenance and quality control of a manufacturing system with deterioration failures. *J. Manuf. Syst.* 2020, 56, 470–483.
36. Liu, Y.H.; Huang, H.P.; Lin, Y.S. Dynamic scheduling of flexible manufacturing system using support vector machines. In *Proceedings of the 2005 IEEE Conference on Automation Science and Engineering, IEEE-CASE 2005*, Edmonton, AB, 1–2 August 2005, 2005, 387–392.
37. Zhou, G.; Chen, Z.; Zhang, C.; Chang, F. An adaptive ensemble deep forest based dynamic scheduling strategy for low carbon flexible job shop under recessive disturbance. *J. Clean. Prod.* 2022, 337, 130541.
38. de la Rosa, F.L.; Gómez-Sirvent, J.L.; Sánchez-Reolid, R.; Morales, R.; Fernández-Caballero, A. Geometric transformation-based data augmentation on defect classification of segmented images of semiconductor materials using a ResNet50 convolutional neural network. *Expert Syst. Appl.* 2022, 206, 117731.
39. Chaudhary, A.; Bodala, D.; Sindhwani, N.; Kumar, A. Analysis of customer loyalty using artificial neural networks. In *2022 International Mobile and Embedded Technology Conference (MECON)*. IEEE, 2022, March, pp. 181–183.
40. Krahe, C.; Marinov, M.; Schmutz, T.; Hermann, Y.; Bonny, M.; May, M.; Lanza, G. AI based geometric similarity search supporting component reuse in engineering design. *Procedia CIRP* 2022, 109, 275–280.
41. Anand, R.; Sindhwani, N.; Juneja, S. Cognitive Internet of things, its applications, and its challenges: A survey. In *Harnessing the Internet of Things (IoT) for a Hyper-Connected Smart World*. Apple Academic Press, 2022, pp. 91–113.
42. Anand, R.; Arora, S.; Sindhwani, N. A miniaturized UWB antenna for high speed applications. In *2022 International Conference on Computing, Communication and Power Technology (IC3P)*. IEEE, 2022, January, pp. 264–267.
43. Anand, R.; Ahamad, S.; Veeraiah, V.; Janardan, S.K.; Dhabliya, D.; Sindhwani, N.; Gupta, A. Optimizing 6G wireless network security for effective communication. In *Innovative Smart Materials Used in Wireless Communication Technology*. IGI Global, 2023, pp. 1–20.

16 Business Intelligence in Action
Way of Successful Implementation of Automated Systems

Venkata Naga Siva Kumar Challa,
K.K. Ramachandran, P.M. Rameshkumar, Luigi Pio
Leonardo Cavaliere, Purnendu Bikash Acharjee,
and Venkata Harshavardhan Reddy Dornadula

16.1 INTRODUCTION

Business Intelligence (BI) is a vital aspect of modern business operations. BI solutions have been available for many years, but the advent of automated BI systems has revolutionized the field. Automated BI systems provide businesses with real-time, enabling them to make data-driven decisions with confidence. In this review chapter, we will explore the successful implementation of automated BI systems through a series of case studies. We will examine the benefits of these systems, the challenges that businesses face when implementing them, and the strategies that successful organizations have used to overcome these challenges. Our aim is to provide insights and recommendations for businesses that are considering implementing automated BI systems. Automated BI systems utilize AI and ML algorithms to collect, process, and analyze data. These systems can analyze large datasets in real-time and generate reports, dashboards, and alerts. Automated BI systems provide businesses with the ability to access critical data and insights quickly and easily, enabling them to make informed decisions in real-time. One of the main advantages of automated BI systems is their ability to reduce the workload on employees. Traditionally, businesses had to rely on their employees to gather, process, and analyze data. This was a time-consuming and error-prone process. Automated BI systems eliminate the need for manual data collection and analysis, enabling employees to focus on higher-value tasks such as decision-making and strategy development. These systems can analyze vast amounts of data from multiple sources, providing businesses with a holistic view of their operations. Automated BI systems can also identify anomalies and

Business Intelligence in Action

outliers, alerting businesses to potential problems before they become major issues. Despite the benefits of automated BI systems, implementing these systems can be challenging. Businesses must have the technical expertise and infrastructure necessary to support these systems. They must also ensure that the data being analyzed is accurate, complete, and up-to-date. In addition, businesses must ensure that the insights generated by the system are communicated effectively to decision-makers. In conclusion, this review chapter will provide insights into the successful implementation of automated BI systems through a series of case studies. We will examine the benefits of these systems, the challenges that businesses face when implementing them, and the strategies that successful organizations have used to overcome these challenges. Our aim is to provide insights and recommendations for businesses that are considering implementing automated BI systems [1].

16.1.1 Importance of Business Intelligence

Business Intelligence (BI) refers to the process of using data analysis tools and techniques to gain insights and make informed decisions. BI solutions involve collecting, integrating, and analyzing data from various sources, including databases, spreadsheets, and enterprise systems. The process of BI typically involves several stages. First, data is collected from various sources and integrated into a central repository. This process involves ensuring data quality and consistency to ensure accurate analysis. Once the data is integrated, it is analyzed using a variety of techniques, including statistical analysis, data mining, and predictive analytics. These techniques can reveal patterns, trends, and insights that are not immediately apparent from the raw data. The insights gained through BI analysis are then presented in a user-friendly format, such as dashboards, reports, or visualizations. These formats allow decision-makers to quickly and easily understand the data and make informed decisions based on the insights gained. BI solutions are used across industries to support a variety of business functions, including finance, marketing, and operations. By using data to drive decision-making, BI solutions can help organizations optimize their performance [2, 3], identify new opportunities, and gain a competitive edge in their respective markets. The data is then transformed into information that can be used for decision-making purposes. BI is essential for businesses because it provides them with a competitive advantage. By using data to make informed decisions, businesses can optimize their operations, identify new opportunities, and mitigate risks. BI enables businesses to gain a holistic view of their operations, which is critical in today's complex business environment. It allows businesses to identify trends and patterns, spot anomalies and outliers, and detect potential problems before they become major issues. BI solutions can be used in various areas of business, including finance, marketing, human resources, supply chain management, and customer relationship management. BI enables businesses to track key performance indicators and metrics, monitor progress, and make adjustments as necessary. By using BI, businesses can identify opportunities to increase efficiency, reduce costs, and improve customer satisfaction. In summary, BI is essential for businesses because it provides them with the insights they need to make informed decisions. By analyzing data, businesses can optimize their operations, identify new opportunities, and mitigate risks. BI allows businesses to gain a holistic view of their operations, which is critical in today's complex business environment [4].

16.1.2 Overview

Automated systems in BI refer to the use of artificial intelligence and machine learning algorithms to automate the data analysis process. These systems are designed to collect, process, and analyze data in real-time, providing businesses with accurate and actionable insights. Automated systems in BI can analyze large amounts of data from various sources, including databases, spreadsheets, and enterprise systems. The systems can generate reports, dashboards, and alerts, which can be customized to meet the specific needs of businesses. These systems can identify patterns, trends, and anomalies in data that might otherwise go unnoticed, enabling businesses to make informed decisions quickly. Automated systems in BI can be used in various areas of business, including finance, marketing, supply chain management, and customer relationship management. The benefits of automated systems in BI include increased efficiency, improved accuracy, and reduced workload on employees. These systems eliminate the need for manual data collection and analysis, enabling employees to focus on higher-value tasks such as decision-making and strategy development. In summary, automated systems in BI use AI and ML algorithms to automate the data analysis process. These systems can analyze large amounts of data in real-time, provide accurate and actionable insights, and be customized to meet the specific needs of businesses. The benefits of these systems include increased efficiency, improved accuracy, and reduced workload on employees [5–9]. Figure 16.1 shows the BI features.

16.2 AUTOMATED BI SYSTEMS

These systems are designed to collect and process large amounts of data from various sources in real-time. Automated BI systems typically work by using algorithms to analyze data, identifying patterns, trends, and anomalies that may not be immediately apparent to humans. The system can then generate reports, dashboards, and alerts based on this analysis, providing customized insights that are relevant to specific business needs. Automated BI systems can be used in a wide range of business applications, including finance, marketing, supply chain management, and customer relationship management. For example, an automated BI system could be used to track customer behavior, monitor inventory levels, analyze financial data, and forecast sales. The benefits of automated BI systems include increased efficiency, improved accuracy, and reduced workload on employees. By automating the process of data collection and analysis, these systems eliminate the need for manual data entry and processing, freeing up employees to focus on higher-value tasks such as decision-making and strategy development. Automated BI systems can also reduce the risk of errors and inconsistencies in data analysis, improving the accuracy and reliability of the insights generated. By leveraging advanced technologies such as AI,

FIGURE 16.1 Business Intelligence features.

ML, and natural language processing (NLP), these systems can automate the process of data analysis and provide customized insights that can help businesses optimize their operations, identify new opportunities, and stay ahead of the competition [10].

16.2.1 BENEFITS OF AUTOMATED BI SYSTEMS

The various advantages of the automated BI systems are [11–13]:

- Automated BI systems offer several advantages for businesses. One of the most significant benefits is increased efficiency. These systems can process large volumes of data quickly and accurately, eliminating the need for manual data entry and analysis. This can save time and resources, enabling employees to focus on higher-value tasks. With automated BI systems in place, businesses can analyze data more efficiently and generate insights faster, which can be particularly beneficial for larger organizations with vast amounts of data to analyze.
- Another advantage of automated BI systems is improved accuracy. Businesses can benefit from using algorithms to analyze data, as it minimizes the risk of human error and ensures that insights are based on accurate and reliable data. By incorporating BI solutions into their decision-making processes, businesses can gain a competitive advantage and avoid making costly mistakes resulting from inaccurate data analysis. By

leveraging these insights, businesses can make better-informed decisions that improve their operations, enhance their customer experience, and drive growth. The implementation of BI solutions requires careful planning and management, including the selection of appropriate tools and the training of staff to use the systems effectively. Overall, BI solutions have become an essential component of modern business operations, enabling organizations to make data-driven decisions and stay ahead of the competition.

- Automated BI systems also offer customized insights based on specific business needs and requirements. These systems can generate customized reports and insights that are tailored to the unique needs of the business, providing more targeted and relevant information than traditional BI systems. This can be particularly valuable for businesses operating in highly competitive industries can make all the difference.
- In addition to providing customized insights, automated BI systems can also provide real-time insights into business operations. By analyzing data in real-time, these systems can help businesses respond quickly to changing circumstances and make informed decisions faster. This can be particularly beneficial for businesses operating in fast-paced industries where timely decision-making is critical to success.
- Despite the many advantages of automated BI systems, there are also some potential disadvantages to consider. These include challenges related to technical expertise, integration, data quality, security, and cost. It's important for businesses to carefully weigh the pros and cons of automated BI systems before making a decision about whether or not to adopt them.

16.2.2 Limitations

While automated BI systems offer a number of advantages, there are also several limitations that should be considered [14–16].

- One major limitation of automated BI systems is the potential for errors in data analysis. While automated systems can reduce the risk of human error, they may still be subject to biases or errors in programming. If the algorithms used by the system are flawed or the data input into the system is incomplete or inaccurate, the insights generated by the system may be flawed as well. Therefore, it is important to ensure that the data being used is accurate and that the algorithms are properly designed and maintained.
- Another limitation of automated BI systems is that they may not be able to capture all the relevant data or account for all the factors that could impact business performance. Automated systems rely on data that is structured and stored in a certain way, which may not capture all of the nuances or context surrounding business operations. Additionally, automated systems may not be able to take into account unstructured data sources, such as social media or customer feedback, which can be valuable sources of information for business decision-making.

- Automated BI systems may also be limited by the types of insights they are able to generate. While these systems are adept at analyzing data and identifying trends, they may not be able to provide the same level of nuanced analysis that a human analyst could. Human analysts are often better able to interpret data in the context of broader business goals and strategies, and may be better equipped to identify opportunities or challenges that may not be immediately apparent from the data alone.
- Another limitation of automated BI systems is the potential for security breaches or data privacy issues. Automated systems rely on large amounts of data, much of which may be sensitive or confidential. If the system is not properly secured, it may be vulnerable to cyber-attacks or data breaches, which could have serious consequences for the business. Additionally, there may be concerns around data privacy, particularly if the system is collecting data on customers or employees.
- Finally, there is the potential for over-reliance on automated BI systems, which could lead to a lack of critical thinking or analysis by human decision-makers. While automated systems can provide valuable insights and help inform business decisions, they should not be seen as a replacement for human analysis and decision-making. Human judgment and intuition are still valuable assets in the decision-making process, and should be used in conjunction with automated systems to ensure the best possible outcomes.

Automated BI systems offer a number of advantages, they are not without their limitations. By leveraging the strengths of both automated and human analysis, businesses can make more informed and effective decisions that drive success and growth.

16.2.3 Challenges

Automated BI systems can offer significant benefits to organizations in terms of efficiency, accuracy, and real-time insights. However, there are also several challenges that organizations may face when implementing and using these systems. Here are some of the key challenges of automated BI systems as shown in Figure 16.2.

- *Data Integration*: one of the main challenges of automated BI systems is integrating data from different sources. Organizations often have data stored in various systems, databases, and applications, and integrating this data into a single system can be complex and time-consuming. Data integration is a critical step in automated BI, and any issues with data quality or consistency can lead to inaccurate insights.
- *Data Quality*: another challenge with automated BI systems is ensuring the quality of data. Automated BI systems rely on accurate and up-to-date data to generate insights, and any errors or inconsistencies in the data can lead to incorrect conclusions. Organizations need to have effective data governance practices in place to ensure the quality and consistency of their data.

FIGURE 16.2 Challenges for automated BI systems.

- *Technical Expertise*: automated BI systems require technical expertise to implement and maintain. Organizations may need to hire additional IT staff or work with external consultants to ensure that the system is set up correctly and functioning properly. This can be a barrier to adoption for smaller organizations with limited resources.
- *Security*: automated BI systems can pose security risks if they are not properly secured or if sensitive data is accessed by unauthorized parties. Organizations need to ensure that their systems are secure and that access to data is restricted to authorized users.
- *Cost*: implementing and maintaining an automated BI system can be costly, particularly for smaller organizations with limited budgets. Organizations need to consider the cost of hardware, software, and personnel when evaluating the potential benefits of automated BI.
- *Change Management*: introducing a new automated BI system can be a significant change for an organization. Employees may need to learn new skills and workflows, and there may be resistance to change. Organizations need to have effective change management processes in place to ensure that employees are prepared for the transition and that the system is adopted successfully.
- *Scalability*: as organizations grow, their data volumes and processing requirements may increase. Automated BI systems need to be scalable

to accommodate these changes and ensure that insights can be generated quickly and accurately.

Automated BI systems offer many benefits to organizations, but they also present several challenges. By addressing these challenges, organizations can unlock the full potential of automated BI and gain valuable insights into their operations and performance [17–19].

16.2.4 Identification of Common Success Factors

Successful implementation of automated BI systems can be attributed to a combination of various factors. However, there are certain common success factors that have been identified in multiple case studies of businesses that have successfully implemented automated BI systems. One common success factor is a clear understanding of business needs and goals. Successful businesses typically have a clear understanding of the specific business needs and goals that they want to achieve with automated BI systems. This understanding helps them to identify the specific features and functionalities that are required in the system to achieve their goals. Another common success factor is a well-defined data strategy. A well-defined data strategy involves identifying the sources of data that will be used in the system and the types of data that will be analyzed. A well-defined data strategy helps to ensure that the automated BI system is designed to provide relevant and actionable insights that support business objectives. Effective collaboration between IT and business teams is another important success factor. The IT team is responsible for implementing and maintaining the automated BI system, while the business team is responsible for identifying business needs and defining data requirements.

Effective collaboration between these two teams is essential to ensure that the automated BI system meets the specific needs of the business and provides actionable insights that support business objectives. Providing ongoing support helps to ensure that the system remains effective and relevant over time. Another important success factor is a focus on data quality. Automated BI systems rely on high-quality data to generate accurate insights. Ensuring that the data used in the system is accurate, complete, and up-to-date is essential to ensure that the insights generated by the system are reliable. Finally, a culture of data-driven decision-making is a key success factor. Encouraging a culture of data-driven decision-making involves promoting the use of data to inform decision-making at all levels of the organization. This helps to ensure that insights generated by the automated BI system are used to inform business decisions and drive business performance. In conclusion, the successful implementation of automated BI systems can be attributed to a combination of various factors. There are certain common success factors that have been identified in multiple case studies of businesses that have successfully implemented automated BI systems. These common success factors include a clear understanding of business needs and goals, a well-defined data strategy, effective collaboration between IT and business teams, adequate training and support, a focus on data quality, and a culture of data-driven decision-making. Businesses that focus on these success factors are more likely to achieve their goals and realize the full benefits of automated BI systems [20, 21].

16.3 TYPES OF AUTOMATED SYSTEMS IN BI

Automated systems in business intelligence (BI) can be categorized based on the specific functions they perform within an organization. Here are some of the most common types of automated systems used in BI [22, 23]:

1. *Reporting Systems*: these automated systems generate reports that provide information about key performance indicators (KPIs) and other metrics relevant to an organization's business objectives. Reporting systems allow organizations to track progress towards goals and make data-driven decisions based on the insights provided by the reports.
2. *Dashboards:* dashboards can be customized to show data relevant to specific roles within an organization, providing each user with the information they need to make informed decisions.
3. *Predictive Analytics Systems*: these automated systems use ML algorithms to identify patterns and predict future outcomes based on historical data. Predictive analytics systems are used to forecast trends, identify potential risks, and inform strategic decision-making.
4. *Process Automation Systems*: these automated systems streamline business processes by automating repetitive tasks such as data entry, report generation, and workflow management.
5. *Data Visualization Systems*: these automated systems use graphical representations to convey complex data in a simple and intuitive way. Data visualization systems are designed to present data in a way that is easy to understand and visually appealing. These systems use graphs, charts, and other visual aids to help users explore and interact with data, making it easier to identify trends, patterns, and relationships. By presenting data in a visual format, these systems can help users quickly understand complex datasets and draw insights that might not be apparent from raw data alone.

Business Intelligence systems integrate various automated processes to provide a comprehensive overview of an organization's performance, enabling data-driven decision-making across all levels of the organization. BI solutions typically involve the automated collection, integration, analysis, and reporting of data from various sources such as databases, spreadsheets, and enterprise systems.

The automation of these processes reduces the risk of human error, ensuring that insights are based on accurate and reliable data. BI systems also provide data visualization capabilities, allowing users to explore and interact with data, making it easier to identify trends and patterns. These systems use various data visualization techniques, such as charts, graphs, and dashboards, to present data in a format that is easy to understand. The integration of automated systems in BI can also help organizations achieve their strategic goals

Business Intelligence in Action

by providing them with accurate, timely, and relevant information. By automating data collection, analysis, and reporting processes, organizations can gain valuable insights into their operations, leading to informed decision-making. This can ultimately give organizations a competitive edge in the market. In summary, the integration of various automated systems in BI can provide organizations with a holistic view of their performance, enabling data-driven decision-making across all levels of the organization.

16.3.1 Predictive Analytics

This is a technique that allows organizations to use past data to predict future events and behaviors. It involves the application of statistical algorithms, data mining, and machine learning techniques to large datasets to uncover patterns, relationships, and hidden insights that may be useful in predicting future outcomes. The data is then preprocessed to ensure that it is clean, complete, and ready for analysis. By using predictive models to identify potential risks and predict the likelihood of events such as fraud, defaults, and other types of losses, businesses can take proactive steps to mitigate these risks and minimize their impact. This can help businesses to protect their assets, maintain regulatory compliance, and maintain the trust of their stakeholders. Another important application of predictive analytics is in customer behavior analysis. By analyzing customer data and identifying patterns in customer behavior, businesses can gain insights into customer preferences, buying patterns, and other factors that influence customer behavior. This can help businesses to optimize their marketing campaigns, improve customer engagement, and increase customer retention rates. Predictive analytics can also be used to optimize business operations by identifying areas of inefficiency and areas where improvements can be made. For example, predictive analytics can be used to identify bottlenecks in supply chain operations, forecast demand for products, and optimize inventory management. One of the key benefits is its ability to provide real-time insights into business operations. By using predictive models to analyze data in real-time, businesses can respond quickly to changing circumstances and make informed decisions faster. This can help businesses to stay ahead of their competition, taking advantage of emerging trends. As businesses continue to generate and collect more data, the importance of predictive analytics in driving business success is only set to increase [24].

16.3.2 Machine Learning

Machine learning algorithms are a key component of predictive analytics, allowing businesses to analyze large sets of historical data and identify patterns and correlations between variables. Through this analysis, businesses can make accurate predictions about future events and behaviors, which can inform decision-making and improve business outcomes. By leveraging the power of machine learning algorithms, businesses can gain insights into customer behavior, market trends, and operational inefficiencies, among other things. This can help them identify potential risks and opportunities, optimize their operations, and gain a competitive edge in the market. Overall, the use of machine learning algorithms in predictive analytics has become increasingly popular across a variety of industries, driving innovation and growth in businesses around the world. This can help businesses to optimize their operations, improve customer engagement, and capitalize on emerging trends. By analyzing data in real-time and adjusting their algorithms accordingly, machine learning systems can respond quickly to changing circumstances and make accurate predictions in real-time. This can help businesses to stay ahead of their competition, capitalize on emerging trends, and make data-driven decisions faster. By using algorithms that can recognize patterns and make predictions based on data, machine learning enables businesses to optimize their operations, improve customer engagement, and capitalize on emerging trends. As businesses continue to generate and collect more data, the importance of machine learning in driving business success is only set to increase [25].

16.3.3 Natural Language Processing

Natural Language Processing (NLP) is a subfield of artificial intelligence that focuses on enabling machines to understand and interpret human language. NLP involves using computer algorithms and statistical models to analyze, process, and generate natural language text and speech. One of the most common applications of NLP is in the development of chatbots and virtual assistants. By analyzing the language used by users, NLP algorithms can understand the intent of the user and provide relevant responses or take specific actions. This has become increasingly important in customer service, where chatbots can provide instant support and resolve issues without the need for human intervention. By analyzing large volumes of text, NLP algorithms can identify patterns and determine the overall sentiment of the text. This can be useful for businesses looking to understand customer feedback or sentiment towards their products or services [26–29].

NLP can also be used for text mining and information extraction. By analyzing large volumes of text data, NLP algorithms can identify key phrases, entities, and relationships between different pieces of information. This can be useful for businesses looking to extract insights from unstructured data sources such as social media or customer feedback forms. Words and phrases can have multiple meanings and interpretations, making it difficult for NLP algorithms to accurately understand the intent of the user. Additionally, NLP algorithms can struggle with understanding language that contains sarcasm, irony, or other forms of figurative language. Another challenge is the need for large volumes of high-quality training data. Despite these challenges, NLP has become an increasingly important tool in a wide range of applications, from virtual assistants to social media analytics. As advances in NLP continue, it is likely that we will see even more innovative applications of this technology in the future [30, 31].

16.3.4 Robotics Process Automation

Robotics Process Automation (RPA) is a type of automated system in business intelligence that involves using software robots to automate repetitive, rule-based tasks.

Business Intelligence in Action

RPA is designed to mimic human actions and interactions with software systems, enabling it to perform tasks such as data entry, report generation, and customer service interactions. RPA has several benefits for organizations, including increased efficiency, cost savings, and improved accuracy. By automating routine tasks, RPA frees up employees to focus on higher-value activities that require human expertise, such as decision-making and problem-solving. RPA can also reduce errors associated with manual data entry and improve overall accuracy. One key advantage of RPA is its ability to integrate with existing systems and processes. RPA can be used to automate tasks across a range of business functions, including finance, human resources, and customer service. By integrating with existing systems, RPA can streamline processes and improve overall efficiency. However, there are also some challenges associated with implementing RPA. One key challenge is the need for technical expertise to design and implement RPA systems. Organizations must have staff with the necessary skills and knowledge to develop and maintain RPA systems, which can be a barrier for smaller organizations. Another challenge is the need to ensure that RPA systems are secure and compliant with data protection regulations. RPA involves the handling of sensitive data, and organizations must take steps to ensure that data is protected from unauthorized access or misuse. Finally, there is the potential for job displacement as a result of RPA implementation. While RPA can help to improve efficiency and reduce costs, it may also lead to the displacement of workers who perform tasks that are automated by RPA. Organizations must carefully consider the impact of RPA on their workforce and take steps to mitigate any negative effects. In conclusion, RPA is a powerful tool for organizations looking to automate repetitive, rule-based tasks. While it has many benefits, it also presents challenges that organizations must carefully consider before implementing RPA systems. By understanding the potential benefits and challenges of RPA, organizations can make informed decisions about whether to adopt this technology and how to effectively implement it [32, 33].

16.4 CONCLUSION

In conclusion, this chapter has provided an overview of automated systems in Business Intelligence (BI) and examined their benefits, limitations, and challenges. The chapter has also explored different types of automated systems, including predictive analytics, machine learning, natural language processing, and robotics process automation. Automated BI systems offer several advantages over traditional manual systems, such as increased efficiency, accuracy, and customized insights. However, these systems also have several limitations, including technical expertise requirements and integration issues. The challenges associated with the implementation of automated BI systems can be addressed through careful planning, collaboration, and ongoing monitoring. To ensure the successful implementation of automated BI systems, businesses must identify and leverage common success factors, such as setting clear goals and objectives, effective change management strategies, collaboration between IT and business teams, and ongoing monitoring and improvement. The use of automated BI systems has become increasingly popular in modern businesses due to the benefits they offer in decision-making processes. The different types of automated systems can help organizations analyze large amounts of data more quickly and accurately [34, 35]. Nevertheless, businesses should carefully evaluate the potential benefits and limitations of these systems before adopting them and develop a comprehensive plan to ensure successful implementation. In summary, this chapter provides valuable insights for businesses looking to implement automated BI systems and emphasizes the importance of careful planning, collaboration, and ongoing monitoring for successful implementation.

REFERENCES

1. Maune, A. Competitive intelligence as a game changer for Africa's competitiveness in the global economy. *J. Intell. Stud. Bus.* 2019, 9(3), 24–38.
2. Anand, R.; Chawla, P. Bandwidth optimization of a novel slotted fractal antenna using modified lightning attachment procedure optimization. In *Smart Antennas: Latest Trends in Design and Application*. Springer International Publishing, 2022, pp. 379–392.
3. Chibber, A.; Anand, R.; Arora, S. A staircase microstrip patch antenna for UWB applications. In *2021 9th International Conference on Reliability, Infocom Technologies and Optimization (Trends and Future Directions)(ICRITO)*. IEEE, 2021, September, pp. 1–5.
4. Jalil, N.A.; Prapinit, P.; Melan, M.; Mustaffa, A.B. Adoption of business intelligence-technological, individual and supply chain efficiency. In *Proceedings of the International Conference on Machine Learning, Big Data and Business Intelligence (MLBDBI)*, Taiyuan, China, 15–17 October 2021, pp. 1245–1267.
5. Dadkhah, M.; Lagzian, M.; Rahimnia, F.; Kimiafar, K. The potential of business intelligence tools for expert finding. *J. Intell. Stud. Bus.* 2019, 9, 82–95.
6. Pandey, D.; Wairya, S.. An optimization of target classification tracking and mathematical modelling for control of autopilot. *The Imaging Science Journal* 2022, 70(6), 371-386.
7. Kumar, M.S.; Sankar, S.; Nassa, V.K.; Pandey, D.; Pandey, B.K.; Enbeyle, W. Innovation and creativity for data mining using computational statistics. In *Methodologies and Applications of Computational Statistics for Machine Intelligence*. IGI Global, 2021, pp. 223–240.
8. Pandey, D.; Pandey, B.K.; Wariya, S. An approach to text extraction from complex degraded scene. *IJCBS* 2020, 1(2), 4–10.
9. Pandey, D.; Pandey, B.K. An efficient deep neural network with adaptive galactic swarm optimization for complex image text extraction. In *Process Mining Techniques for Pattern Recognition*. CRC Press, 2022, pp. 121–137.
10. Soni, N.; Sharma, E.K.; Singh, N.; Kapoor, A. Artificial Intelligence in business: From research and innovation to market deployment. In *Proceedings of the International Conference on Computational Intelligence and Data Science (ICCIDS)*, Taiyuan, China, 8–10 November 2019, pp. 1834–1865.
11. Martynov, V.V.; Shavaleeva, D.N.; Zaytseva, A.A. Information technology as the basis for transformation into a digital society and Industry 5.0. In *Proceedings of the International Conference Quality Management, Transport and Information Security, Information Technologies (IT&QM&IS)*, Sochi, Russia, 23–27 September 2019, pp. 1654–1663.
12. Afandi, M.I.; Wahyuni, E.D. Mobile Business Intelligence assistant (m-BELA) for higher education executives. In *Proceedings of the 4th International Conference on*

Information Technology, Information Systems and Electrical Engineering (ICITISEE). Yogyakarta, Indonesia, 20 November 2019, pp. 1345–1376.
13. Murmura, F.; Bravi, L.; Santos, G. Sustainable process and product innovation in the eyewear sector: The role of Industry 4.0 enabling technologies. *Sustainability* 2021, 13(1), 365–375.
14. Vrchota, J.; Řehoř, P.; Maříková, M.; Pech, M. Critical success factors of the project management in relation to Industry 4.0 for sustainability of projects. *Sustainability* 2021, 13(1), 281–301.
15. Kłobukowski, P.; Pasieczny, J. Impact of resources on the development of local entrepreneurship in Industry 4.0. *Sustainability* 2020, 12(24), 10272.
16. Szabo, R.Z.; Vuksanović Herceg, I.; Hanák, R.; Hortovanyi, L.; Romanová, A.; Mocan, M.; Djuričin, D. Industry 4.0 implementation in B2B companies: Cross-country empirical evidence on digital transformation in the CEE region. *Sustainability* 2020, 12(22), 9538.
17. Michna, A.; Kmieciak, R. Open-mindedness culture, knowledge-sharing, financial performance, and Industry 4.0 in SMEs. *Sustainability* 2020, 12(21), 9041.
18. Kayes, A.S.M.; Kalaria, R.; Sarker, I.H.; Islam, M.S.; Watters, P.A.; Ng, A.; Hammoudeh, M.; Badsha, S.; Kumara, I. A survey of context-aware access control mechanisms for cloud and fog networks: Taxonomy and open research issues. *Sensors (Basel)* 2020, 20(9), 2464.
19. Reyes, P.M.; Visich, J.K.; Jaska, P. Managing the dynamics of new technologies in the global supply chain. *IEEE Eng. Manag. Rev.* 2019, 8(1), 156–162.
20. Luberecki, B. Going after the government. In *Proceedings of the Conference Business Intelligence*, Houston, TX, 1–2 March 2019, pp. 91–92.
21. Gebhardt, G.F.; Farrelly, F.J.; Conduit, J. Market intelligence dissemination practices. *J. Mark.* 2019, 83(3), 72–90.
22. Gounder, M.S.; Iyer, V.V.; Mazyad, A.A. A survey on Business Intelligence tools for university dashboard development. In *Proceedings of the 3rd MEC International Conference on Big Data and Smart City*, Muscat, Oman, 15–16 March 2016, pp. 1452–1509.
23. Priyadarshni, D. Intelligence software mining with Business Intelligence tool for automation of micro services in SOA: A use case for analytics. In *Proceedings of the 7th International Conference on Computing for Sustainable Global Development (INDIACom)*, New Delhi, India, 12–14 March 2020, pp. 1287–1325.
24. Cao, J.; Wang, S.; Wang, B.; Ding, Z.; Wang, F.W. Integrating multisource texts in online Business Intelligence systems. *IEEE Trans. Syst. Man Cybern. Syst.* 2018, 50, 1638–1648.
25. Zhao, L. Business Intelligence application of enhanced learning in Big Data scenario. In *Proceedings of the International Joint Conference on Information, Media and Engineering (IJCIME)*, Osaka, Japan, 19 December 2019, pp. 875–907.
26. Passlick, J.; Guhr, N.; Lebek, B.; Breitner, M.H. Encouraging the use of self-service business intelligence—An examination of employee-related influencing factors. *J. Decis. Syst.* 2020, 29(1), 1–26.
27. Shabib-Ahmed, S.; Tarun, S. Study on the various intellectual property management strategies used and implemented by ICT firms for business intelligence. *J. Intell. Stud. Bus.* 2019, 9, 30–42.
28. Singh, P.; Sindhwani, N.; Tiwari, S.; Jangra, V. Automatic candidature selection by artificial natural language processing. In *Mobile Radio Communications and 5G Networks: Proceedings of Third MRCN 2022*. Singapore: Springer Nature Singapore, 2023, pp. 471–482.

29. Kaura, C.; Sindhwani, N.; Chaudhary, A. Analysing the impact of cyber-threat to ICS and SCADA systems. In *2022 International Mobile and Embedded Technology Conference (MECON)*. IEEE, 2022, March, pp. 466–470.
30. Capinzaiki, S.L.; Pomim, M.L.; Mosconi, E. A competitive intelligence model based on information literacy: Organizational competitiveness in the context of the 4th Industrial Revolution. *J. Intell. Stud. Bus.* 2018, 8, 55–65.
31. Bordeleau, F.E.; Mosconi, E.; De Santa-Eulalia, L.A. Business intelligence and analytics value creation in Industry 4.0: A multiple case study in manufacturing medium enterprises. *Prod. Plan. Control* 2019, 31(2–3), 173–185.
32. Elordi, G.; De la Calle, A.; Gil, M.J.; Errasti, A.; Uradnicek, J. Aplicación de un sistema Business Intelligence en un contexto Big Data de una empresa industrial alimentaria. *Ind. Inf. Conoc.* 2017, 92, 347–353.
33. Herli, M.; Tjahjadi, B.; Hafidhah, L. Effectiveness of the Business Intelligence system in the manufacturing decision-making process: The case in fertilizer companies. *Talent Dev. Excell.* 2020, 12, 1814–1820.
34. Anand, R.; Sindhwani, N.; Dahiya, A. Design of a high directivity slotted fractal antenna for C-band, X-band and Ku-band applications. In *9th International Conference on Computing for Sustainable Global Development (INDIACom)*. IEEE, 2022, March, pp. 727–730.
35. Gupta, A.; Anand, R.; Pandey, D.; Sindhwani, N.; Wairya, S.; Pandey, B.K.; Sharma, M. Prediction of breast cancer using extremely randomized clustering forests (ERCF) technique: Prediction of breast cancer. *Int. J. Distrib. Syst. Technol. (IJDST)* 2021, 12(4), 1–15.

17 A Review of Dielectric Resonator Antennas (DRA)-Based RFID Technology for Industry 4.0

Manvinder Sharma, Rajneesh Talwar,
Digvijay Pandey, Vinay Kumar Nassa,
Binay Kumar Pandey, and Pankaj Dadheech

17.1 INTRODUCTION

The fourth-generation industrial revolution, often known as Industry 4.0 or I4.0 in the manufacturing sector, requires digitization [1]. The development of sensing and communication technologies has made it possible to have high visibility in industrial operations [2]. Automation that is IoT-enabled makes it easier to collect data from sensors in real-time about goods, processes, and stages. The digital twin of this data is what is used to manage industrial operations with high transparency, efficiency, accuracy, and risk reduction [3].

A broad variety of embedded computer systems, sensors, including wireless sensor networks and control systems, and automation, with automatic identification and data capture (AIDC) as a crucial element, are all included in the Internet of Things (IoT) [4–13]. AIDC sensors (RFID) include wireless sensors like radio-frequency identification, digital sensors like webcams, and optical sensors like barcodes. The I4.0's objective is achieved through passive radio frequency identification (RFID) technology, such as Ultra high frequency (UHF), by enabling a common object to be recognized, found, authenticated, and engaged with other IoT devices [14, 15]. RFID encourages global adoption of UHF RFID in the same way that other industry organizations like the WiFi Alliance, Bluetooth Special Interest Group (SIG), and others do [16].

A reader and reading antenna make up a typical passive UHF RFID system, which transmits and receives radio-frequency signals from passive RFID tags as shown in Figure 17.1. The passive RFID tag consists of a tiny chip connected to a printed antenna that collects energy and transmits it back to the antenna of the reader. Almost anything in our everyday lives can have this tag attached to it. Reader antennas are positioned

FIGURE 17.1 A typical UHF/RAIN RFID system.

throughout corridors, entranceways, and warehouses to follow the movements of the assets [2]. A server receives tag data through Wi-Fi or Bluetooth when one or more readers activate reader antennas. Rugged and tough antennas are needed since these reader antennas can get knocked around and damaged from forklift movements. Durable antennas can withstand collisions, mechanical shocks, thermal shocks, and vibrations. In general, RFID tags are built to last for a long period across the supply chain, but reader antennas need extra protection, such as an enclosure, to survive blows and thumps in the plant. This chapter offers recommendations for a durable antenna design and production process that does not require the use of an enclosure during installation [17].

17.2 INDUSTRY 4.0

17.2.1 Introduction to Industry 4.0

Since the creation of the first manufacturing plants, new technology and improved processes have fostered the expansion of the manufacturing industry [18]. The term "Industry 4.0," sometimes referred to as the "fourth industrial revolution," is a recent technological development that enables mass manufacturing of more items more simply and quicker. Real-time data may be accessed at any stage of the production process to find any anomalies or failures. RFID technology can let manufacturers immediately address any issues [19]. Connecting several machines may boost output significantly, save costs, and control the items being produced better [20]. Figure 17.2 depicts the progression of the industries towards Industry 4.0.

17.2.2 RFID's Role in Industry 4.0

RFID tags are starting to permeate the industrial sector and are starting to become a significant part of business. As part of Industry 4.0 industrial logistic systems, RFID has been investigated for use on mechanical equipment [21]. RFID sensors function via radio transmissions and real-time data-capable sensors [9]. Numerous options exist for these devices to assist with industrial logistics, and controlling the units needs relatively little manual labor. Without taking costs into account, RFID may dramatically boost the production processes and output of many organizations [22].

17.2.3 RFID Impact

While many firms use a cataloging system of some kind, RFID improves the effectiveness of product management. RFID and manufacturers continue to work together

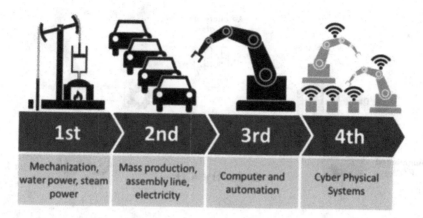

FIGURE 17.2 The evolution of the industries to Industry 4.0.

in a number of ways, such as customizing products, enhancing cataloging, and producing better results [23]. Certain factories can function entirely on machines, and all data collected from the factory floor is under control. Humans may spend more time making sure the plant runs well and preventing faults. The cost for customers to replace damaged goods is rising as more products are upgraded [24].

17.2.4 The RFID–Industry 4.0 Connection

Figure 17.3 depicts the major pillars of Industry 4.0. Many of the technologies that support Industry 4.0 have been around for a while. RFID has been used in industry and the supply chain for a long time, especially for asset management and inventory control [25, 26].

RFID is playing a bigger role in Industry 4.0. The use of data contained in RFID tags by manufacturing applications is evolving, enabling more flexible and affordable manufacture of custom items as well as improved automation and standardization [27–30]. Overall effectiveness and adaptability are increased as a consequence, and expenses are decreased.

Unlike other well-known identification technologies such as active tags or barcodes, passive RFID tags don't need a line of sight to function and don't need a power source [31].

To meet the demands of different applications or environments, contactless technology is available in a broad variety of form factors. They are programmable with several kbits of data and may be attached to practically any object. Data from tags is processed in a matter of milliseconds, and the data may be modified at any moment [32].

17.2.5 Need of RFID to Industry 4.0

From production to retail and beyond, data becomes an advantage with RFID. RFID's contribution to "big data" and the new infrastructure that enables it can be used by

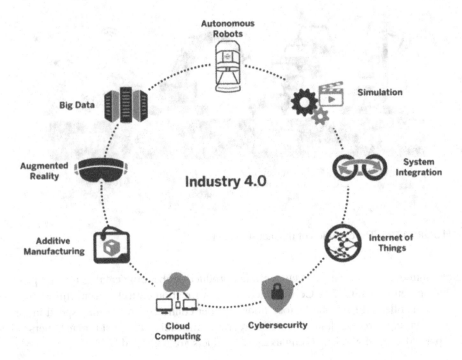

FIGURE 17.3 Key pillars of Industry 4.0.

manufacturers. As a result, manufacturing teams may manage variants, just-in-time and just-in-sequence production processes, and obtain greater insights into production control, process optimization, and quality control using real-time data [33]. Big data tracks the RFID tag throughout the supply chain and can be set up to track customer interactions as well. Having rapid access to such a vast amount of data opens up a whole new world of possibilities for producers and brand owners [34].

17.2.5.1 Manufacturing

RFID turns data into a benefit across industries, from manufacturing to retail and beyond. Manufacturers can benefit from RFID's contribution to "big data" and the new infrastructure that makes it possible. In order to handle variations, just-in-time and just-in-sequence production processes, and gain a deeper understanding of production control, process optimization, and quality control utilizing real-time data [33], manufacturing teams may now do so. Big data may be used to track customer interactions in addition to tracking the RFID tag across the supply chain. For producers and brand owners, having quick access to such a massive amount of data opens up a whole new world of possibilities [34].

The same production equipment may authenticate its own spare parts, ensuring that the correct replacement is installed in the exact location, effectively eliminating production errors [25].

17.2.5.2 Supply Chain

RFID's track-and-trace features may increase efficiency and visibility across the supply chain. Digital identification may be used by brand owners to deter counterfeiting and grey market diversion as well as to facilitate the movement of goods through customs and border checkpoints [28].

17.2.5.3 Pre- and Post-Sale

With the use of RFID, buyers may learn more about a product's contents and intended purpose in retail environments [27]. By offering information on warranty details, replacement parts, and appropriate recycling after the sale, the RFID tag may keep brand owners and consumers interested [23]. By allowing the Internet of Shopping, the latter phase of e-commerce, RFID introduces Industry 4.0 into the market [29].

17.3 RFID

17.3.1 Working Mechanism

Most RFID systems rely on tags affixed to the items being recognized. Each tag has a separate internal "read-only" or "rewrite" memory, depending on the kind and application [34–37]. Products' unique IDs, production dates, and other information are frequently stored in this memory. Through tags, the RFID system can find objects within its range thanks to magnetic fields produced by the RFID reader [35]. The tags answer to the inquiry using the reader's high frequency electromagnetic radiation and query signal; the query frequency can reach 50 times per second [34]. As a result, the fundamental parts of the system, notably tags and readers, are now able to communicate [36].

Massive volumes of data are generated as a result. Supply chain businesses utilize filters that are passed to backend information systems to manage this issue. To put it another way, this issue is managed with software like Savant. Between the information technology and the RFID reader, this software functions as a barrier [36, 37].

Several protocols control the communication between the reader and the tag. These protocols (International Organization for Standardization - ISO 15693 and ISO 18000-3 for HF or ISO 18000-6 and Electronic Product Code (EPC) for UHF) start the identification process when the reader is switched on.

These protocols work on certain frequency ranges, such as HF 13.56 MHz or UHF 860-915 MHz. The reader will instantly wake up, decode the signal, and modulate the reader's field in response if it is turned on and the tag is found in the reader fields [37]. In the event that all tags within the reader's range reply simultaneously, the reader must recognize signal collision (indication of multiple tags) [36]. In order to resolve signal conflicts, anti-collision algorithms are used. This enables the reader to sort tags and select/handle each tag according to the frequency range (50 to 200 tags) and protocol used.

The reader may perform actions on the tags via this connection, such as reading the tag's identification number and adding data to the tag [37].

17.3.2 Components of an RFID System

The many parts that make up the RFID system are linked together in the manner mentioned in the preceding section. The items (tags) may then be removed by the

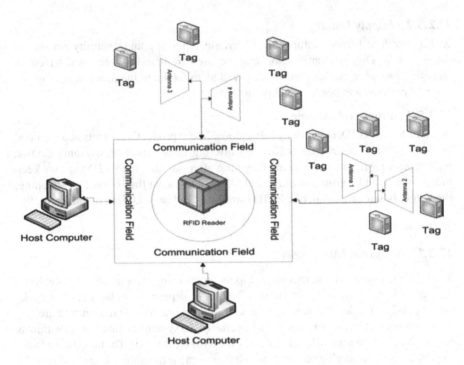

FIGURE 17.4 Components of an RFID system.

RFID system, which may subsequently perform other actions on them. Implementing an RFID solution is made possible by the integration of RFID components [38]. The RFID system is made up of the following parts, as shown in Figure 17.4.

17.3.2.1 Tags (Attached with an Object, Unique Identification)

The particular identification number (ID) for each object is stored on microchips in tags. The ID is recorded as a serial number in the RFID memory. The integrated circuits on the chip are inserted into a silicon chip [27]. An RFID memory chip can be either permanent or removable depending on the read/write capabilities. Because read-only tags have fixed data that cannot be changed without reprogramming, read-only and rewrite circuits are different [35]. Rewrite tags, on the other hand, may be freely and whenever desired programmed through the reader. RFID tags come in a variety of sizes and shapes, depending on the purpose and environment where they will be utilized.

Numerous materials are used in these tags. In the case of credit cards, tiny plastic bits are bonded to labels and other items. Numerous other items, such as apparel, industrial materials, and papers, also include embedded labels [39].

Figure 17.5 displays the various forms and sizes of RFID tags. Based on their ability to read and write data, RFID tags may also be classed [40]. Figure 17.5 shows the five categories of RFID tags.

There are three classifications for tags: passive, semi-active, and active. Active and passive tag characteristics are combined in semi-active tags. As a result, both

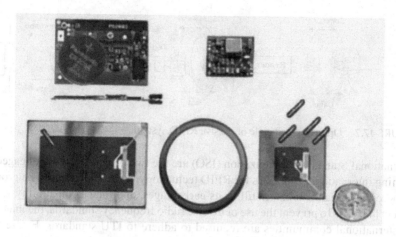

FIGURE 17.5 Variety of RFID tags (various shapes and sizes).

active and passive tags are used by industry and the majority of RFID systems. One of the most crucial characteristics of RFID tags is their function in the RFID system.

Their range, frequency, memory, security, kind of data, and other variables determine this. These characteristics, which differ in their usage and support for RFID system operations, are crucial for RFID performance [34, 41]. Figure 17.6 contrasts the active and passive tags while taking these characteristics into account.

17.3.2.2 Working Frequencies

The range of RFID tags is determined by frequency. This frequency determines interference resistance and other performance traits [42]. The RFID tag to use and the frequency to utilize are decided by the application [40]. EPC Global and the

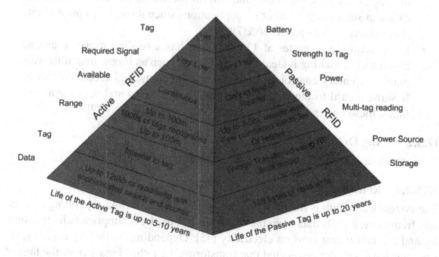

FIGURE 17.6 RFID active and passive tags comparison.

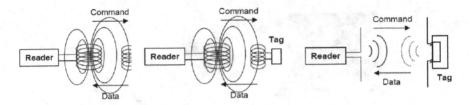

FIGURE 17.7 Operating principle of passive RFID system.

International Standards Organization (ISO) are the two principal groups engaged in defining international standards for RFID technology in the UHF band. The growth of these two organizations is still in its early stages, and their compatibility is not perfect [43, 44]. To prevent the use of diverse radio frequency standards, the majority of international communities are required to adhere to ITU standards. Figure 17.7 depicts RFID active and passive tags.

The most often utilized frequencies are as follows:

- Microwave can read tags more quickly than UHF tags since it runs at 2.45 GHz. With a tag read range of around 1 meter, it performs better in applications like vehicle tracking (in and out with obstacles) even if the reading rate on wet surfaces and close to metals is not the same at this frequency [37].
- With a quick multiple read rate, Ultra High Frequency, which works in the 860–930 MHz band, can identify several tags at once. It has a fast reading speed as a result. It is subject to the same restrictions as a microwave when used on a wet surface or next to metal. It is quicker than high frequency data transfer and has a reading range of 3 meters [37].
- High Frequency operates at 13.56 MHz and has a reading range of less than one meter, but it is affordable and helpful for access control, item identification at sales points, and other applications since it can be implanted into thin objects such as paper [36, 37].
- Low Frequency operates at 125 kHz and has a reading range of around half a meter, making it ideal for applications such as shops, manufacturing plants, inventory control through in and out counts, and access control by flashing a card to the reader. When applied to damp and near metal surfaces, these low frequency tags are mainly unaffected [37, 39].

17.3.2.3 Tag Detector

RFID antennas collect data and are used as a medium for tag reading [37].

17.3.2.4 RFID Reader

The system's core hub is the RFID reader. Using RFID antennas operating at a certain frequency, it pulls data from tags [37, 39]. The reader is simply a radio transmitter and a receiver that runs on electricity [45]. Depending on the tag capacity, the antennas have a reader connected that transforms the radio signals from the tags. A single reader may operate on many frequencies and has anti-collision mechanisms. These readers will thus be required to collect information or, if appropriate, write it onto tags and transmit it to computer systems. Readers, also known as serial readers, can be linked to a computer via wired options such RS-232, RS-485, and USB connections.

17.3.2.5 Tag Standards

Readers use near- and far-field radiation to communicate with the tag through its antennas [37]. A tag must first acquire energy before connecting with a reader if it is to reply to that reader. For instance, passive tags use one of the two methods described below [37, 41, 46–48].

17.4 OPERATING PRINCIPLE

There is no independent power source for the passive tag. As seen in Figure 17.8, the reader powers the chip in this tag. An RF signal is sent to the tag by the reader antenna. LF and HF tags collect energy from RF signals via inductive coupling, whereas UHF tags employ backscatter coupling. Backscatter coupling and inductive coupling both employ electromagnetic waves to transmit data between the tag and the reader.

17.4.1 INDUCTIVE COUPLING

An inductively coupled tag is comprised of an electronic data carrying component, often a single microprocessor, and an antenna consisting of a wide area coil. These are always off. This suggests that the reader is where it gets its power. For this reason, the reader's antenna coil produces a powerful electromagnetic field that permeates both the surrounding region and the cross-section of the coil area. The antenna coil of the transponder receives a small amount of the emitted field. Through induction, the transponder's antenna coil produces a voltage. This voltage is rectified and used as the power source for the microchip. Inductively coupled systems employ a

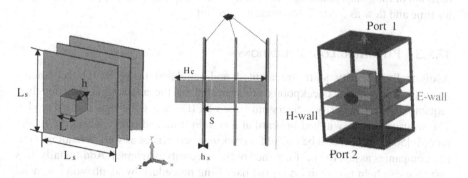

FIGURE 17.8 Multilayer DRA.

transformer-type coupling between the reader's primary coil and the transponder's secondary coil.

This is true when the distance between the coils is less than 0.16 and the transponder is in the transmitter antennas near field.

17.4.2 Backscatter Coupling

As we know from RADAR technology, electromagnetic waves are reflected by objects with diameters greater than half the wavelength of the wave. The efficacy with which an item reflects electromagnetic waves is indicated by its reflection cross-section. Some of the received RF energy is retransmitted by a tuned receiving antenna. Backscattering is the term for retransmission that happens in the same direction as the original transmitter. This backscattering can be detected by another antenna.

17.5 RFID APPLICATIONS

17.5.1 Healthcare Applications

Healthcare RFID solutions have the potential to significantly reduce costs while simultaneously enhancing patient care. RFID tags might be used for medical items used in the healthcare industry, such as patient records and medical equipment tracking, to help cut down on mistakes. RFID improves the scenario for a patient's treatment by including medical items utilized throughout that patient's care. A better patient experience will come from paramedical staff members being more effective and efficient thanks to RFID-based real-time information on the whereabouts of objects [34, 46].

17.5.2 Security and Control Applications

Equipment, user personal and professional goods, and cars can all be fitted with RFID tags. RFID technology allows for the granting and revocation of authorization for users/persons in a secure area as well as the tracking of individual access and the duration of their stay. Auditing can benefit from it as well. These apps closely monitor time and flow as a crucial component [48, 49].

17.5.3 Patrolling Log Applications

Additionally, security staff are audited and regulated using RFID technology. Security guard patrol checkpoints are provided by the application. During their sequential patrol, security officers must scan RFID checkpoint tags using a reader. The security guard switched his card at a certain time and place, which the reader records [50–54]. This will be useful for tracking occurrences as well as helping security companies assess the performance of their security personnel. Additionally, this software can help in streamlining the patrolling procedure by identifying the need for additional patrols or checkpoints in a monitored region [55–57].

Review of DRA-Based RFID Technology for Industry 4.0   313

17.5.4 Baggage Applications

Due to missing or delayed luggage or goods, airlines, package delivery firms, and other organizations suffer significant financial losses. It might be difficult to manage a large number of shipments from different sites to different destinations using different routes. The best resource management, operation, and package transfer in this situation is provided by RFID applications [58–61]. RFID facilitates package identification and produces data that can advise the industry on prospective growth areas [62–64]. Additionally, it updates customers on their orders.

17.5.5 Toll Road Applications

RFID applications increase toll collection and billing while enhancing traffic flow since cars and other vehicles cannot pass past toll booths without stopping to pay. RFID is used to speed up transactions and automatically identify the account holder [65, 66]. In order to ensure smooth traffic flow and identify traffic trends that may be utilized to inform administration or decision support systems, this application makes use of data mining techniques. The information might be used, for example, to report traffic issues or to develop and set future policies [67, 68].

17.5.6 Asset Tracking and Locating Objects

RFID may be used to find items or keep them from becoming lost. A tag containing an RFID chip is placed on an object for physical verification. An online database is used to track the movement of objects [69].

17.5.7 Libraries of RFID Labels

Libraries may control their book inventories using RFID. Tags, readers, self-checkout/in, book drop readers, and middleware are just a few of the RFID components used for this management. These elements assist it in managing the borrowing and returning of books. RFID technology detects when a book is being borrowed and when it has been returned [69].

17.5.8 Animal Identification

One of the original uses for RFID. The RFID tag can be permanently inserted under the animal's skin. This method is less unpleasant since the tag cannot be removed or altered using an identification mark. Data cannot be changed because the RFID chip within the tag is "Read-only." This chip holds a range of information, including the animal's birthdate, most recent vaccination, past illnesses, and distinctive traits [69].

17.5.9 Anti-Theft System

An RFID anti-theft tag, which is attached to the item via a sturdy thread or a plastic band, may safeguard any object. If someone walks to the exit wearing the tag, RFID door antennas placed nearby will detect its presence and send out a warning [69].

17.5.10 WASTE MANAGEMENT

RFID may be used to control rubbish as well. Every garbage truck has an RFID reader, and every trash can has an RFID tag attached to it. The scanner scans the tag as soon as the trash can is dumped into the truck and wirelessly transmits the information to the driver's cabin. At the conclusion of the journey, data is transferred to a central server. These details include the quantity of trash cans and the date and person responsible for collecting them [69].

17.5.11 NATIONAL IDENTIFICATION

National identity has been a significant problem for all nations. Identification may be accomplished using RFID technology. A user can only have one card with an RFID chip. The RFID tag number is then connected to a database on the internet that is accessible by multiple entities. An individual must provide a single ID card for identification purposes [69].

17.6 DIFFERENT DRA STRUCTURES FOR RFID APPLICATIONS

In terms of wireless communication, an antenna is a crucial component [70–74]. There are several different types of antennas available today for sending and receiving electromagnetic waves [75–79].

A multilayer dielectric resonator antenna array transmits fixed radio frequency (RF) presented in the form of RF identification (RFID) reader [80] at 5.8 GHz as shown in Figure 17.9. Three layers of square dielectric resonator antenna (DRA) components are placed in a unit cell of the transmit array on a dielectric substrate. A 909 square DRA transmit array with circular polarization is intended for long-range RFID applications and operates at 5.8 GHz. The transmit array's gain is 20.2 dB at its highest point. The right-hand circular polarization level is less than 31 dB at the required frequency, with an SLL of -22 dB. A 9-near-field focused DRA transmit array is investigated for fixed RFID at 5.8 GHz. At the same operating frequency,

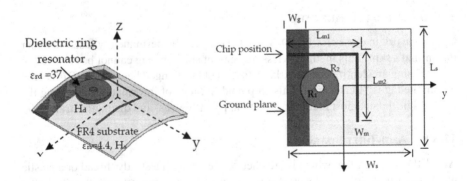

FIGURE 17.9 Curved Dual Band DRA.

Review of DRA-Based RFID Technology for Industry 4.0

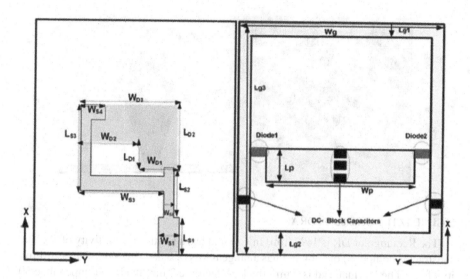

FIGURE 17.10 L-Shaped DRA.

the attributes of the near-field-focused transmit array are compared to those of the far-field transmit array.

The application of a new dielectric resonator antenna in a radio frequency identification system is described in this study [81]. A curved dual band dielectric resonator antenna (shown in Figure 17.10) is suggested for RFID applications. The 2.45 GHz (2.25–2.55 GHz) and 5.8 GHz (5.65–5.95 GHz) bands are covered by the tag antenna. We investigate the free-space radiation properties of the tag antenna. Under different weights, the radar cross-sections are measured. It describes how curvature affects the tag antenna's performance. Two cases should be taken into account. In the first instance, the tag antenna is mounted atop a cylindrical bottle that is filled with polyethylene material. It is investigated how object qualities affect radiation properties and radar cross-section. In the second scenario, a sphere-shaped bottle filled with polyethylene material has a tag antenna attached to it. Calculations are made for the radiation and backscattering characteristics. The Finite Element Method (FEM) is used for the simulation, and the Finite Integration Technique (FIT) is used to validate the results.

Figure 17.11 from [82] illustrates how an L-shaped DRA is excited by an inverted question mark-shaped feed, producing two orthogonal modes inside the DR that produce a circularly polarized pattern. A thin microstrip line is applied to a defective ground plane to provide a wideband response. This technique also aids in the location of two diode switches, which are positioned to the left and right of the microstrip line. The intended structure worked between 2.9 and 3.8 GHz. In switching operation 1, it offers impedance bandwidth of 24%, axial ratio (AR) bandwidth of 42%, and Circular Polarization (CP) 100%. It offers an impedance bandwidth of 60%, AR bandwidth of 58.88%, and CP performance of 76.36% in switching operations. The design provides a radiation efficiency of 93% and a peak gain of 3.4 dBi.

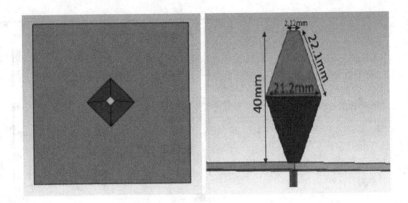

FIGURE 17.11 Octahedron DRA.

The Rectangular DR is 1492 mm³ in size and has a relative permittivity of 30. The substrate is composed of FR4 and has a tangent loss of 0.002 and a relative permittivity of 4.2. The DR material is 2 mm thick. A feedline 3 mm thick and impedance 50 is used to feed the DR, which is then followed by a quarter wave transform 1.5 mm thick. It is faulty if a rectangular area of the ground plane is etched. The DR is rectangularly carved to create an L-shaped pattern. Additionally, a feedline that resembles an inverted question mark is extended to excite CP fields. Placing a parasitic strip below the bottom side of the DR increases the axial ratio and impedance bandwidth.

In [83], two cone-shaped elements are inverted (as seen in Figure 17.12) to create a three-dimensional rhombus-shaped DRA with a ground plane that is 100 mm by

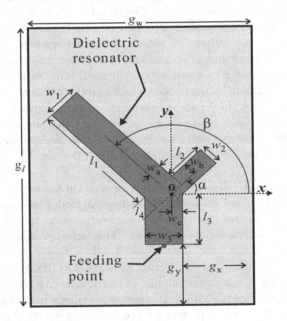

FIGURE 17.12 Y-Shaped DRA.

Review of DRA-Based RFID Technology for Industry 4.0

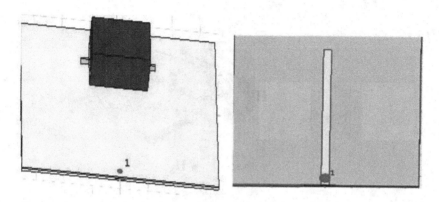

FIGURE 17.13 Rectangular DRA.

100 mm and is 2 mm thick. To excite the DRA structure's rhombus form, a dielectric feed line is utilized. PEC and Rogers RT 6010 are the materials utilized in the design of the ground plane and DRA, respectively. The Rogers RT 6010 has an epsilon value of 10.2. The antenna has a 2 GHz bandwidth and offers gains of 2.7, 3.4, and 4.6 dB at 2.3, 2.9, and 3.7 GHz, respectively. The design's impedance is 50. The design operates in the S band and offers excellent efficiency, a consistent radiation pattern, and outstanding gain.

Here in [84], a substrate with a thickness of 1.52 mm made of taconic RF-35 is employed, on top of which a Y-shaped DR with relative permittivity of 10 is positioned (see Figure 17.13). A Y-shaped DR is created by combining three rectangular arms – long, medium, and short – into one unit. A vertical strip is linked as feedline to the DR's short arm. Dual band response and a 61.80% impedance bandwidth (2.18–4.13 GHz) are provided by the antenna. Axial ratios of 5.27% (2.4–2.53 GHz) and 15.83% (3.2–3.75 GHz) are reached in the lower band. 4.11 dBi and 6.48 dBi of peak gain are obtained. At 2.46 GHz and 3.5 GHz, the antenna offers co-polarized gains of 3.86 dBi and 5.23 dBi. The antenna's frequency range is compatible with WLAN and WiMAX.

In [85], a micro strip line is used to feed slot coupled RDRA (as shown in Figure 17.14). FR4 substrate used has a thickness and permittivity of 0.8 mm and 4.4 respectively. At the bottom of the substrate a microstrip line of thickness 0.4 mm is placed. A slot of thickness 0.4 mm is introduced on the ground. To maximize the coupling and to serve as a matching circuit, the microstrip line is extended beyond the slot center. DRA is placed on the slot and has a permittivity of 30. It is observed that permittivity increases as resonance frequency decreases. Also return loss increases as slot length decreases. Antennas are used for direct-to-home satellite television applications and space-to-earth communication in frequency from 2.712–2.7621 GHz and 2.1849–2.2312 GHz. The antenna provides a gain of 4.64 dBi and directivity 6.437 dBi with a desirable radiation pattern.

For WiMAX and S band applications, the Dual Band Integrated DRA in [86] combines a ring with a modified polygon-shaped slot antenna; excitation is given via

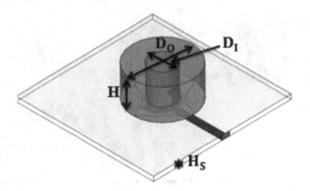

FIGURE 17.14 Dual Band Integrated DRA,

a circular-shaped aperture. This is depicted in Figure 17.14. The developed antenna has a 1.2 GHz bandwidth in the 2.75–3.96 GHz frequency range, a return loss of -40 dB and -32 dB, and gains of 4.9 dB and 5 dB at 2.9 GHz and 3.7 GHz. The antenna offers 5 dB & 6.4 dB gain at 2.9 GHz & 3.7 GHz respectively. The substrate is FR4, and its dimensions are 50 mm by 50 mm. Its thickness, loss tangent, and relative permittivity are 1.6 mm, 0.02 mm, and 4.4 mm, respectively. In order to excite a polygon-shaped slot antenna, a microstrip feed line is employed. Permittivity and loss tangent of alumina, which is employed as a dielectric material, are 9.8 and 0.02 respectively [87–91].

17.7 CONCLUSION

The usage of DRA for RFID applications in Industry 4.0 is covered in this chapter. We need Internet of Things (IoT) enabled automation tools for real-time information gathering on goods, processes, and stages via sensors if we want to improve the performance of industries in terms of efficiency, accuracy, and risk reduction. Tags, antennas, and RFID readers are necessary for RFID systems. Antennas are crucial since they are both a means of data collection and a means of reading tags. Antennas should therefore be extremely resistant to heat shocks and vibrations without sacrificing RF performance. DRA antennas studied in this work have emerged as a good candidate for all these requirements. Different DRA designs have been studied in terms of parameters like shape, size, substrate material, bandwidth, feeding mechanism, and peak gain.

REFERENCES

1. J. Qin, Y. Liu, and R. Grosvenor, A categorical framework of manufacturing for Industry 4.0 and beyond, changeable, agile, reconfigurable & virtual production. *Procedia CIRP*, 2016, 52, 173–178.

2. P. Parthiban, IoT antennas for Industry 4.0 – Design and manufacturing with an example. In *IEEE International IOT, Electronics and Mechatronics Conference (IEMTRONICS)*, 2020.
3. C. Occhiuzzi, S. Caizzone, and G. Marrocco, Passive UHF RFID antennas for sensing applications: Principles, methods, and classifications. *Antennasand Propagation Magazine, IEEE*, 2013, 55(6), 14–34.
4. R. Meelu, and R. Anand, Energy efficiency of cluster-based routing protocols used in wireless sensor networks. In *AIP Conference Proceedings*, 2010, November, 1324(1), 109–113. American Institute of Physics.
5. R. Anand, J. Singh, D. Pandey, B.K. Pandey, V.K. Nassa, and S. Pramanik, Modern technique for interactive communication in LEACH-based ad hoc wireless sensor network. In *Software Defined Networking for Ad Hoc Networks*. Springer International Publishing, Cham, 2022, pp. 55–73.
6. R. Anand, N. Sindhwani, and S. Juneja, Cognitive Internet of things, its applications, and its challenges: A survey. In *Harnessing the Internet of Things (IoT) for a Hyper-Connected Smart World*. Apple Academic Press, 2022, pp. 91–113.
7. N. Sindhwani, R. Anand, M. Niranjanamurthy, D.C. Verma, and E.B. Valentina (Eds.). *IoT Based Smart Applications*. Springer Nature, 2022.
8. M. Sharma, and H. Singh, Substrate integrated waveguide based leaky wave antenna for high frequency applications and IoT. *International Journal of Sensors, Wireless Communications and Control*, 2021, 11(1), 5–13.
9. S. Bommareddy, J.A. Khan, and R. Anand, A review on healthcare data privacy and security. *Networking Technologies in Smart Healthcare*, 2022, 165–187.
10. R. Gupta, G. Shrivastava, R. Anand, and T. Tomažič, IoT-based privacy control system through android. In *Handbook of E-business Security*. Auerbach Publications, 2018, pp. 341–363.
11. N. Jain, A. Chaudhary, N. Sindhwani, and A. Rana, Applications of Wearable devices in IoT. In *9th International Conference on Reliability, Infocom Technologies and Optimization (Trends and Future Directions)(ICRITO)*. IEEE, 2021, September, pp. 1–4.
12. B. Gupta, A. Chaudhary, N. Sindhwani, and A. Rana, Smart shoe for detection of electrocution using Internet of things (IoT). In *9th International Conference on Reliability, Infocom Technologies and Optimization (Trends and Future Directions)(ICRITO)*. IEEE, 2021, September, pp. 1–3.
13. M. Sharma, and H. Singh, Contactless methods for respiration monitoring and design of SIW-LWA for real-time respiratory rate monitoring. *IETE Journal of Research*, 2022, 1–11.
14. M. Toivonen, T. Björninen, L. Sydänheimo, L. Ukkonen, and Y. Rahmat-Samii, Impact of moisture and washing on the performance of embroidered UHF RFID tags. *IEEE Antennas and Wireless Propagation Letters*, 2013, 12, 1590–1593.
15. G. Marrocco, C. Occhiuzzi, and F. Amato, Sensor-oriented passiverfid. In *The Internet of Things*, D. Giusto, A. Iera, G. Morabito, and L. Atzori, Eds. Springer, New York, 2010, pp. 273–282.
16. K. Koski *et al.*, Durability of embroidered antennas in wireless body-centric healthcare applications. In *2013 7th European Conference on Antennas and Propagation (EuCAP)*, Gothenburg, 2013, pp. 565–569.
17. T. Phatarachaisakul, T. Pumpoung, C. Phongcharoenpanich, and S. Kosulvit, Tag antenna using printed dipole with H-slot for UHF RFID applications. In *2014 International Electrical Engineering Congress (iEECON)*. Chonburi, 2014, pp. 1–4.
18. P. Parthiban, Fixed UHF RFID reader antenna design for practical applications: A guide for antenna engineers with examples. *IEEE Journal of Radio Frequency Identification*, September 2019, 3(3), 191–204.

19. B. Fady, J. Terhzaz, A. Tribak, F. Riouch, and A. Mediavilla, Novel miniaturized planar low-cost multiband Antennafor Industry 4.0 communications. *Progress in Electromagnetics Research C*, 2019, 93, 29–38.
20. A. H. George M. Fernando, A. S. George, T. Baskar, and D. Pandey, Metaverse: The next stage of human culture and the internet. *International Journal of Advanced Research Trends in Engineering and Technology (IJARTET)*, (2021). 8(12), 1–10.
21. K.D. Thoben, S. Wiesner, and T. Wuest, Industrie 4.0 and smart manufacturing- A review of research issues and application examples. *International Journal of Automation Technology*, 2017, 11(1), 4–16.
22. Industrial internet of things: Unleashing the potential of connectedproducts and services. *World Economic Forum*, Tech. Rep., 2015.
23. A. Hakeema, D.Solyalia, M.Asmaela, and Q.Zeeshana, Smart manufacturing for Industry 4.0 using radio frequency identification (RFID) Technology, 2020. https://doi.org/10.17576/jkukm-2020-32(1)-05.
24. Y.Y. Fu et al., Experimental study on the washing durability of electro-textile UHF RFID tags. *IEEE Antennas and Wireless Propagation Letters*, 2015, 14, 466–469.
25. M. Sharma, S. Singh, S.G. Dishant Khosla, and A. Gupta, Waveguide diplexer: Design and analysis for 5G communication. In *2018 Fifth International Conference on Parallel, Distributed and Grid Computing (PDGC)*. IEEE, 2018, pp. 586–590.
26. M. Sharma, and H. Singh, SIW based leaky wave antenna with semi C-shaped slots and its modeling, design and parametric considerations for different materials of dielectric. In *Fifth International Conference on Parallel, Distributed and Grid Computing (PDGC)*. IEEE, 2018, pp. 252–258.
27. S.H. Zaiund-Deen, S.M. Gaber, A.M. Abd-Elhady, A.A. Kishk, and K.H. Awadalla, Wideband perforated rectangular dielectric resonator antenna reflectarray. *Proceedings of IEEE Antennas and Propagation Society International Symposium*, June 2011, 1, 113–116.
28. A. Kumar, D. Parkash, and M.V. Kartikeyan, Planer antennas for passive UHF RFID tag. *Progress in Electromagnetics Research*, 2009, 91, 95–212.
29. S.H. Zainud-Deen, H.A. Malhat, and K.H.Awadalla, Fractal antenna for passive UHF RFID applications. *Progress in Electromagnetics Research B, PIER B*, 2009, 16, 209–228.
30. Z.N. Chen, *Antennas for Portable Devices*. John Wiley& Sons Ltd, 2007.
31. L.W. Myer, and A.L. Scholtz, A Dual-band HF/ UHFAntennas for RFID Tags. *Proceedings of IEEE Antennas and Propagation Society International Symposium*, June 2008, 1, 1–5.
32. M. Au, and M.K. Rahim, Triple-band printed dipole TagAntenna for RFID. *Progress in Electromagnetics Research C, PIER C*, 2009, 9, 145–153.
33. S. Singh, B.S. Singla, M. Sharma, S. Goyal, and A. Sabo, Comprehensive study on Internet of things (IoT) and design considerations of various microstrip patch antennas for IoT applications. In *Mobile Radio Communications and 5G Networks, Proceedings of MRCN 2020*. Springer, Singapore, 2021, pp. 19–30.
34. K. Ahsan, H. Shah, and P. Kingston, Context based knowledge management in healthcare: An EA approach. AMCIS 2009, Available at AIS Library.
35. M. Sharma, D. Pandey, S.G. Dishant Khosla, B.K. Pandey, and A.K. Gupta, Design of a GaN-based flip chip light emitting diode (FC-LED) with Au bumps & thermal analysis with different sizes and adhesive materials for performance considerations. *Silicon*, 2022, 14(12), 7109–7120.
36. L. Srivastava, RFID: Technology, applications and policy implications, presentation, international telecommunication union, Kenya, 2005.

37. Application notes, introduction to RFID technology CAENRFID: The art of identification (2008).
38. L. Sandip, *RFID Sourcebook*. IBM Press, 2005, ISBN: 0-13-185137-3.
39. M. Sharma, B. Sharma, A.K. Gupta, and B.S. Singla, Design of 7 GHz microstrip patch antenna for satellite IoT-and IoE-based devices. In *Recent Innovations in Computing: Proceedings of ICRIC* Springer Singapore, 2020, pp. 627–637.
40. A. Narayanan, S. Singh, and M. Somasekharan, Implementing RFID in library: Methodologies, advantages and disadvantages, 2005.
41. Intermec, ABCs of RFID: Understanding and using radio frequency identification. White Paper, 2009.
42. E. Zeisel, and R. Sabella, RFID+. *Exam. Cram*, 2006. ISBN: 0-7897-3504-0.
43. US Department of Homeland Security, Additional guidance and security controls are needed over systems using RFID and DHS. Department of Homeland Security (Office of Inspector General), 2006, OIG-06–53.
44. US Department of Homeland Security, Enhanced security controls needed for us-visit's system using RFID technology. Department of Homeland Security (Office of Inspector General), 2006, OIG-06–39.
45. US Government Accountability Office, Information security: Radio frequency identification technology in the federal government, 2005, Report to Congressional Requesters, GAO-05-551.
46. K. Ahsan, H. Shah, and P. Kingston, Role of enterprise architecture in healthcare IT. *Proceeding ITNG2009*, IEEE, 2009.
47. Y. Meiller, and S. Bureau, Logistics projects: How to assess the right system? The case of RFID solutions in healthcare. *Americas Conference on Information Systems (AMCIS) 2009 Proceedings, Association for Information Systems Year*, 2009.
48. R. Parks, W. Yao, and C.H. Chu, RFID privacy concerns: A conceptual analysis in the healthcare sector. *Americas Conference on Information Systems (AMCIS) 2009 Proceedings, Association for Information Systems Year*, 2009.
49. M. Sharma, H. Singh, and D. Pandey, Parametric considerations and dielectric materials impacts on the performance of 10 GHzSIW-LWA for respiration monitoring. *Journal of Electronic Materials*, 2022, 51(5), 2131–2141.
50. S.H. Zainud-Deen, S.M. Gaber, H.A. Malhat, and K.H. Awadalla, Multilayer dielectric resonator antenna TRANSMITARRAY for near-field and far-fieldfixed RFID reader. *Progress in Electromagnetics Research C*, 2012, 27, 129–142.
51. M. Sharma, A.K. Gupta, T. Arora, D. Pandey, and S. Vats. Comprehensive Analysis of Multiband Microstrip Patch Antennas used in IoT-based Networks. In *2023 10th International Conference on Computing for Sustainable Global Development (INDIACom)*, (2023, March). (pp. 1424–1429). IEEE.
52. N. Sindhwani, R. Anand, G. Nageswara Rao, S. Chauhan, A. Chaudhary, A. Gupta, and D. Pandey, Comparative Analysis of Optimization Algorithms for Antenna Selection in MIMO Systems. In *Advances in Signal Processing, Embedded Systems and IoT: Proceedings of Seventh ICMEET-2022*, 2023, (pp. 607–617). Singapore: Springer Nature Singapore.
53. B.S. Deepak, G.M. Reddy, M.Y. Varma, S. Bharani, and B.K. Subhash, A compact octahedron dielectric resonator antenna for S-Band wireless applications. *Turkish Journal of Computer and Mathematics Education*, 2021, 12(12), 2675–2680.
54. A. Altaf, and M. Seo, Dual-band circularly polarized dielectric Resonator Antenna for WLAN and WiMAX applications. *Sensors*, 2020.20(4), 1137.

55. M. Sharma, A.K. Gupta, J. Singh, R. Mittal, H. Singh, and D. Pandey, Effects of slot shape in performance of SIW based Leaky Wave Antenna. In *2022 International Conference on Computational Modelling, Simulation and Optimization (ICCMSO)*, 2022, December, (pp. 268–273). IEEE.
56. D. Pathak, S.K. Sharma, and V.S.Kushwah, Dual band integrated dielectric resonator antenna for S-Band and WiMAX applications. *International Journal of Applied Engineering Research*, 2017, 12(24), 13995–13999, ISSN 0973-4562.
57. S. Minhas, and D. Khosla, Compact size and slotted patch antenna for WiMAX and WLAN. *Indian Journal of Science and Technology*, 2017, 10(16), 1–5.
58. N. Nasimuddin, and K.P. Esselle, Antennas with dielectric resonators and surface mounted short horn for high gain and wide bandwidth. *IET Microwaves, Antennas and Propagation*, 2007, 1(3), 723–728.
59. N. Kumar, G. Saini, P. Sahni, and D. Khosla, A compact multiband PIFA for personal communication handheld devices. *International Conference on Engineering, Technology and Management*, 2016, 4(4), 13–15.
60. R. Raghavan, D.C. Verma, D. Pandey, R. Anand, B.K. Pandey, and H. Singh. Optimized building extraction from high-resolution satellite imagery using deep learning. *Multimedia Tools and Applications*, 2022, 81(29), 42309–42323.
61. R. Kaur, and D. Khosla, A wideband antenna for wireless communication devices. *International Journal of Modern Computer Science*, 2015a, 3(3), 129–132.
62. S. Khan, X.C. Ren, H. Ali, C. Tanougast, A. Rauf, S.N.K. Marwat, and M.R. Anjum, Reconfigurable compact wideband circularly polarised dielectric resonator antenna for wireless applications. *Computers, Materials & Contin*, 2021, 68 2095–2109.
63. R. Kaur, and D. Khosla, Study of planar inverted-F antenna structures for various wireless devices. *IJEEE*, 2015b, 2(3), 31–34.
64. B. S. Deepak, G.M. Reddy, M.Y. Varma, S. Bharani, and B.K. Subhash, A compact octahedron dielectric resonator antenna for S-B and wireless applications. *Turkish Journal of Computer and Mathematics Education*, 2021, 12(12), 2675–2680.
65. M. Sharma, and H. Singh, SIW based leaky wave antenna with semi C-shaped slots and its modeling, design and parametric considerations for different materials of dielectric. *2018 Fifth International Conference on Parallel, Distributed and Grid Computing (PDGC)*. IEEE, pp. 252–258, 2018.
66. R. Anand, B. Khan, V.K. Nassa, D. Pandey, D. Dhabliya, B.K. Pandey, and P. Dadheech, Hybrid convolutional neural network (CNN) for Kennedy Space Center hyperspectral image. *Aerospace Systems*, 2023, 6(1), 71–78.
67. S.P. Kaur, and M. Sharma, Radially optimized zone-divided energy-aware wireless sensor networks (WSN) protocol using BA (bat algorithm). *IETE Journal of Research*, 2015, 61(2), 170–179.
68. V. Govindaraj, S. Dhanasekar, K. Martinsagayam, D. Pandey, B.K. Pandey, and V.K. Nassa, Low-power test pattern generator using modified LFSR. *Aerospace Systems*, 2023, 1–8.
69. M. Sharma, S. Singh, D. Khosla, S. Goyal, and A. Gupta, Waveguide diplexer: Design and analysis for 5G communication. *Fifth International Conference on Parallel, Distributed and Grid Computing (PDGC)*. IEEE, 2018, pp. 586–590.
70. D. Pathak, S.K. Sharma, and V.S. Kushwah, Dual band integrated dielectric resonator antenna for S-Band and WiMAX applications. *International Journal of Applied Engineering Research*, 2017, 12(24), 13995–13999, ISSN 0973-4562.
71. A. Kumar, and M. Sharma, Designing of ultra wide band microstrip antenna using triple slotted patch and DGS. *Journal of Telecommunication, Switching Systems and Networks*, 2019, 5(3), 12–19.

72. M. Sharma, and H. Singh, Substrate integrated waveguide based Leaky wave antenna for high frequency applications and IoT. *International Journal of Sensors, Wireless Communications and Control*, 2021., 11(1), 5–13.
73. M. Sharma, and H. Singh, A review on substrate integrated waveguide for mmW. Circ. Computer Science, ICIC 2018, 137–138.
74. S.H. Zainud-Deen, S.M. Gaber, H.A. Malhat, and K.H. Awadalla, Multilayer dielectric resonator antenna TRANSMITARRAY for near-field and far-fieldfixed RFID reader. *Progress in Electromagnetics Research C*, 2012, 27, 129–142.
75. P. Iyyanar, R. Anand, T. Shanthi, V.K. Nassa, B.K. Pandey, A.S. George, and D. Pandey, A Real-Time Smart Sewage Cleaning UAV Assistance System Using IoT. In *Handbook of Research on Data-Driven Mathematical Modeling in Smart Cities*, 2023, (pp. 24–39). IGI Global.
76. S. Polniak, RFID case study book: RFID application stories from around the globe. *Abhisam Software*, 2007.
77. D. Khosla, and K.S. Malhi, Investigations on designs of dielectric resonator antennas for WiMax&WLAN applications. In *Proceedings of Fifth International Conference on Parallel, Distributed and Grid Computing (PDGC)*, December 2018, pp. 646–651.
78. R.K. Mongia, and A. Ittipiboon, Theoretical and experimental investigation on rectangular dielectric resonator antenna. *IEEE Transaction on Antennas and Propagation*, 1997, 45(9), 1348–1359.
79. D. Khosla, and K.S. Malhi, Investigations on designs of dielectric resonator antennas for WiMax&WLAN applications. In *Proceedings of Fifth International Conference on Parallel, Distributed and Grid Computing (PDGC)*, December 2018, pp. 646–651.
80. A.K. Okaya, and L.F. Barash, The dielectric microwave resonator. *Proceedings of the IRE*, 1962, 50(10), 2081–2092.
81. P. Palta, M. Sharma, D. Khosla, andS. Goyal, SIW-based leaky wave antenna: Design and analysis for silicon. *International Journal of Mobile Computer Devices*, 2018, 4(2), 20–25.
82. G.D. Makwana, and K.J. Vinoy, A microstrip line fed rectangular dielectric resonator antenna for WLAN application. *Proceeding of IEEE International Symposium on Microwave (ISM-08)*, 2008, pp. 299–303.
83. N. Sharma, and D. Khosla, A compact two element U shaped MIMO planar inverted-F antenna(PIFA) for 4G LTE mobile devices. *5th IEEE International Conference on Parallel, Distributed and Grid Computing (PDGC-2018)*, December 2018, pp. 838–841.
84. G.D. Makwana, and K.J. Vinoy, Design of a compact rectangular dielectric resonator antenna at 2.4 GHz. *Progress in Electromagnetics Research C*, 2009, 11, 69–79.
85. M. Sharma, S. Singh, D. Khosla, and S. Goyal, Waveguide diplexer: Design and analysis for 5G communication. *5th IEEE International Conference on Parallel, Distributed and Grid Computing (PDGC-2018)*, December 2018, pp. 586–590.
86. N. Nasimuddin, and K.P. Esselle, A low profile compact microwave antenna with high gain and wide bandwidth. *IEEE Transactions on Antennas and Propagation*, 2007, 55(6), 1880–1883.
87. D. Pandey, and S. Wairya, An optimization of target classification tracking and mathematical modelling for control of autopilot. *The Imaging Science Journal*, 2022, 70(6), 371–386.
88. B. Kumar Pandey, D. Pandey, V.K. Nassa, T. Ahmad, C. Singh, A.S. George, and M.A. Wakchaure, Encryption and steganography-based text extraction in IoT using the EWCTS optimizer. *The Imaging Science Journal*, 2021, 69(1–4), 38–56.
89. D. Pandey, S. Wairya, M. Sharma, A.K. Gupta, R. Kakkar, and B.K. Pandey, An approach for object tracking, categorization, and autopilot guidance for passive homing missiles. *Aerospace Systems*, 2022, 5(4), 553–566.

90. M.S. Kumar, S. Sankar, V.K. Nassa, D. Pandey, B.K. Pandey, and W. Enbeyle, Innovation and creativity for data mining using computational statistics. In *Methodologies and Applications of Computational Statistics for Machine Intelligence*. IGI Global, 2021, pp. 223–240.
91. D. Pandey, B.K. Pandey, and S. Wariya, An approach to text extraction from complex degraded scene. *IJCBS*, 2020, 1(2), 4–10.

18 Leveraging Blockchain for Improved Supply Chain Management and Traceability in Industry 4.0

P. Balaji, Luigi Pio Leonardo Cavaliere,
B. Nagarjuna, S. Ramesh Babu,
M. Kavitha, and Barinderjit Singh

18.1 INTRODUCTION

In the contemporary world, businesses must prioritize the efficient and effective management of their supply chain to achieve success in most industries. While the traditional challenges of SCM such as timely delivery, cost minimization, and quality maintenance remain significant, the need for traceability and transparency has become increasingly important. Stakeholders including consumers, regulators, and others demand more visibility into supply chain operations to ensure ethical and sustainable practices and mitigate the risks associated with product safety and security [1].

Moreover, I-4.0, which is the fourth industrial revolution, is driven by advanced technologies like the Internet of Things (IoT), artificial intelligence, and big data. Although I-4.0 brings the promise of greater efficiency and productivity, it also introduces new complexities and risks to SCM. As supply chains become more interconnected and globalized, the need for secure and transparent systems for SCM and traceability becomes even more crucial. In summary, businesses need to prioritize efficient and effective SCM, which includes cost minimization, timely delivery, and quality maintenance, while also ensuring traceability and transparency to meet stakeholder demands for ethical and sustainable practices and to mitigate the risks associated with product safety and security. Additionally, with the emergence of I-4.0 and its advanced technologies, it has become even more critical to have secure and transparent systems. The chapter will also present examples of blockchain-based traceability systems and examine the potential benefits they offer. Traceability offers a substantial potential for improving efficiency, transparency, and trust in I-4.0. The benefits of BT include improved traceability, reduced costs, enhanced security, and increased transparency. As BT continues to evolve and develop, it is likely to play an increasingly significant role in SCM and traceability, providing new opportunities for organizations to optimize their operations and meet the needs of their stakeholders [2].

18.1.1 Definition of I-4.0

I-4.0 aims to create "smart factories" that are highly connected and automated, allowing for greater efficiency, flexibility, and customization in production processes. It represents a shift towards digitalization and the use of real-time data to optimize operations and decision-making, leading to increased productivity and competitiveness [3].

18.1.2 Importance of SCM and Traceability

Traceability enables companies to identify and address issues such as contamination, product recalls, and counterfeiting, which can have serious consequences for consumers and the company's reputation. Additionally, traceability can help companies comply with regulations related to product safety and ethical sourcing. In recent years, consumer and regulatory demands for greater transparency and accountability in supply chains have increased, further highlighting the importance of effective SCM and traceability. Companies that can demonstrate transparency and traceability in their supply chains are likely to have a competitive advantage, as they can build trust with customers, investors, and other stakeholders. Furthermore, effective SCM and traceability can help companies identify areas for improvement and reduce their environmental impact, leading to a more sustainable and responsible business model [4, 5].

18.2 OVERVIEW OF SCM AND TRACEABILITY

Traceability, on the other hand, refers to the ability to track and trace products or components throughout the supply chain. It involves documenting the movement of products from one stage of the supply chain to another and keeping records of the products' origin, destination, and any processing or handling they undergo. Traceability is critical to ensuring that products meet quality and safety standards, and can help to identify and address any issues that arise in the supply chain.

As the global economy becomes increasingly interconnected, traceability is becoming more important, with consumers and regulators demanding greater transparency and accountability in SCM. Traceability can help prevent fraud, ensure compliance with regulations, and reduce the risk of supply chain disruptions. With the emergence of I-4.0, new technologies such as blockchain are offering innovative solutions to improve SCM and traceability. This presents an exciting opportunity for companies to explore new approaches to these critical business processes, ultimately leading to better outcomes for all stakeholders involved [6, 7].

18.2.1 CHALLENGES IN SCM AND TRACEABILITY

While SCM and traceability are essential for business success, there are several challenges that companies face in implementing and maintaining effective SCM and traceability systems. Some of these challenges include [8–10]:

1. *Complexity*: modern supply chains can be incredibly complex, with multiple suppliers, distributors, and other stakeholders involved in the process. This complexity can make it difficult to track products and ensure that they are meeting quality and safety standards.
2. *Data management*: effective SCM and traceability require accurate and timely data. However, collecting and managing this data can be a significant challenge, particularly in large, global supply chains.
3. *Integration*: many companies use a variety of different systems and processes to manage their supply chain, which can make it challenging to integrate these systems and ensure that data is flowing smoothly between them.
4. *Cost*: implementing and maintaining effective SCM and traceability systems can be costly, particularly for smaller companies with limited resources.
5. *Regulatory compliance*: companies operating in regulated industries, such as food and pharmaceuticals, must comply with strict regulations around SCM and traceability. Failure to comply with these regulations can result in costly fines and reputational damage.
6. *Cybersecurity*: as supply chains become increasingly digital, cybersecurity threats are becoming a growing concern. Hackers can target supply chain systems to steal sensitive data, disrupt operations, or introduce counterfeit products into the supply chain.
7. *Human error*: despite the use of technology, human error can still occur in SCM and traceability. This can result in inaccurate data, delays, or other issues that can impact the supply chain.

Addressing these challenges requires a comprehensive approach that includes the use of innovative technologies such as blockchain, as well as effective data management, process integration, and cybersecurity measures. By addressing these challenges, companies can improve SCM and traceability, reduce costs, and enhance customer satisfaction.

18.2.2 OVERVIEW OF BT

The use of advanced cryptographic algorithms ensures the security of transactions, making them resistant to tampering or fraud. Transactions are validated through a consensus mechanism, making them more secure and resistant to manipulation. Smart Contracts (SC) can lead to faster and more cost-effective transactions, particularly in industries such as finance and SCM. Furthermore, BT eliminates the need for intermediaries or central authorities, allowing transactions to be processed quickly and efficiently. This can provide significant benefits in today's fast-paced business environment. Its decentralized structure and advanced cryptographic algorithms make it resistant to tampering and fraud, while its immutability ensures that all transactions are permanently recorded. The ability to support SC and streamline business processes can lead to faster and more cost-effective transactions, providing a competitive advantage in today's business environment [11].

18.3 BENEFITS AND APPLICATIONS

Table 18.1 shows the benefits and applications of blockchain in SCM and traceability. The utilization of BT in SCM and traceability can offer various advantages. By promoting transparency and traceability, blockchain can help track and trace products throughout the supply chain, enhance compliance with regulations and standards, and provide better insight into the supply chain. As a result, it can lead to improved quality control, customer satisfaction, and brand reputation. Furthermore, blockchain can enhance the efficiency of supply chain processes by eliminating intermediaries and automating transactions through the use of SC. This can result in cost savings and increased efficiency. Moreover, BT can optimize inventory management and logistics, which can enhance overall supply chain performance while reducing waste. Another significant benefit of blockchain is its ability to provide enhanced security. The use of advanced cryptography ensures the security of transactions and prevents tampering or fraud, making it an ideal solution for ensuring the security of sensitive data such as financial information and intellectual property. By utilizing BT, stakeholders can reduce the risk of cyber-attacks and data breaches while increasing trust and accountability. Lastly, BT can promote greater collaboration and coordination between stakeholders in the supply chain. By offering a shared, secure,

TABLE 18.1
Benefits and Applications of Blockchain in SCM and Traceability [12, 13]

Benefits	Applications
Enhanced transparency and traceability	Tracking and tracing products from the source to destination
	Verifying the authenticity of products and materials
	Monitoring compliance with regulations and standards
	Providing greater visibility into the supply chain
Increased efficiency	Streamlining transactions and reducing the need for intermediaries
	Automating supply chain processes with SC
	Improving inventory management and reducing waste
	Optimizing logistics and transportation
Improved security	Ensuring the security of sensitive data such as financial information and intellectual property
	Preventing fraud and counterfeiting
	Protecting against cyber-attacks and data breaches
	Enhancing trust

and transparent platform for communication and data sharing, blockchain can break down silos and enhance communication between stakeholders. As a result, it can lead to increased efficiency and effectiveness in the supply chain. Ultimately, blockchain can support the development of decentralized and democratic supply chains that prioritize transparency, sustainability, and social responsibility.

18.3.1 BLOCKCHAIN-BASED SOLUTIONS FOR SCM

BT has made significant strides in revolutionizing the way supply chains are managed, with the potential to bring greater transparency, efficiency, and security to the process. One of the main advantages of BT is the ability to create an immutable and tamper-proof record of all transactions in the supply chain, providing a high degree of transparency and accountability. This can be particularly beneficial in industries where traceability is critical, such as food and pharmaceuticals. Additionally, BT can reduce fraud and counterfeiting in the supply chain, automate many processes, and create a more decentralized and democratic supply chain. One of the most significant advancements in blockchain-based solutions for SCM is the development of enterprise blockchain solutions. These solutions are specifically designed for businesses and provide a range of features and functions to help manage the supply chain more effectively. Some of the key features of enterprise blockchain solutions include scalability, interoperability, security, customizability, and analytics. These solutions can handle large amounts of data, integrate with existing systems, provide a high degree of security and privacy, are customized to meet specific needs, and provide detailed analytics and insights into the supply chain. Overall, the advancements in blockchain-based solutions for SCM in I-4.0 are significant and have the potential to transform the way businesses manage their supply chains. BT provides greater transparency, efficiency, and security, helping businesses to reduce costs, improve customer satisfaction, and stay ahead of the competition in the era of I-4.0 [14, 15]. Figure 18.1 shows SCM Stages.

18.3.2 EXAMPLES OF SUCCESSFUL BLOCKCHAIN-BASED SCM SYSTEMS

There are several successful blockchain-based SCM systems that have been implemented in various industries [16, 17] including:

1. *IBM Food Trust*: a platform designed to enhance food safety and traceability in the food industry.
2. *Walmart's Blockchain Traceability System*: a mechanism put in place by Walmart to increase the food products' quality and safety. The system enables the monitoring and logging of data regarding the production, distribution, and point of origin of food goods.
3. *Maersk and IBM's Trade Lens*: a shipping platform that provides end-to-end supply chain visibility and transparency. It enables all participants in the supply chain to track and trace shipments in real-time, reducing work and increasing efficiency.

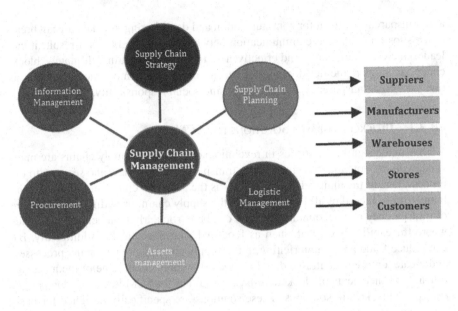

FIGURE 18.1 SCM stages.

4. *Provenance*: a system created to offer the fashion industry transparency and traceability. It gives customers the ability to trace the origin and manufacturing of textile products, guaranteeing that they are sustainably and ethically sourced.

These examples demonstrate the potential of BT to improve SCM and traceability in various industries. By enabling end-to-end transparency and traceability, blockchain-based systems can help to increase efficiency, reduce costs, and enhance sustainability and social responsibility in the supply chain.

18.4 BLOCKCHAIN-ENABLED TRACEABILITY

Supply chain traceability has become increasingly important due to the growing consumer demand for transparency and accountability from companies. BT has emerged as a promising solution for improving supply chain traceability. Blockchain-enabled traceability can help to track and trace products from source to destination. By using BT, stakeholders can access a permanent and tamper-proof record of all transactions, ensuring that products have not been altered or tampered with during transport [18–22].

Blockchain-enabled traceability can also improve supplier management as this technology helps to reduce the risk of fraud, errors, and delays in the supply chain, providing stakeholders with real-time information on the quality and availability of goods. Moreover, blockchain-enabled traceability can also improve sustainability and social responsibility in the supply chain by providing a transparent and verifiable record of environmental and social impact data. By using BT, stakeholders can access real-time information on the environmental and social impact of products, helping to ensure that products meet the required standards for sustainability and social responsibility. One of the main challenges is the complexity of supply chains, which can involve multiple parties and transactions. Ensuring that all stakeholders are on board and using the same blockchain platform can be a significant challenge. Another challenge is the cost of implementing and maintaining BT. While the benefits of blockchain-enabled traceability can be significant, the initial investment in technology and infrastructure can be high, especially for small and medium-sized enterprises. By using BT, stakeholders can improve product traceability, supplier management, payment and financing, and sustainability and social responsibility [23–26].

18.4.1 Challenges in Traceability in SCM

Traceability in SCM is critical for ensuring transparency and accountability. However, there are several challenges associated with implementing effective traceability systems in supply chains [27–30].

- One of the main challenges is the complexity of supply chains, which can involve multiple parties and transactions. Ensuring that all stakeholders are on board and using the same traceability system can be a significant challenge.
- Another challenge is the lack of standardized data and labeling systems. Without a standardized approach to labeling and data collection, it can be difficult to track and trace products effectively. This can lead to errors and delays in the supply chain, as well as increased risk of fraud and counterfeiting.
- The cost of implementing and maintaining traceability systems can also be a challenge, especially for small and medium-sized enterprises. The investment required in technology and infrastructure can be high, which can be a barrier to entry for some companies.
- There may also be resistance from some stakeholders in the supply chain who may be concerned about sharing data or revealing information about their operations. Ensuring that all stakeholders are comfortable with the level of transparency required for effective traceability can be a significant challenge.
- Finally, there is a need for effective governance and regulation to ensure that traceability systems are implemented and used effectively. This requires collaboration between governments, industry associations, and other stakeholders to establish and enforce standards and guidelines for traceability in supply chains.

In conclusion, while traceability is critical for improving transparency and accountability in supply chains, there are several challenges associated with implementing effective traceability systems. Addressing these challenges will require collaboration between stakeholders in the supply chain, as well as effective governance and regulation to ensure that standards and guidelines are established and enforced effectively.

18.4.2 Applicability of BT to Enhance Traceability

In recent years, the use of BT has emerged as a promising solution for enhancing traceability in SCM. One of the key features of BT is its immutable and transparent ledger. This decentralized and tamper-proof ledger records all transactions and movements of goods creating a permanent and reliable record of all activities. This makes it easier to trace and track the products throughout the supply chain, helping to prevent issues such as counterfeiting, theft, or product recalls. In addition to the immutable and transparent ledger, BT enables the creation of unique digital identities for products. These digital identities can be linked to the blockchain ledger, making it easier to trace the products' origin, quality, and location. By using these digital identities, stakeholders can have access to real-time information on the product's location, ensuring that products have not been altered or tampered with during transport. SC is another feature of BT that can enhance traceability. SC are self-executing contracts that can automate certain processes in the supply chain, such as triggering a payment when certain conditions are met. This can help to improve the efficiency of transactions and reduce the risk of errors and fraud. By using SC, stakeholders can ensure that all transactions in the supply chain are transparent and traceable. Real-time tracking is another key feature of BT that can enhance traceability in SCM. The features of BT, including its immutable and transparent ledger, digital identities for products, SC, and real-time tracking, can help to prevent issues and ensure that all transactions in the supply chain are transparent and traceable. Finally, BT can improve transparency in the supply chain by enabling stakeholders to access real-time information on the environmental and social impact of products. By using BT, stakeholders can have access to information on the production processes used to create products, helping to ensure that products are produced in a responsible and ethical manner [31, 32].

18.5 EMERGING TRENDS AND TECHNOLOGIES

I-4.0 has brought significant advancements and changes to the field of SCM and traceability. Several trends and technologies have emerged in recent years that are helping manufacturers optimize their supply chains and improve traceability [33].

One of the key trends in SCM is the use of real-time data and analytics to improve visibility and control over the supply chain. With the help of IoT devices, manufacturers can collect real-time data on inventory levels, production processes, and transportation. This data can be analyzed using advanced analytics and AI algorithms to optimize the supply chain and improve overall efficiency [34–37].

Another important trend is the use of blockchain-based solutions for SCM. BT provides a secure and transparent way of recording transactions and sharing information across multiple parties. This can help manufacturers improve traceability and reduce the risk of fraud or errors in the supply chain. By using blockchain-based solutions, manufacturers can create a digital record of every transaction in the supply chain, from raw material sourcing to the delivery of finished goods [38–40].

In addition to these trends, there are several emerging technologies that are transforming SCM and traceability in I-4.0. These include:

1. RFID (Radio-Frequency Identification) technology: RFID tags can be attached to products or packaging to track their movement through the supply chain. This technology provides real-time visibility into inventory levels and can help to reduce the risk of product loss or theft.
2. Autonomous vehicles: self-driving vehicles are being used to transport goods between warehouses and distribution centers. This technology can help to reduce transportation costs and improve delivery times.
3. Drones: drones are being used to monitor inventory levels, inspect warehouses, and deliver goods in hard-to-reach areas. This technology can help to improve efficiency and reduce the risk of accidents in the supply chain.

Another application of AI in SCM is predictive maintenance. By analyzing sensor data from equipment and machinery, AI algorithms can identify signs of impending equipment failure and alert maintenance teams before a breakdown occurs. This can help companies reduce downtime and avoid costly repairs.

Emerging trends and technologies in SCM and traceability in I-4.0 are revolutionizing the way companies manage their supply chains. From advanced analytics and AI to blockchain-based solutions, these technologies are enabling companies to optimize their operations, reduce waste, and improve overall efficiency. As the pace of technological change continues to accelerate, it will be crucial for companies to stay up-to-date with the latest developments in SCM and traceability in order to remain competitive in the era of I-4.0 [41, 42].

18.6 CONCLUSION

In conclusion, BT offers significant potential for improving SCM and traceability in the context of I-4.0. Stakeholders may benefit from the immutability and transparency of blockchain ledgers. Furthermore, blockchain-based traceability systems can improve supply chain transparency, increase operational efficiency, reduce fraud and errors, and improve customer trust and loyalty. The chapter has highlighted various use cases and applications of BT in SCM, including digital identities, SC, real-time tracking, and interoperability. These use cases and applications have the potential to enable stakeholders to access real-time information on the location, condition, and quality of products throughout the supply chain, enabling them to optimize their operations and improve efficiency. The chapter has also discussed emerging trends and technologies in blockchain, including IoT, AI, and sustainability [43–49]. These emerging trends and technologies have the potential to further enhance the impact of blockchain on SCM and traceability, enabling stakeholders to automate certain processes, improve sustainability and social responsibility, and optimize their operations even further. Overall, the chapter has shown that leveraging blockchain for improved SCM and traceability is an important area of research and development, with significant potential for enhancing supply chain efficiency, transparency, and sustainability. However, there are still challenges to be addressed, including interoperability, scalability, and regulatory compliance. Further research is needed to explore the potential of BT in SCM and to address these challenges, paving the way for the adoption of blockchain-based solutions in I-4.0.

REFERENCES

1. Cao, Y., Yi, C., Wan, G., Hu, H., Li, Q., Wang, S. An Analysis on the Role of Blockchain-Based Platforms in Agricultural Supply Chains. *Transp. Res. Part E-Logist. Transp. Rev.* 2022, *163*, 102731.
2. Tseng, C.-T., Shang, S.S.C. Exploring the Sustainability of the Intermediary Role in Blockchain. *Sustain. Switz.* 2021, *13*, 1936.
3. Khanna, A., Jain, S., Burgio, A., Bolshev, V., Panchenko, V. Blockchain-Enabled Supply Chain Platform for Indian Dairy Industry: Safety and Traceability. *Foods* 2022, *11*, 2716.
4. Varriale, V., Cammarano, A., Michelino, F., Caputo, M. Sustainable Supply Chains with Blockchain, IoT and RFID: A Simulation on Order Management. *Sustain. Switz.* 2021, *13*, 6372.
5. Lezoche, M., Hernandez, J.E., Díaz, M.D.M.E.A., Panetto, H., Kacprzyk, J. Agri-Food 4.0: A Survey of the Supply Chains and Technologies for the Future Agriculture. *Comput. Ind.* 2020, *117*, 103187.
6. Dadi, V., Nikla, S.R., Moe, R.S., Agarwal, T., Arora, S. Agri-Food 4.0 and Innovations: Revamping the Supply Chain Operations. *Prod. Eng. Arch.* 2021, *27*, 75–89.
7. Song, L., Wang, X., Merveille, N. Research on Blockchain for Sustainable E-Agriculture. In *Proceedings of the 2020 IEEE Technology and Engineering Management Conference, TEMSCON 2020*, Novi, MI, 3–6 June 2020.
8. Kumar, S., Raut, R.D., Nayal, K., Kraus, S., Yadav, V.S., Narkhede, B.E. To Identify Industry 4.0 and Circular Economy Adoption Barriers in the Agriculture Supply Chain by Using ISM-ANP. *J. Clean. Prod.* 2021, *293*, 126023.
9. Sodamin, D., Vaněk, J., Ulman, M., Šimek, P. Fair Label versus BT from the Consumer Perspective: Towards a Comprehensive Research Agenda. *AGRIS -Line Pap. Econ. Inform.* 2022, *14*, 111–119.
10. Scuderi, A., La Via, G., Timpanaro, G., Sturiale, L. The Digital Applications of "Agriculture 4.0": Strategic Opportunity for the Development of the Italian Citrus Chain. *Agric.-Basel* 2022, *12*, 400.
11. Singh, V., Sharma, S.K. Application of BT in Shaping the Future of Food Industry Based on Transparency and Consumer Trust. *J. Food Sci. Technol.* 2022, *246*, 6–17.
12. Singh, S., Madaan, G., Kaur, J., Swapna, H.R., Pandey, D., Singh, A., Pandey, B.K.. Bibliometric Review on Healthcare Sustainability. *Handbook of Research on Safe Disposal Methods of Municipal Solid Wastes for a Sustainable Environment* 2023, 142–161.
13. Dey, S., Saha, S., Singh, A.K., McDonald-Maier, K. SmartNoshWaste: Using Blockchain, Machine Learning, Cloud Computing and QR Code to Reduce Food Waste in Decentralized Web 3.0 Enabled Smart Cities. *Smart Cities* 2022, *5*, 162–176.
14. Dong, S., Yang, L., Shao, X., Zhong, Y., Li, Y., Qiao, P. How Can Channel Information Strategy Promote Sales by Combining ICT and Blockchain? Evidence from the Agricultural Sector. *J. Clean. Prod.* 2021, *299*, 126857.

15. Alkahtani, M., Khalid, Q.S., Jalees, M., Omair, M., Hussain, G., Pruncu, C.I. E-Agricultural Supply Chain Management Coupled with Blockchain Effect and Cooperative Strategies. *Sustainability* 2021, *13*, 816.
16. Vodenicharova, M.S. Supply Chain Study in Food Industry in Bulgaria. *Int. J. Retail Distrib. Manag.* 2020, *48*, 921–938.
17. Nurgazina, J., Pakdeetrakulwong, U., Moser, T., Reiner, G. Distributed Ledger Technology Applications in Food Supply Chains: A Review of Challenges and Future Research Directions. *Sustainability* 2021, *13*, 4206.
18. Rahman, L.F., Alam, L., Marufuzzaman, M., Sumaila, U.R. Traceability of Sustainability and Safety in Fishery Supply Chain Management Systems Using Radio Frequency Identification Technology. *Foods* 2021, *10*, 2265.
19. Swapna, H.R., Singh, S., Madaan, G., Mishra, S., Pandey, D., Kanike, U. K.. Globalization and Emerging Opportunities and Challenges in Sustainable Environment in Industry 4.0. *Handbook of Research on Safe Disposal Methods of Municipal Solid Wastes for a Sustainable Environment* 2023, 48–68.
20. Adams, D., Donovan, J., Topple, C. Achieving Sustainability in Food Manufacturing Operations and Their Supply Chains: Key Insights from a Systematic Literature Review. *Sustain. Prod. Consum.* 2021, *28*, 1491–1499.
21. Haji, M., Kerbache, L., Muhammad, M., Al-Ansari, T. Roles of Technology in Improving Perishable Food Supply Chains. *Logist.-Basel* 2020, *4*, 33.
22. Qian, J., Ruiz-Garcia, L., Fan, B., Villalba, J.I.R., McCarthy, U., Zhang, B., Yu, Q., Wu, W. Food Traceability System from Governmental, Corporate, and Consumer Perspectives in the European Union and China: A Comparative Review. *Trends Food Sci. Technol.* 2020, *99*, 402–412.
23. Oruma, S.O., Misra, S., Fernandez-Sanz, L. Agriculture 4.0: An Implementation Framework for Food Security Attainment in Nigeria's Post-Covid-19 Era. *IEEE Access* 2021, *9*, 83592–83627.
24. Iyyanar, P., Anand, R., Shanthi, T., Nassa, V.K., Pandey, B.K., George, A. S., Pandey, D. A Real-Time Smart Sewage Cleaning UAV Assistance System Using IoT. In *Handbook of Research on Data-Driven Mathematical Modeling in Smart Cities* (pp. 24–39). IGI Global, 2023.
25. Barbosa, M.W. Uncovering Research Streams on Agri-Food Supply Chain Management: A Bibliometric Study. *Glob. Food Secur.* 2021, *28*, 100517.
26. Mendi, A.F. Blockchain for Food Tracking. *Electronics* 2022, *11*, 2491.
27. Wünsche, J.F., Fernqvist, F. The Potential of BT in the Transition Towards Sustainable Food Systems. *Sustain. Switz.* 2022, *14*, 7739.
28. Dayioglu, M.A., Turker, U. Digital Transformation for Sustainable Future—Agriculture 4.0: A Review. *J. Agric. Sci.-Tarim Bilim. Derg.* 2021, *27*, 373–399.
29. Lin, S.-Y., Zhang, L., Li, J., Ji, L., Sun, Y. A Survey of Application Research Based on Blockchain Smart Contract. *Wirel. Netw.* 2022, *28*, 635–690.
30. Rejeb, A., Keogh, J.G., Zailani, S., Treiblmaier, H., Rejeb, K. BT in the Food Industry: A Review of Potentials, Challenges and Future Research Directions. *Logistics* 2020, *4*, 27.
31. Ekawati, R., Arkeman, Y., Suprihatin, Sunarti, T.C. Proposed Design of White Sugar Industrial Supply Chain System Based on BT. *Int. J. Adv. Comput. Sci. Appl.* 2021, *12*, 459–465.
32. Ali, M.H., Chung, L., Kumar, A., Zailani, S., Tan, K.H. A Sustainable Blockchain Framework for the Halal Food Supply Chain: Lessons from Malaysia. *Technol. Forecast. Soc. Chang.* 2021, *170*, 120870.
33. Bechtsis, D., Tsolakis, N., Iakovou, E., Vlachos, D. Data-Driven Secure, Resilient and Sustainable Supply Chains: Gaps, Opportunities, and a New Generalised Data Sharing and Data Monetisation Framework. *Int. J. Prod. Res.* 2022, *60*, 4397–4417.

34. Kayikci, Y., Usar, D.D., Aylak, B.L. Using BT to Drive Operational Excellence in Perishable Food Supply Chains during Outbreaks. *Int. J. Logist. Manag.* 2022, *33*, 836–876.
35. Rana, R.L., Tricase, C., De Cesare, L. BT for a Sustainable Agri-Food Supply Chain. *Br. Food J.* 2021, *123*, 3471–3485.
36. Kumar, A., Srivastava, S.K., Singh, S. How BT Can Be a Sustainable Infrastructure for the Agrifood Supply Chain in Developing Countries. *J. Glob. Oper. Strateg. Sourc.* 2022, *15*, 380–405.
37. Joo, J., Han, Y. An Evidence of Distributed Trust in Blockchain-Based Sustainable Food Supply Chain. *Sustainability* 2021, *13*, 10980.
38. Tsolakis, N., Niedenzu, D., Simonetto, M., Dora, M., Kumar, M. Supply Network Design to Address United Nations Sustainable Development Goals: A Case Study of Blockchain Implementation in Thai Fish Industry. *J. Bus. Res.* 2021, *131*, 495–519.
39. Kramer, M.P., Bitsch, L., Hanf, J. Blockchain and Its Impacts on Agri-Food Supply Chain Network Management. *Sustainability* 2021, *13*, 2168.
40. Phua, C., Andradi-Brown, D.A., Mangubhai, S., Ahmadia, G.N., Mahajan, S.L., Larsen, K., Friel, S., Reichelt, R., Hockings, M., Gill, D. Marine Protected and Conserved Areas in the Time of COVID. *Parks* 2021, *27*, 85–102.
41. Pal, K., Yasar, A.U.H. Internet of Things and Blockchain Technology in Apparel Manufacturing Supply Chain Data Management. *Procedia Comput. Sci.* 2020, *170*, 450–457. https://doi.org/10.1016/j.procs.2020.03.088
42. Peña, M., Llivisaca, J., Siguenza-Guzman, L. Blockchain and Its Potential Applications in Food Supply Chain Management in Ecuador. *Adv. Intell. Syst. Comput.* 2020, *1066*, 101–112. https://doi.org/10.1007/978-3-030-32022-5_10
43. Sindhwani, N., Anand, R., Niranjanamurthy, M., Verma, D. C., Valentina, E. B. (Eds.). *IoT Based Smart Applications*. Springer Nature, 2022.
44. Singh, P., Kaiwartya, O., Sindhwani, N., Jain, V., Anand, R. (Eds.). *Networking Technologies in Smart Healthcare: Innovations and Analytical Approaches.* CRC Press, 2022.
45. Pandey, D., George, S., Aremu, B., Wariya, S., Pandey, B.K. Critical Review on Integration of Encryption, Steganography, IOT and Artificial Intelligence for the Secure Transmission of Stego Images. 2021.
46. Kumar Pandey, B., Pandey, D., Nassa, V.K., Ahmad, T., Singh, C., George, A.S., Wakchaure, M.A. Encryption and steganography-based text extraction in IoT using the EWCTS optimizer. *The Imaging Science Journal* 2021, 69(1–4), 38–56.
47. Pandey, D., Wairya, S. An optimization of target classification tracking and mathematical modelling for control of autopilot. *The Imaging Science Journal* 2022, 70(6), 371–386.
48. Sindhwani, N., Anand, R., Vashisth, R., Chauhan, S., Talukdar, V., Dhabliya, D. Thingspeak-Based Environmental Monitoring System Using IoT. In *2022 Seventh International Conference on Parallel, Distributed and Grid Computing (PDGC)* (pp. 675–680). IEEE, 2022, November.
49. Singh, H., Pandey, B. K., George, S., Pandey, D., Anand, R., Sindhwani, N., Dadheech, P. Effective Overview of Different ML Models Used for Prediction of COVID-19 Patients. In *Artificial Intelligence on Medical Data: Proceedings of International Symposium, ISCMM 2021* (pp. 185–192). Springer Nature, 2022, July.

19 Securing Automated Systems with BT Opportunities and Challenges

Luigi Pio Leonardo Cavaliere, Swati Rawat, Neeru Sidana, Purnendu Bikash Acharjee, Latika Kharb, and Venkateswararao Podile

19.1 INTRODUCTION

With the rise of automation in various industries, the need for securing automated systems has become increasingly important. Automated systems, such as those in finance, healthcare, and transportation, process large volumes of sensitive data and transactions, making them vulnerable to cyber-attacks and data breaches. In particular, we examine how BT can be used to enhance the security of automated systems, the potential opportunities of using BT in securing automated systems, the challenges that come with implementing BT, and the use cases of BT in securing automated systems. The role of BT in securing automated systems is particularly significant because of the nature of BT. BT is inherently decentralized, meaning that it does not rely on a central authority to manage data or transactions. Instead, it relies on a distributed network of nodes to verify and validate transactions. This decentralization feature of BT can help address the security issues associated with centralized systems. Furthermore, the immutable and transparent nature of the BT ledger makes it possible to maintain a tamper-proof record of all transactions. This can help in preventing fraudulent activities and cyber-attacks, as well as ensuring that all transactions are transparent and traceable. Despite the potential benefits of using BT to secure automated systems, there are also significant challenges that need to be addressed. These challenges include the high computational power required for BT transactions, integration challenges, BT scalability, and regulatory challenges. High computational power is required to process BT transactions, which can result in increased energy consumption and operational costs. Integration challenges arise when trying to integrate BT into existing automated systems, which may require significant changes to the system architecture. BT scalability is another challenge that needs to be addressed, as BT systems currently have limited capacity for processing large volumes of transactions. Finally, regulatory challenges arise when trying to implement BT in regulated industries, where compliance with existing regulations may be challenging. Despite these challenges, the potential of BT in securing automated systems cannot be ignored. Use cases of BT in securing automated systems include SC and automation, SCM, identity management, and IoT security. In conclusion, securing automated systems is crucial for maintaining the integrity and security of sensitive data and transactions. BT offers significant opportunities for enhancing the security of automated systems, but also comes with significant challenges that need to be addressed. This chapter aims to explore these opportunities and challenges, and provide insights into the potential use cases of BT in securing automated systems [1–3].

19.1.1 SIGNIFICANCE

Automated systems are used in various industries, including finance, healthcare, transportation, and manufacturing, to streamline processes, increase efficiency, and reduce costs. However, with the increasing reliance on automated systems, the security risks associated with these systems have also increased. Automated systems process large volumes of sensitive data and transactions, making them vulnerable to cyber-attacks and data breaches. These security risks can result in financial losses, damage to reputation, and even loss of life in some cases. Securing automated systems is, therefore, crucial for maintaining the integrity and security of sensitive data and transactions. It is necessary to ensure that automated systems are protected from cyber-attacks and data breaches, and that any vulnerabilities are identified and addressed in a timely manner. In addition to the security risks associated with automated systems, there are also regulatory and compliance requirements that need to be addressed. Overall, securing automated systems is critical for ensuring the integrity and security of sensitive data and transactions, complying with regulatory requirements, and maintaining the trust of customers and stakeholders [4].

19.2 THE ROLE OF BT IN SECURING AUTOMATED SYSTEMS

The role of BT in securing automated systems is significant due to its inherent properties that address many of the security issues associated with centralized systems. BT's decentralized architecture makes it possible to maintain a tamper-proof record of all transactions, preventing fraudulent activities and cyber-attacks, and ensuring transparency and traceability. Moreover, the immutable and transparent nature of the BT ledger can help in enhancing the security of automated systems. By using BT to record and verify all transactions, automated systems can reduce the risk of data tampering and unauthorized access. One of the ways BT can enhance the security of automated systems is through the use of SC. In conclusion, the role of BT in securing automated systems is significant due to its inherent properties, which address many of the security issues associated with centralized systems. By using BT to enhance the security of automated systems, organizations can ensure the integrity and security of sensitive data and transactions, comply with regulatory requirements, and maintain the trust of customers and stakeholders [5].

19.2.1 BLOCKCHAIN-BASED SYSTEMS WITH ENHANCED SECURITY

Traditional financial systems are often centralized and controlled by a single authority, which can make them vulnerable to attacks and fraud. BT-based systems, on the other hand, are decentralized and utilize advanced cryptography to secure transactions. This makes it virtually impossible for anyone to tamper with the system or steal funds. Another way in which BT-based systems can enhance security is by creating a secure and transparent SCM system. By using BT, it is possible to track the movement of goods from the source to the destination and ensure that all parties involved in the supply chain are held accountable for their actions. This can help to prevent counterfeit goods and other forms of supply chain fraud. In addition, BT-based systems can be used to enhance the security of data storage and transfer. By using BT, it is possible to create a decentralized system for storing and transferring data, which is much more secure than traditional centralized systems. This can help to prevent data breaches, hacking, and other forms of cyber-attacks.

BT-based systems offer enhanced security and transparency compared to traditional centralized systems. By utilizing advanced cryptography and a decentralized ledger, it is possible to create a secure and transparent system that can help to prevent fraud, hacking, and other forms of cyber-attacks. As such, BT-based systems are increasingly being adopted by various industries to enhance security and improve efficiency [6, 7].

19.3 OPPORTUNITIES OF SECURING AUTOMATED SYSTEMS WITH BLOCKCHAIN

Table 19.1 shows opportunities for securing automated systems with BT. Blockchain offers numerous opportunities to enhance the security of automated systems in various industries and applications. The table provides an overview of seven key opportunities, along with examples and explanations. These opportunities include decentralization, immutable record keeping, SC, SCM, identity management, IoT security, and cryptocurrency. Decentralization eliminates the need for centralized authorities in managing transactions, and it is useful in decentralized cryptocurrency exchanges. Immutable record keeping ensures accurate records, and it can be beneficial in medical record keeping. SC can be used to automate processes in the insurance industry. SCM can be improved by tracking the movement of goods, reducing the risk of fraud and counterfeiting. Identity management can be secured by managing identities through BT, which can be useful in government-issued digital identities. IoT security can be enhanced by securing devices and data with BT, reducing the risk of hacking and data breaches. Finally, cryptocurrency offers a decentralized and secure platform for peer-to-peer transactions, increasing transaction efficiency and reducing the need for intermediaries. Overall, BT offers several opportunities to enhance the security of automated systems, as demonstrated by the table's contents.

TABLE 19.1
Opportunities of Securing Automated Systems with BT [8–10]

Opportunities	Explanation	Example
Decentralization	Eliminates the need for a central authority or intermediary to manage transactions, reducing the risk of data tampering and unauthorized access	Decentralized cryptocurrency exchanges
Immutable Record	Ensures all transactions are recorded and verified, making it impossible to tamper with or alter any previous transaction without invalidating the entire chain	Medical record keeping
SC	Self-executing contracts stored on the BT that automatically execute when certain conditions are met, reducing the risk of human error and increasing transaction efficiency	Insurance claims processing
SCM	BT can be used to track the movement of goods and products throughout the supply chain, reducing the risk of counterfeiting, theft, and fraud	Tracking the origin of diamonds in the jewelry industry
Identity Management	BT can be used to manage identities, ensuring that only authorized parties have access to sensitive data and transactions	Government-issued digital identities
Internet of Things (IoT) Security	BT can be used to secure IoT devices and data, reducing the risk of hacking and data breaches	Smart home devices
Cryptocurrency	Creating a decentralized and secure platform for peer-to-peer transactions, reducing the need for intermediaries and increasing the efficiency of transactions	Bitcoin

19.3.1 Decentralized Data Management

Decentralized data management refers to a data storage system that is distributed across a network of computers rather than being centralized in one location. This is achieved by using BT, which ensures that data is verified and recorded in an immutable ledger. Decentralized data management offers several benefits. Firstly, it improves data security by reducing the risk of data breaches and cyber-attacks. Since the data is distributed across multiple locations, even if one node is compromised, the rest of the network remains secure. Secondly, it increases data accessibility, as data can be accessed from any node in the network. This is particularly useful for large organizations with multiple offices or remote workers who need access to the same data. In finance, decentralized data management can improve the security and efficiency of payment systems by eliminating intermediaries. In SCM, decentralized data management can improve transparency by providing a clear and auditable record of the movement of goods and products. Decentralized data management using BT offers numerous benefits such as improved security, accessibility, and privacy. Its applications are vast and varied, making it a promising solution for many industries seeking to improve their data management and security practices [11].

19.3.2 Immutable and Transparent Ledger

An immutable and transparent ledger is a key feature of BT that ensures that all transactions recorded on the BT are permanent and cannot be altered or deleted. Once a transaction is recorded on the BT, it becomes a permanent part of the ledger and is viewable to anyone with access to the BT network. This ensures that all transactions are transparent and can be verified by anyone on the network. The immutable and transparent ledger offers several benefits. Firstly, it improves data security by making it difficult for hackers to manipulate or destroy data since any attempt to do so would require consensus from the network participants. Secondly, it enhances transparency by providing a clear and auditable record of all transactions on the BT. This is particularly useful in industries such as finance, where transparency is crucial in maintaining trust and integrity in the system. Thirdly, it reduces the risk of fraud and corruption since any attempts to manipulate the ledger would be immediately detected by the network participants. The immutable and transparent ledger has numerous applications in various industries. In finance, it can be used to improve the security and efficiency of payment systems, reducing the need for intermediaries and increasing transaction speed. In SCM, it can be used to track the movement of goods and products, improving transparency and reducing the risk of fraud and counterfeiting. In the healthcare industry, it can be used to securely store and share patient data between healthcare providers, improving data security and privacy. In conclusion, the immutable and transparent ledger is a fundamental feature of BT that provides numerous benefits, including improved security, transparency, and reduced risk of fraud. Its applications are vast and varied, making it a promising solution for many industries seeking to improve their data management and security practices [12, 13].

19.3.3 Reduced Cyber-Attacks

Reducing cyber-attacks is another potential benefit of using BT. Cyber-attacks, such as hacking and phishing, are major threats to automated systems, and they can cause significant financial losses, reputation damage, and legal liabilities. BT's decentralized architecture, cryptographic security, and immutable records can help reduce the risk of cyber-attacks in several ways.

Firstly, BT's decentralized architecture eliminates the need for a central authority or intermediary to manage transactions, making it difficult for hackers to compromise the system. Since BT operates on a network of distributed nodes, no single point of failure exists. This reduces the chances of a successful cyber-attack since attackers would need to compromise a significant portion of the network to take control. Secondly, BT's cryptographic security helps protect against attacks such as data tampering and identity theft. The use of public and private keys to authenticate transactions ensures that only authorized parties can access sensitive data and make transactions on the BT. BT presents an opportunity to reduce cyber-attacks and improve the security of automated systems. Its decentralized architecture, cryptographic security, and immutable record offer several benefits in terms of reducing the risk of cyber-attacks, detecting fraudulent activities, and ensuring the integrity of data. As such, BT has the potential to revolutionize the way we secure automated systems and protect against cyber threats [14, 15].

19.3.4 Secure Data Sharing

One of the significant opportunities of using BT in securing automated systems is the ability to facilitate secure data sharing. BT's decentralized architecture and immutable record make it possible for different parties to share information without the need for intermediaries while ensuring data integrity and confidentiality. With BT, data can be encrypted and shared among authorized parties, providing a secure and efficient way of exchanging information. This opportunity presents itself in industries such as healthcare, where patients' medical records can be securely shared among healthcare providers, improving patient care while maintaining data privacy. For instance, Medrek is a BT-based system that allows patients to control access to their medical records and share them with healthcare providers securely. The system uses SC to facilitate secure data sharing, reducing the risk of data breaches and unauthorized access to patient records. Another example is the financial services industry, where BT is being used to facilitate secure and efficient cross-border payments. Ripple, a BT-based payment protocol, enables banks and financial institutions to make secure and instant cross-border transactions, reducing transaction costs and settlement times. In summary, BT presents an opportunity to enhance secure data sharing across different industries, providing a secure and efficient way of exchanging information while maintaining data privacy and integrity [16, 17].

19.4 CHALLENGES OF SECURING AUTOMATED SYSTEMS WITH BT

While BT presents several opportunities to secure automated systems, it also poses some challenges. In this section, we will discuss some of the challenges of securing automated systems with BT [18–22].

- Firstly, scalability is one of the significant challenges of BT. As the number of transactions increases, the size of the BT grows, making it difficult for nodes to store and validate transactions. This challenge is particularly significant in public BTs like Bitcoin, where the number of transactions can reach thousands per second.
- Secondly, interoperability is another challenge of BT. Different BT platforms operate on different protocols, making it challenging to transfer assets and data across different BT networks. This challenge makes it difficult for businesses to adopt BT, as they may need to use different BT networks to meet their specific business needs.

Securing Automated Systems with BT

- Thirdly, regulatory challenges are also a significant barrier to the adoption of BT. Different countries have different regulations governing the use of BT, making it challenging for businesses to operate across borders. Additionally, the lack of clear regulatory frameworks can lead to legal and regulatory uncertainty, hindering the adoption of BT.
- Fourthly, security risks associated with SC are another challenge of BT. SC can contain bugs and vulnerabilities that can be exploited by hackers to steal funds or disrupt transactions. Additionally, SC are irreversible once deployed, making it difficult to fix any bugs or vulnerabilities

BT presents several opportunities to secure automated systems; it also poses several challenges, including scalability, interoperability, regulatory challenges, security risks associated with SC, and energy consumption. Addressing these challenges will be crucial in ensuring the widespread adoption of BT in securing automated systems.

19.4.1 High Computational Power Requirement

One of the challenges of securing automated systems with BT is the high computational power requirement. The process of verifying and adding transactions to the BT, known as mining, is a computationally intensive process that requires significant processing power and energy consumption. This can limit the scalability of BT and make it difficult for smaller organizations or individuals to participate in the network. Additionally, the increasing demand for computing power can lead to a concentration of mining power in the hands of a few powerful entities, leading to centralization and reducing the security of the network. To address this challenge, alternative consensus algorithms such as proof-of-stake and proof-of-authority have been developed, which require less computational power and are more energy-efficient. However, these algorithms have their own trade-offs and limitations, and more research is needed to determine their effectiveness in securing automated systems with BT [23].

19.4.2 Integration Challenges

Another challenge of securing automated systems with BT is the integration of existing systems with BT networks. Integrating BT with legacy systems can also lead to compatibility issues and may require significant changes to existing systems. Additionally, BT networks may have different security protocols and requirements than existing systems, which can pose a challenge for integration. To address this challenge, organizations need to carefully consider the benefits and risks of integrating BT with existing systems and develop a comprehensive integration plan that addresses compatibility issues and security requirements. This may involve collaborating with BT experts and investing in specialized training and development programs [24].

19.4.3 BT Scalability

BT scalability is another challenge that needs to be addressed when securing automated systems with BT. Currently, most BT networks have limited capacity and can only process a small number of transactions per second. This is due to the computational power required to validate and add transactions to the BT, as well as the need for consensus among network participants. As BT networks become more widely adopted, the number of transactions they need to process will increase, putting pressure on the network's capacity. This can lead to slower transaction times and higher fees, which can impact the user experience and limit the scalability of BT-based systems. To address this challenge, several approaches have been proposed, including shredding, which involves breaking up the BT into smaller parts, and off-chain solutions like payment channels, which can reduce the number of transactions that need to be processed on the main BT. However, implementing these solutions requires significant technical expertise and resources, and there is still a need for further research and development in this area. As BT continues to evolve, it is likely that new solutions will emerge to address the scalability challenge and enable BT-based systems to process a higher volume of transactions more efficiently [25].

19.4.4 Regulatory Challenges

Regulatory challenges are another important aspect of securing automated systems with BT. BT-based systems often operate across multiple jurisdictions, and different countries have different regulations regarding data privacy, security, and financial transactions. In some cases, existing regulations may not be well-suited to the unique characteristics of BT, such as its decentralized architecture and immutable record keeping. This can create uncertainty and make it difficult for organizations to comply with relevant laws and regulations. Moreover, the decentralized nature of BT means that there is no central authority to regulate or oversee transactions. This can create a challenge for regulatory bodies, as they may not have clear lines of authority or the necessary tools to monitor and enforce compliance with relevant laws and regulations. To address these challenges, several initiatives have been proposed, including the development of new regulatory frameworks specifically designed for BT-based systems, as well as collaborations between industry stakeholders and regulatory bodies to identify and address potential regulatory issues. Overall, regulatory challenges are an important consideration when securing automated systems with BT, and there is a need for ongoing dialogue and collaboration between industry stakeholders and regulatory bodies to ensure that BT-based systems can operate in a compliant and secure manner [26].

19.5 BT FOR SECURING AUTOMATED SYSTEM

The use of automated systems is becoming increasingly prevalent in various industries, from manufacturing to healthcare. While automated systems offer numerous benefits, including increased efficiency and reduced costs, they also pose significant security risks. Cyber-attacks, hacking, and other forms of fraud can compromise the security of automated systems and result in financial losses, reputational damage, and even physical harm. In order to enhance the security of automated systems, BT can be utilized. This can help to prevent counterfeit goods, fraud, and other forms of supply chain fraud. Another use case for BT in automated systems is in the area of financial transactions. Traditional financial systems are often centralized and controlled by a single authority, which can make them vulnerable to attacks and fraud. BT-based financial systems, on the other hand, are decentralized and utilize advanced cryptography to secure transactions. This makes it virtually impossible for anyone to tamper with the system or steal funds. Additionally, by using a consensus mechanism such as proof-of-work or proof-of-stake, it is possible to ensure that only valid transactions are added to the BT. In addition, BT can be used to enhance the security of data storage and transfer in automated systems. By using BT, it is possible to create a decentralized system for storing and transferring data, which is much more secure than traditional centralized systems. This can help to prevent data breaches, hacking, and other forms of cyber-attacks. Moreover, BT allows for the creation of SC, which are self-executing contracts that can be programmed to trigger specific actions based on predefined conditions. This can help to automate various processes in automated systems while maintaining security and transparency. The integration of BT can significantly improve the security of automated systems [27–30]. An example of this can be seen in the use of BT to secure the supply chain logistics industry. In a traditional supply chain logistics system, data is often siloed and difficult to access. This lack of transparency can create opportunities for fraud and corruption. However, by implementing BT, the entire supply chain can be secured using an immutable and transparent ledger that records all transactions and movements of goods. This means that all parties involved in the supply chain, from manufacturers to distributors, can access a single source of truth for all transactions. The transparency and immutability of the BT make it nearly impossible for bad actors to tamper with the data or commit fraud.

Preventing fraud, BT can also enhance the security of the supply chain by making it easier to track goods and monitor their movements. This level of transparency allows for early detection of any issues or delays in the supply chain, enabling stakeholders to quickly address any problems before they become bigger issues. Integrating BT into automated systems, such as those in the supply chain logistics industry, can significantly enhance their security. By creating an immutable and transparent ledger, BT enables all parties involved to access a single source of truth, preventing fraud and enhancing transparency. This technology has the potential to revolutionize the way automated systems operate, increasing efficiency and improving security across a range of industries as shown in Figure 19.1.

19.6 CONCLUSION

The rapid growth of automated systems in various industries has brought about significant benefits, but it has also created security risks. In this chapter, we have explored the opportunities and challenges associated with using BT to enhance the security of

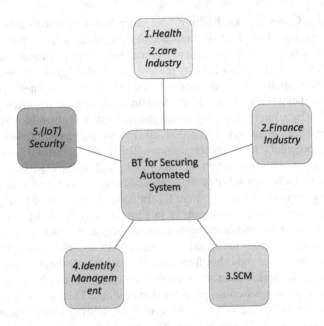

FIGURE 19.1   BT for securing automated system.

automated systems. We have discussed the role of BT in securing automated systems, emphasizing how it can enhance security and transparency. We have also examined BT-based systems with enhanced security that can be utilized to secure automated systems. Furthermore, we have explored the opportunities of securing automated systems with BT, including decentralized data management, immutable and transparent ledger, reduced cyber-attacks, and secure data sharing. These opportunities can help to improve the security of automated systems while enhancing efficiency. Nevertheless, challenges such as high computational power requirements, integration challenges, BT scalability, and regulatory challenges must be addressed to fully realize the benefits of using BT to secure automated systems. By utilizing BT, a more secure and transparent system can be created, ultimately enhancing the reliability and safety of automated systems [31–34]. In conclusion, BT provides a promising solution for enhancing the security of automated systems, but it also presents challenges that must be overcome through continued research and development.

REFERENCES

1. Pandey, J.K.; Jain, R.; Dilip, R.; Kumbhkar, M., Jaiswal, S.; Pandey, B.K.; ... Pandey, D. Investigating Role of IoT in the Development of Smart Application for Security Enhancement. In *IoT Based Smart Applications*. Cham: Springer International Publishing, 2022 (pp. 219–243).
2. Iyyanar, P.; Anand, R.; Shanthi, T.; Nassa, V. K.; Pandey, B. K.; George, A. S.; Pandey, D. A Real-Time Smart Sewage Cleaning UAV Assistance System Using IoT. In *Handbook of Research on Data-Driven Mathematical Modeling in Smart Cities*, IGI Global. 2023, (pp. 24–39).

3. George, A.S.; George, A.H.; Baskar, T.; Pandey, D. The Transformation of the workspace using Multigigabit Ethernet. *Partners Universal International Research Journal* 2022, 1(3), 34–43.
4. Ravidas, S.; Lekidis, A.; Paci, F.; Zannone, N. Access control in Internet-of-Things: A survey. *Journal of Network and Computer Applications* 2019, *144*, 79–101.
5. Raghavan, R.; Verma, D.C.; Pandey, D.; Anand, R.; Pandey, B. K.; Singh, H. Optimized building extraction from high-resolution satellite imagery using deep learning. *Multimedia Tools and Applications* 2022, 81(29), 42309–42323.
6. Jia, J.; Qiu, X.; Cheng, C. Access control method for web of things based on role and SNS. In *Proceedings of the 2012 IEEE 12th International Conference on Computer and Information Technology*, Chengdu, China, 27–29 October 2012, pp. 316–321.
7. Xu, R.; Chen, Y.; Blasch, E.; Chen, G. Blendcac: A BT-enabled decentralized capability-based access control for iots. In *Proceedings of the 2018 IEEE International Conference on Internet of Things (iThings) and IEEE Green Computing and Communications (GreenCom) and IEEE Cyber, Physical and Social Computing (CPSCom) and IEEE Smart Data (SmartData)*, Halifax, NS, 30 July–3 August 2018, pp. 1027–1034.
8. Novo, O. BT Meets IoT: An architecture for scalable access management in IoT. *IEEE Internet of Things Journal* 2018, 5(2), 1184–1195.
9. Dorri, A.; Kanhere, S.S.; Jurdak, R. BT in Internet of Things: Challenges and solutions. *Yingyong Kexue Xuebao/Journal of Applied Sciences* 2020, *38*, 22–33.
10. Gao, W.; Hatcher, W.G.; Yu, W. A survey of BT: Techniques, applications, and challenges. In *Proceedings of the 2018 27th International Conference on Computer Communication and Networks (ICCCN)*, Hangzhou, China, 30 July–2 August 2018.
11. Dorri, A.; Kanhere, S.S.; Jurdak, R.; Gauravaram, P. BT for IoT security and privacy: The case study of a smart home. In *Proceedings of the 2017 IEEE International Conference on Pervasive Computing and Communications Workshops (PerCom Workshops)*, Kona, HI, 13–17 March 2017, pp. 618–623.
12. Ouaddah, A.; Elkalam, A.A.; Ouahman, A.A. Towards a novel privacy-preserving access control model based on BT technology in IoT. In *Europe and MENA Cooperation Advances in Information and Communication Technologies*. Springer, 2017, pp. 523–533.
13. Liu, J.; Xiao, Y.; Chen, C.L.P. Authentication and access control in the Internet of Things. In *Proceedings of the 2012 32nd International Conference on Distributed Computing Systems Workshops*, Macau, China, 18–21 June 2012, pp. 588–592.
14. Ndibanje, B.; Lee, H.J.; Lee, S.G. Security analysis and improvements of authentication and access control in the internet of things. *Sensors* 2014, *14*(8), 14786–14805.
15. Kaiwen, S.; Lihua, Y. role-based hybrid access control in the internet of things. In *Asia-Pacific Web Conference*. Springer, 2014, *8710*, pp. 333–343.
16. Touati, L.; Challal, Y. Poster: Activity-based access control for IoT. In *Proceedings of the 1st International Workshop on Experiences with the Design and Implementation of Smart Objects*, Paris, France, 7–11 September 2015, pp. 29–30.
17. Touati, L.; Challal, Y. Batch-based CP-ABE with attribute revocation mechanism for the Internet of Things. In *Proceedings of the 2015 International Conference on Computing, Networking and Communications (ICNC)*, Garden Grove, CA, 16–19 February 2015, pp. 1044–1049.
18. Sicari, S.; Rizzardi, A.; Coen-Porisini, A.; Cappiello, C. A NFP model for Internet of Things applications. In *Proceedings of the 2014 IEEE 10th International Conference on Wireless and Mobile Computing, Networking and Communications (WiMob)*, Larnaca, Cyprus, 8–10 October 2014, pp. 265–272.
19. Goncalves, F.; Macedo, J.; Nicolau, M.J.; Santos, A. Security architecture for mobile e-health applications in medication control. In *Proceedings of the 2013 21st*

International Conference on Software, Telecommunications and Computer Networks-(SoftCOM 2013), Primosten, Croatia, 18–20 September 2013.
20. Gusmeroli, S.; Piccione, S.; Rotondi, D. IoT access control issues: A capability based approach. In *Proceedings of the 2012 Sixth International Conference on Innovative Mobile and Internet Services in Ubiquitous Computing*, Palermo, Italy, 4–6 July 2012, pp. 787–792.
21. Bernabe, J.B.; Ramos, J.L.H.; Gomez, A.F.S. TACIoT: Multidimensional trust-aware access control system for the Internet of Things. *Soft Computing* 2016, 20(5), 1763–1779.
22. Mahalle, P.N.; Thakre, P.A.; Prasad, N.R.; Prasad, R. A fuzzy approach to trust based access control in internet of things. In *Proceedings of the Wireless VITAE 2013*, Atlantic City, NJ, 24–27 June 2013, pp. 1–5.
23. NTT Innovation Institute. Mandatory access control over IoT communications. Available online: https://labevent.ecl.ntt.co.jp/forum2017/elements/pdf_eng/03/C-8_e.pdf.
24. Seitz, L.; Selander, G.; Gehrmann, C. Authorization framework for the Internet-of-Things. In *Proceedings of the 2013 IEEE 14th International Symposium on "A World of Wireless, Mobile and Multimedia Networks"(WoWMoM)*, Madrid, Spain, 4–7 June 2013.
25. Naedele, M. An access control protocol for embedded devices. In *Proceedings of the 2006 4th IEEE International Conference on Industrial Informatics*, Singapore, 16–18 August 2006, pp. 565–569.
26. Zhang, R.; Zhang, Y.; Ren, K. Distributed privacy-preserving access control in sensor networks. *IEEE Transactions on Parallel and Distributed Systems* 2012, 23, 1427–1438.
27. Pandey, D.; Wairya, S. An optimization of target classification tracking and mathematical modelling for control of autopilot. *The Imaging Science Journal* 2022, 70(6), 371–386.
28. Pandey, D.; Wairya, S.; Sharma, M.; Gupta, A.K.; Kakkar, R.; Pandey, B.K. An approach for object tracking, categorization, and autopilot guidance for passive homing missiles. *Aerospace Systems* 2022, 5(4), 553–566.
29. Kumar, M.S.; Sankar, S.; Nassa, V.K.; Pandey, D.; Pandey, B.K.; Enbeyle, W. Innovation and creativity for data mining using computational statistics. In *Methodologies and Applications of Computational Statistics for Machine Intelligence*. IGI Global, 2021, pp. 223–240.
30. Gupta, A.; Srivastava, A.; Anand, R. Cost-effective smart home automation using Internet of Things. *Journal of Communication Engineering and Systems* 2019, 9(2), 1–6.
31. Gupta, A.; Asad, A.; Meena, L.; Anand, R. IoT and RFID-based smart card system integrated with health care, electricity, QR and banking sectors. In: *Artificial Intelligence on Medical Data: Proceedings of International Symposium, ISCMM*. Springer Nature Singapore, 2021, pp. 253–265.
32. Gupta, A.; Anand, R.; Pandey, D.; Sindhwani, N.; Wairya, S.; Pandey, B.K.; Sharma, M. Prediction of breast cancer using extremely randomized clustering forests (ERCF) technique: Prediction of breast cancer. *International Journal of Distributed Systems and Technologies (IJDST)* 2021, 12(4), 1–15.
33. Pandey, D.; Wairya, S.; Sharma, M.; Gupta, A.K.; Kakkar, R.; Pandey, B.K. An approach for object tracking, categorization, and autopilot guidance for passive homing missiles. *Aerospace Systems* 2022, 5(4), 553–566.
34. Jain, S.; Sindhwani, N.; Anand, R.; Kannan, R. COVID detection using chest X-ray and transfer learning. In *Intelligent Systems Design and Applications: 21st International Conference on Intelligent Systems Design and Applications (ISDA 2021) Held During December 13–15*. Springer International Publishing, 2021, pp. 933–943.

20 Transforming Healthcare with Industry 4.0
The Impact of Social Media on Mental Health

Shashidhar Sonnad, Luigi Pio Leonardo Cavaliere, H.V. Vinay, Sonali Vyas, Lims Thomas, and S. Durga

20.1 INTRODUCTION

SM has become an integral part of modern society, and it has a profound impact on the mental state of the community. The influence of online platforms on people's MH and well-being has been the subject of numerous studies, and the results are mixed. While these platforms offer many benefits, such as facilitating communication, providing access to information, and connecting people, they can also have some negative effects on the mental state of the community. One of the most significant effects of online platforms on the mental state of the community is social comparison. The culture of comparison fostered by SM encourages people to compare their lives with others. They compare their physical appearance, lifestyle, job, and relationships with those of others on digital platforms. This can lead to feelings of inadequacy, anxiety, and depression, particularly in young people who may be more susceptible to the pressure of social comparison [1].

Another significant effect of digital platforms on the mental state of the community is increased social isolation. Despite their name, these platforms can actually lead to social isolation as people spend more time online and less time interacting with others in-person. This can lead to loneliness, depression, and anxiety, particularly in those who are more vulnerable, such as older adults and people with MH issues. Online platforms can also have a detrimental impact on self-esteem. In addition to the negative effects on MH, online platforms can also have a significant impact on physical health. Spending too much time online can lead to poor sleep quality, eye strain, and a sedentary lifestyle. These can all have negative consequences for physical health, including weight gain, diabetes, and cardiovascular disease. While online platforms offer many benefits, they can also have negative effects on the mental state of the community. It is important to use them in moderation and be mindful of their impact on mental and physical health. By taking steps to mitigate these negative effects, we can continue to enjoy the benefits of online platforms while minimizing their potential harm. It is essential to be mindful of the potential negative effects of SM and take steps to mitigate them. This includes limiting screen time, engaging in offline activities, seeking support from friends and family, and prioritizing mental and physical health. By doing so, individuals can continue to enjoy the benefits of SM while minimizing its potential harm. Ultimately, it is up to each person to find a balance between their online and offline lives and take responsibility for their mental and physical well-being [2, 3].

20.1.1 Overview of the Healthcare Industry

The healthcare industry is a vast and complex system that encompasses a wide range of medical services, facilities, and professionals. It includes hospitals, clinics, primary care providers, specialists, pharmacists, medical equipment manufacturers, insurance companies, and government agencies, among others. The healthcare industry plays a critical role in promoting and maintaining the health and well-being of individuals and populations [4–8]. One of the primary functions of the healthcare industry is the provision of medical care and treatment for individuals who are sick or injured. This includes preventive care, such as vaccinations and screenings, as well as diagnostic and therapeutic services, such as surgeries, medications, and rehabilitation. In addition to medical care, the healthcare industry also plays a vital role in promoting health and wellness through health education and public health initiatives. This includes health promotion campaigns, disease prevention efforts, and community outreach programs designed to raise awareness about healthy lifestyles and behaviors. The healthcare industry is subject to extensive regulation and oversight from government agencies, such as the Food and Drug Administration and the Centres for Medicare and Medicaid Services. These agencies are responsible for ensuring the safety and efficacy of medical products and services, as well as monitoring the quality and cost of healthcare delivery. The healthcare industry is also undergoing significant technological transformation, with the rise of digital health technologies, such as electronic health records, telemedicine, and health apps. These technologies have the potential to improve the efficiency and effectiveness of healthcare delivery, as well as enhance patient engagement and empowerment. Despite the many benefits of the healthcare industry, it also faces significant challenges, including rising costs, workforce shortages, and disparities in access to care. Addressing these challenges will require innovative solutions and collaborations across the healthcare ecosystem. Overall, the healthcare industry is a vital component of modern society, providing essential medical services, promoting health and wellness, and contributing to economic growth and development [9, 10].

20.1.2 Introduction to Industry 4.0 in Healthcare

One benefit of I-4.0 technologies in healthcare is the ability to create smart, connected medical devices and equipment that can improve patient outcomes and reduce healthcare costs. For example, IoT-enabled medical devices can monitor patients remotely and

Transforming Healthcare with Industry 4.0

transmit real-time data to healthcare providers, enabling early detection of potential health problems and more timely interventions. Despite the many benefits of I-4.0 technologies in healthcare, there are also challenges and concerns that need to be addressed. These include issues related to data privacy and security, ethical considerations around the use of AI and automation in healthcare, and the need for workforce training and development to support the adoption of these technologies [11]. In conclusion, I-4.0 technologies are transforming the healthcare industry, providing new opportunities for improving patient outcomes, enhancing efficiency, and promoting innovation. As the healthcare industry continues to evolve and adopt these technologies, it will be essential to ensure that they are used in a responsible and ethical manner, and that patients and healthcare providers are empowered to leverage the benefits of these technologies while addressing the challenges and concerns [12].

20.1.3 Importance

Leveraging I-4.0 technologies in healthcare (Figure 20.1) can have a significant impact on patient outcomes, improving both the quality and efficiency of healthcare delivery. Some of the key benefits of leveraging I-4.0 technologies for improved patient outcomes include:

I-4.0 technologies have transformed healthcare, with their potential to significantly improve patient care and outcomes. These technologies leverage data, automation, and artificial intelligence (AI) to provide accurate and efficient healthcare services. One of the most prominent I-4.0 technologies used in healthcare is the Internet of Things (IoT). IoT devices, such as wearables and sensors, are used to monitor patient health remotely. This technology allows for real-time data collection, analysis, and feedback, which can improve patient outcomes. For example, IoT devices can monitor blood sugar levels in patients with diabetes, allowing for early detection of any issues and better disease management [13].

FIGURE 20.1 I-4.0 technologies for betterment.

Another technology that is being used in healthcare is AI. AI is particularly useful for identifying patterns and making predictions based on large datasets. This can help healthcare professionals to identify and diagnose diseases more accurately, and to develop more effective treatments. For example, AI algorithms can analyze medical images, such as X-rays and magnetic resonance imaging, to detect early signs of cancer. Robotics is another I-4.0 technology that is being used in healthcare. Robotics can be used for surgical procedures, drug delivery, and physical therapy. This technology allows for precise and efficient healthcare services, reducing the risk of human error and improving patient outcomes. For example, robotic surgery can reduce the risk of complications during surgery and speed up recovery times.

For example, big data analytics can be used to monitor patient responses to medications, allowing healthcare professionals to adjust dosages and treatment plans as needed. Finally, blockchain technology is being explored in healthcare as a way to improve data security and privacy. Blockchain can be used to store and share medical records securely, ensuring that patient data is kept confidential and secure. This technology can also be used to track and manage the supply chain of pharmaceuticals, reducing the risk of counterfeit drugs entering the market [14–16].

I-4.0 technologies offer many benefits for effective patient care. These technologies have the potential to improve accuracy, efficiency, and accessibility in healthcare services. By leveraging data, automation, and AI, healthcare professionals can provide more precise and personalized care to their patients. While these technologies are still in the early stages of development, their potential to revolutionize healthcare is immense. As these technologies continue to evolve, we can expect to see even greater improvements in patient outcomes and quality of care.

20.2 TECHNOLOGIES IN HEALTHCARE

I-4.0 technologies have been rapidly advancing and are being leveraged in healthcare to improve patient outcomes. Virtual Reality (VR) technology, for example, has been used for medical training and to provide distraction therapy for patients. Medical students can use VR to practice procedures in a safe environment before performing them on real patients. Surgeons can also use VR to plan and rehearse surgeries, which can improve accuracy and reduce complications. Moreover, VR can be used to provide a calming and immersive environment for patients to help them relax and focus their attention away from their symptoms. Table 20.1 shows I-4.0 Technologies in Healthcare [17, 18].

Augmented Reality (AR) technology can also enhance medical education and patient care [19, 20]. Medical students can visualize complex medical concepts and procedures in a more interactive and immersive way using AR. In the operating room, AR can be used to overlay critical information, such as the location of blood vessels and nerves, onto the patient's body. This can help improve surgical accuracy and reduce complications. Additionally, AR can be used in patient education to help patients better understand their medical conditions and treatment options.

Another I-4.0 technology being utilized in healthcare is 3D printing. 3D printing technology has been used to create customized prosthetics, implants, and surgical

TABLE 20.1
Industry 4.0 Technologies in Healthcare

Technology	Description	Examples
Virtual Reality (VR)	Computer-generated simulations of three-dimensional environments that can be experienced through a headset or other devices.	Simulation training for surgical procedures, patient therapy, and pain management.
Augmented Reality (AR)	Digital information overlaid onto real-world environments through a device, such as a smartphone or smart glasses.	Overlaying patient information onto their physical body during surgery, medical education, and patient education.
3D Printing	The creation of physical objects through the layer-by-layer deposition of material.	Customized prosthetics, implants, and surgical tools.
Nanotechnology	Manipulating matter at the molecular or atomic level to create new materials or devices.	Smart pills for targeted drug delivery, nanorobots for disease detection and treatment.

tools that are tailored to the specific needs of each patient. 3D-printed prosthetics can be made quickly and at a lower cost than traditional prosthetics, making them more accessible to patients. Furthermore, 3D printing allows for rapid prototyping of medical devices, making it easier to develop and test new products. Additionally, 3D printing can be used to create models of organs and tissues, allowing surgeons to practice and plan surgeries before performing them [21].

Nanotechnology is another I-4.0 technology that has the potential to revolutionize healthcare. It allows for targeted drug delivery and disease detection and treatment at the molecular or atomic level. Smart pills, for example, can be designed to release medication at specific locations in the body, reducing side effects and improving efficacy. Nanorobots can also be used to detect and treat diseases at the cellular level, providing more precise and effective interventions. Additionally, nanotechnology can be used to develop new materials for medical devices, such as artificial organs and tissues. However, the use of nanotechnology in healthcare is still in its early stages, and there are still many challenges to overcome, such as ensuring the safety and efficacy of these technologies [22].

20.3 IMPROVED PATIENT OUTCOMES THROUGH INDUSTRY 4.0

I-4.0 technologies have the potential to significantly improve patient outcomes in healthcare. By leveraging these technologies, healthcare professionals can provide more personalized, efficient, and effective care to their patients. For example, telemedicine and remote monitoring technologies can help patients receive care in the comfort of their own homes, reducing the need for hospital visits and improving access to care, particularly for patients who live in rural or remote areas. These technologies can also help patients manage chronic conditions by providing continuous monitoring and feedback, allowing healthcare professionals to intervene early and prevent complications. Virtual and augmented reality technologies can improve patient outcomes by providing more accurate and personalized treatment plans. Medical professionals can use these technologies to visualize and plan surgeries, reducing the risk of complications and improving outcomes. These technologies can also be used to provide distraction therapy for patients with chronic pain or anxiety, improving their quality of life. 3D printing and nanotechnology can also improve patient outcomes by providing more personalized and effective treatments. For example, 3D printing can be used to create customized prosthetics and implants that fit perfectly and are more comfortable for patients. Nanotechnology can be used to deliver medications more precisely and effectively, reducing side effects and improving efficacy [23].

20.3.1 Enhanced Diagnosis and Treatment

I-4.0 technologies have the potential to enhance diagnosis and treatment in healthcare, leading to better patient outcomes. These technologies can improve the accuracy and speed of diagnoses, allowing for earlier and more effective treatment. One example of how these technologies can improve diagnosis is through the use of artificial intelligence (AI) and machine learning algorithms [24, 25]. These tools can analyze large amounts of patient data, such as medical histories and imaging scans, to identify patterns and make accurate diagnoses. This can lead to earlier detection and treatment of diseases, reducing the risk of complications and improving outcomes. Another way that these technologies can enhance diagnosis is through the use of wearable devices and sensors. These devices can continuously monitor patient health metrics, such as heart rate and blood pressure, and alert healthcare professionals to any changes or anomalies. This can help healthcare professionals detect conditions early and intervene before they become more serious. Additionally, these technologies can improve treatment in healthcare. For example, robotic surgery can provide more precise and less invasive surgeries, reducing the risk of complications and improving recovery times. Personalized medicine, which uses patient-specific data to develop tailored treatment plans, can improve treatment efficacy and reduce side effects. Moreover, smart medication systems can improve medication management by reminding patients to take their medications, monitoring their compliance, and alerting healthcare professionals to any issues. This can improve medication adherence and prevent adverse outcomes. This technology has the potential to significantly enhance diagnosis and treatment in healthcare, improving patient outcomes and reducing costs. However, successful implementation requires careful planning and investment in infrastructure, training, and education to ensure that healthcare professionals have the skills and resources they need to use these technologies effectively [26].

20.3.2 Personalized Medicine

Personalized medicine is an approach to healthcare that uses patient-specific data, such as genetic information and medical histories, to develop tailored treatment plans. This approach recognizes that each patient is unique and that treatments that work for one patient may not work for another. The use of personalized medicine can improve treatment efficacy and reduce the risk of side effects. For example, genetic testing can identify patients who are more likely to have adverse reactions to certain medications, allowing healthcare professionals to choose alternative treatments. The development of personalized medicine has been made possible by advances in genomics, data analytics, and digital health technologies. These technologies allow for the collection, analysis, and interpretation of large amounts of patient data, providing insights into individual health and disease risk. However, the adoption of personalized medicine faces challenges, such as the high cost of genetic testing and the need for better education and training of healthcare professionals [27].

20.3.3 Remote Patient Monitoring

Patients with diabetes can use connected devices to monitor their blood glucose levels and receive real-time feedback on their diet, exercise, and medication management. Healthcare professionals can use this data to make personalized treatment decisions, adjust medications, and provide ongoing support and education. Remote patient monitoring (RPM) can also be used for postoperative care, allowing patients to recover at home while being monitored remotely by healthcare professionals. This approach can reduce the risk of complications, improve patient comfort and satisfaction, and reduce the length of hospital stays. The use of RPM has been accelerated by the COVID-19 pandemic, which has highlighted the importance of remote healthcare technologies. RPM can help reduce the risk of exposure to infectious diseases, minimize the need for in-person healthcare visits, and improve the continuity of care. However, the adoption of RPM faces challenges, such as the need for reliable and secure connectivity, interoperability with existing healthcare systems, and regulatory barriers. Additionally, the technology must be designed with the patient in mind to ensure ease of use and patient engagement. Despite these challenges, the use of RPM is expected to grow in the coming years, driven by advances in digital health technologies and the increasing demand for remote healthcare services. As such, RPM represents a promising approach to improving patient outcomes and transforming healthcare delivery [28].

20.3.4 Telemedicine

Telemedicine is the use of telecommunications technology to provide remote healthcare services and consultations between healthcare providers and patients. Telemedicine services can include virtual appointments with healthcare providers, remote monitoring of patient health data, and remote diagnostic tests such as imaging and lab tests. These services can be delivered through a variety of mediums such as video conferencing, messaging apps, and mobile apps. The use of telemedicine has increased significantly in recent years, particularly during the COVID-19 pandemic when in-person visits were limited. However, there are still some challenges to be addressed, such as ensuring the security and privacy of patient information and ensuring equitable access to telemedicine services for all patients. Overall, telemedicine has great potential to improve healthcare access and outcomes, especially in the context of I-4.0 technologies [29].

20.3.5 PREVENTIVE HEALTHCARE

Preventive healthcare is an approach to healthcare that focuses on preventing diseases or illnesses before they occur. This is achieved through a combination of education, lifestyle modifications, and early detection and treatment of risk factors or conditions that can lead to more serious health problems. I-4.0 technologies have the potential to enhance preventive healthcare through improved data analytics and monitoring. For example, wearable devices can collect and transmit real-time data on a patient's activity levels, heart rate, and other vital signs, allowing healthcare providers to detect potential health issues early on and provide targeted interventions to prevent the development of more serious conditions. Another example of I-4.0 technology in preventive healthcare is the use of predictive analytics to identify individuals at high risk for certain diseases based on their genetic makeup, lifestyle factors, and medical history. This can allow healthcare providers to implement early screening and intervention strategies to prevent the onset of disease or catch it at an early stage when it is more treatable. Preventive healthcare can also be improved through the use of telemedicine and remote monitoring technologies, allowing healthcare providers to monitor and educate patients on lifestyle modifications and provide early interventions as needed. Overall, the integration of I-4.0 technologies in preventive healthcare has the potential to improve patient outcomes, reduce healthcare costs, and promote healthier lifestyles and behaviors. However, it is important to ensure that these technologies are implemented in an ethical and equitable manner to ensure that all patients have access to the benefits of preventive healthcare [30–32].

20.4 CHALLENGES IN HEALTHCARE

While I-4.0 technologies offer numerous benefits and opportunities for improving healthcare, there are also several challenges and limitations that must be considered [33].

- *Cost*: the adoption of I-4.0 technologies in healthcare can be expensive, requiring significant investments in infrastructure, hardware, and software. In addition, there may be ongoing maintenance costs associated with these technologies, which can make them unaffordable for some healthcare organizations.
- *Implementation Challenges*: the implementation of I-4.0 technologies in healthcare can be complex and require significant changes to existing workflows and processes. Healthcare organizations must be prepared to invest time and resources into training staff and ensuring a smooth transition to these new technologies.
- *Interoperability*: With the use of multiple digital systems and technologies, there is a need for interoperability to ensure that patient data can be shared seamlessly across different platforms. This requires standardization of data formats and protocols, which can be challenging to achieve.

- *Ethics and Governance*: the use of I-4.0 technologies in healthcare raises ethical and governance issues, such as how to ensure that the technologies are used in a responsible and ethical manner, how to protect patient autonomy and privacy, and how to ensure that the technologies are used to benefit all patients, regardless of their socio-economic status.
- *Patient Acceptance*: patients may be hesitant to adopt new technologies, particularly if they are unfamiliar with them or perceive them as intrusive. Healthcare organizations must engage patients and provide education and support to ensure that patients are comfortable with the use of these technologies and understand the benefits they offer.
- *Regulatory and Legal Issues*: the use of I-4.0 technologies in healthcare is subject to regulatory and legal requirements, such as data protection laws, medical device regulations, and intellectual property laws. Healthcare organizations must ensure that they comply with these requirements to avoid legal and financial liabilities.

I-4.0 technologies offer numerous opportunities for improving healthcare; they also present several challenges and limitations that must be addressed to ensure their successful adoption and implementation in healthcare settings. Healthcare organizations must carefully consider these challenges and develop strategies to address them to fully leverage the benefits of these technologies for improved patient outcomes [33].

20.5 THE EVIDENCE OF THE IMPACT OF SM ON MH

The use of technology to communicate and interact with others has revolutionized human behavior. However, overindulgence in such technology can have negative impacts on MH, particularly on anxiety and depression. One of the reasons why technology use can be linked to anxiety and depression is the constant comparison to others. Users may compare their lives, appearance, and achievements to others, leading to feelings of inadequacy and low self-esteem. Furthermore, users may become overly reliant on validation and feedback from others, which can contribute to anxiety and stress. The constant need to be online and connected can also lead to sleep deprivation, which can further contribute to anxiety and depression [34]. Moreover, technology platforms can provide an environment for cyberbullying, which can cause psychological harm, especially to vulnerable individuals. The anonymity and ease of access make it easier for bullies to target individuals and make hurtful comments, leading to depression and social anxiety. Victims of cyberbullying may feel helpless and isolated, and the negative impact on their MH can be long-lasting. However, technology can also provide a source of social support and positive interaction. Additionally, it is crucial to prioritize positive interactions, such as those with family and friends, and seek help when experiencing MH issues.

20.5.1 Positive and Negative Effects

The impact of technology on MH, particularly in terms of its effects on anxiety and depression, can be both positive and negative. One of the positive effects of

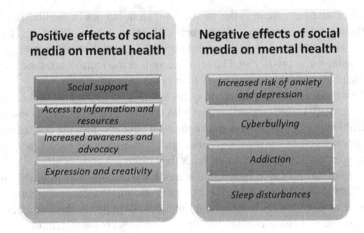

FIGURE 20.2 Positive and negative effects of SM on MH.

technology is its ability to connect people and promote social support. Through various online platforms, individuals can connect with others, share experiences, and find support. This can be particularly beneficial for those who may be geographically isolated or have difficulty accessing traditional forms of social support. However, the overuse of technology, especially through SM, can have negative impacts on MH. The constant comparison to others, the need for validation, and the pressure to present a perfect life can lead to feelings of inadequacy, low self-esteem, and anxiety. Furthermore, online platforms can be a breeding ground for cyberbullying, which can cause psychological harm, particularly to vulnerable individuals.

Additionally, the constant need to be online and connected can lead to sleep deprivation, which can exacerbate existing MH issues or contribute to the development of new ones. The constant bombardment of information, often with negative or distressing content, can also contribute to stress and anxiety. However, it is important to note that the impact of technology on MH is not uniform across all individuals or contexts. Some people may be more resilient to the negative effects of technology, while others may be more susceptible. Additionally, the way technology is used and the content that is consumed can have a significant impact on its effects on MH. In conclusion, while technology has provided many benefits, its overuse, particularly through SM, can have negative impacts on MH, including anxiety and depression. It is important to be mindful of our technology use and to take breaks when necessary to protect our MH. Seeking support from online resources, MH professionals, or in-person social support can also help promote positive MH outcomes [35]. Figure 20.2 shows the positive and negative effects of SM on Mental Health.

20.6 CONCLUSION

In conclusion, the evidence suggests that SM has both positive and negative effects on MH. While it has the potential to promote social support and reduce stigma, it also poses risks such as cyberbullying, social comparison, and addiction. The impact of SM on MH is complex and multifaceted, and it is influenced by individual differences and contextual factors. The emergence of I-4.0 technologies has provided new opportunities to improve healthcare outcomes, including MH outcomes. However, their implementation must be carefully managed to avoid exacerbating existing inequalities and to protect patient privacy. Additionally, the potential for these technologies to transform MH care delivery is significant, but further research is needed to understand their optimal use and effectiveness. Future research should focus on longitudinal studies that assess the long-term impact of SM on MH, as well as more nuanced approaches that take into account individual differences and contextual factors. This will require collaboration between researchers, clinicians, and technology developers to develop effective strategies for promoting positive MH outcomes in the digital age [36]. Overall, it is clear that SM is a powerful tool with both potential benefits and risks for MH. It is important that we continue to monitor and understand its impact, and develop evidence-based strategies to promote positive MH outcomes in the digital age. By doing so, we can harness the potential of technology to transform MH care and improve the lives of individuals and communities around the world.

REFERENCES

1. Zhao, J. Neural Network-Based Optimal Tracking Control of Continuous-Time Uncertain Nonlinear System via Reinforcement Learning. *Neural Process. Lett.* 2020, *51*(3), 2513–2530.
2. Xiong, L.; Zhang, H.; Li, Y.; Liu, Z. Improved Stability and H∞ Performance for Neutral Systems with Uncertain Markovian Jump. *Nonlinear Anal. Hybrid Syst.* 2016, *19*, 13–25.
3. Kamble, S. S.; Gunasekaran, A.; Gawankar, S. A. Sustainable Industry 4.0 Framework: A Systematic Literature Review Identifying the Current Trends and Future Perspectives. *Process Saf. Environ. Prot.* 2018, *117*, 408–425.
4. Bommareddy, S.; Khan, J. A.; Anand, R. A Review on Healthcare Data Privacy and Security. *Netw. Technol. Smart Healthc.* 2022, 165–187.
5. Pandey, B. K.; Pandey, D.; Anand, R.; Singh, H.; Sindhwani, N.; Sharma, Y. (2022) The Impact of Digital Change on Student Learning and Mental Anguish in the COVID Era. *An Interdisciplinary Approach in the Post-COVID-19 Pandemic Era*, 197–206.
6. Pandey, D.; Pandey, B. K.; Sindhwani, N.; Anand, R.; Nassa, V. K.; Dadheech, P. (2022) An Interdisciplinary Approach in the Post-COVID-19 Pandemic Era. *An Interdisciplinary Approach in the Post-COVID-19 Pandemic Era*, 1–290.
7. Sindhwani, N.; Rana, A.; Chaudhary, A. Breast Cancer Detection Using Machine Learning Algorithms. In *2021 9th International Conference on Reliability, Infocom Technologies and Optimization (Trends and Future Directions)(ICRITO) 2021*. IEEE, 2021, September, pp. 1–5.
8. Gupta, A.; Goyal, B.; Dogra, A.; Anand, R. Proximity Coupled Antenna with Stable Performance and High Body Antenna Isolation for IoT-Based Devices. In *Communication, Software and Networks: Proceedings of India*. Springer Nature, 2022, pp. 591–600.
9. Lv, Z.; Song, H. Mobile Internet of Things under Data Physical Fusion Technology. *IEEE Internet Things J.* 2020, *7*(5), 4616–4624.

10. Boyes, H.; Hallaq, B.; Cunningham, J.; Watson, T. The Industrial Internet of Things (IIoT): An Analysis Framework. *Comput. Ind.* 2018, *101*, 6–18.
11. Anand, R.; Shrivastava, G.; Gupta, S.; Peng, S. L.; Sindhwani, N. Audio Watermarking with Reduced Number of Random Samples. In *Handbook of Research on Network Forensics and Analysis Techniques.* IGI Global, 2018, pp. 372–394.
12. Qadri, Y. A.; Nauman, A.; Bin Zikria, Y.; Vasilakos, A. V.; Kim, S. W. The Future of Healthcare Internet of Things: A Survey of Emerging Technologies. *IEEE Commun. Surv. Tutor.* 2020, *22*(2), 1121–1167.
13. Qiu, T.; Shi, X.; Wang, J.; Li, Y.; Qu, S.; Cheng, Q.; Cui, T.; Sui, S. Deep Learning: A Rapid and Efficient Route to Automatic Metasurface Design. *Adv. Sci.* 2019, *6*, 1900128.
14. Li, T.; Xu, M.; Zhu, C.; Yang, R.; Wang, Z.; Guan, Z. A Deep Learning Approach for Multi-Frame In-Loop Filter of HEVC. *IEEE Trans. Image Process.* 2019, *28*(11), 5663–5678.
15. Qian, J.; Feng, S.; Tao, T.; Hu, Y.; Li, Y.; Chen, Q.; Zuo, C. Deep-Learning-Enabled Geometric Constraints and Phase Unwrapping for Single-Shot Absolute 3D Shape Measurement. *APL Photonics* 2020, *5*(4), 046105.
16. Yang, J.; Li, S.; Wang, Z.; Dong, H.; Wang, J.; Tang, S. Using Deep Learning to Detect Defects in Manufacturing: A Comprehensive Survey and Current Challenges. *Materials (Basel)* 2020, *13*(24), 5755.
17. Pandey, D., Ogunmola, G. A., Enbeyle, W., Abdullahi, M., Pandey, B. K., & Pramanik, S. (2021). COVID-19: A framework for effective delivering of online classes during lockdown. *Human Arenas*, 1-15.
18. Theodoridis, S. Chapter 1—Introduction. In *Machine Learning: A Bayesian and Optimization Perspective*, 2nd ed.; Theodoridis, S., Ed. Academic Press, 2020, pp. 1–17. ISBN 978-0-12-818803-3.
19. Badotra, S.; Tanwar, S.; Rana, A.; Sindhwani, N.; Kannan, R. (Eds.). *Handbook of Augmented and Virtual Reality* (Vol. 1). Walter de Gruyter GmbH & Co KG., 2023
20. Nijhawan, M.; Sindhwani, N.; Tanwar, S.; Kumar, S. Role of Augmented Reality and Internet of Things in Education Sector. In *IoT Based Smart Applications.* Springer International Publishing, 2022, pp. 245–259.
21. He, S.; Guo, F.; Zou, Q.; Ding, H. MRMD2.0: A Python Tool for Machine Learning with Feature Ranking and Reduction. *Curr. Bioinform.* 2020, *15*(10), 1213–1221.
22. Gao, N.; Luo, D.; Cheng, B.; Hou, H. Teaching-Learning-Based Optimization of a Composite Metastructure in the 0–10 kHz Broadband Sound Absorption Range. *J. Acoust. Soc. Am.* 2020, *148*(2), EL125–EL129.
23. El-Rashidy, N.; El-Sappagh, S.; Islam, S.; El-Bakry, H.; Abdelrazek, S. End-to-End Deep Learning Framework for Coronavirus (COVID-19) Detection and Monitoring. *Electronics* 2020, *9*(9), 1439.
24. Kumar, M. S.; Sankar, S.; Nassa, V. K.; Pandey, D.; Pandey, B. K.; Enbeyle, W. Innovation and Creativity for Data Mining Using Computational Statistics. In *Methodologies and Applications of Computational Statistics for Machine Intelligence*. IGI Global, 2021, pp. 223–240.
25. Anand, R.; Nirmal, V.; Chauhan, Y.; Sharma, T. An Image-Based Deep Learning Approach for Personalized Outfit Selection. In *2023 10th International Conference on Computing for Sustainable Global Development (INDIACom) 2023.* IEEE, 2023, March, pp. 1050–1054.
26. Bhavsar, K. A.; Singla, J.; Al-Otaibi, Y. D.; Song, O.-Y.; Bin Zikriya, Y.; Bashir, A. K. Medical Diagnosis Using Machine Learning: A Statistical Review. *Comput. Mater. Contin.* 2021, *67*, 107–125.

27. Cao, B.; Wang, X.; Zhang, W.; Song, H.; Lv, Z. A Many-Objective Optimization Model of Industrial Internet of Things Based on Private Blockchain. *IEEE Netw.* 2020, *34*(5), 78–83.
28. Li, B.-H.; Liu, Y.; Zhang, A.-M.; Wang, W.-H.; Wan, S. A Survey on Blocking Technology of Entity Resolution. *J. Comput. Sci. Technol.* 2020, *35*(4), 769–793.
29. Kurdi, H.; Alsalamah, S.; Alatawi, A.; Alfaraj, S.; Altoaimy, L.; Ahmed, S. H. HealthyBroker: A Trustworthy Blockchain-Based Multi-Cloud Broker for Patient-Centered eHealth Services. *Electronics* 2019, *8*(6), 602.
30. Carvalho, N.; Chaim, O.; Cazarini, E.; Gerolamo, M. Manufacturing in the Fourth Industrial Revolution: A Positive Prospect in Sustainable Manufacturing. *Procedia Manuf.* 2018, *21*, 671–678.
31. Lelisho, M.E.; Pandey, D.; Alemu, B.D.; Pandey, B.K.; Tareke, S.A. The negative impact of social media during COVID-19 pandemic. *Trends in Psychology* 2023, 31(1), 123–142.
32. Meslie, Y.; Enbeyle, W.; Pandey, B. K.; Pramanik, S.; Pandey, D.; Dadeech, P.; Belay, A.; Saini, A. Machine Intelligence-Based Trend Analysis of COVID-19 for Total Daily Confirmed Cases in Asia and Africa. In: *Methodologies and Applications of Computational Statistics for Machine Intelligence.* IGI Global, 2021, pp. 164–185.
33. Chen, C.; Wang, X.; Wang, Y.; Yang, D.; Yao, F.; Zhang, W.; Wang, B.; Sewvandi, G. A.; Yang, D.; Hu, D. Additive Manufacturing of Piezoelectric Materials. *Adv. Funct. Mater.* 2020, *30*(52), 2005141.
34. Popov, V.; Grilli, M.; Koptyug, A.; Jaworska, L.; Katz-Demyanetz, A.; Klobčar, D.; Balos, S.; Postolnyi, B. O.; Goel, S. Powder Bed Fusion Additive Manufacturing Using Critical Raw Materials: A Review. *Materials (Basel)* 2021, *14*(4), 909.
35. Popov, V.; Fleisher, A.; Muller-Kamskii, G.; Avraham, S.; Shishkin, A.; Katz-Demyanetz, A.; Travitzky, N.; Yacobi, Y.; Goel, S. Novel Hybrid Method to Additively Manufacture Denser Graphite Structures Using Binder Jetting. *Sci. Rep.* 2021, *11*(1), 2438.
36. Kumar Pandey, B.; Pandey, D.; Nassa, V. K.; Ahmad, T.; Singh, C.; George, A.S.; Wakchaure, M.A. Encryption and steganography-based text extraction in IoT using the EWCTS optimizer. *The Imaging Science Journal* 2021, 69(1–4), 38–56..

21 Multiband Antenna Design for Internet of Things (IoT) Applications

Aikjot Kaur Narula and Amandeep Singh Sappal

21.1 INTRODUCTION

The fourth industrial revolution (industry 4.0 or I4.0) necessitates digitalization as a critical component in the manufacturing sector. The advancement of sensing and communication technology has resulted in increased visibility in production processes. Automation facilitated by the Internet of Things (IoT) aids in the real-time collection of information about products, processes, and stages via sensors. This data is referred to as the digital twin, and it allows industrial processes to be handled efficiently, with high transparency, efficiency, accuracy, and decreased risks. IoT is an umbrella term for numerous embedded computer systems, sensors such as wireless sensor networks, control systems, and automation, all of which include automated identification and data-capture (AIDC) [1–6]. Optical sensors such as barcodes, digital sensors such as cameras, and wireless sensors such as radio-frequency identification (RFID) are examples of typical AIDC sensors. Passive RFID technology, such as Ultra-high frequency (UHF), makes the I4.0 dream a reality by allowing an everyday object to be identified, located, authenticated, and engaged with other IoT devices [7–9]. Society is on the verge of a new industrial revolution, in which self-organizing factories will function autonomously with minimum human interference and products will optimize their own manufacturing cycle based on real-time demand of clients. This concept, known as industry 4.0, is reshaping the global industrial environment. To realize Industry 4.0's full potential, it combines three essential technical developments: the Internet of Things, sophisticated wireless technology, and artificial intelligence. Manufacturers are striving to integrate digital technologies that boost flexibility and efficiency, eliminate unplanned downtimes, and speed up production. Interoperability, a chance to generate customer value by allowing dependable machine-to-machine (M2M) capabilities, lies at the heart of this development, known as Industry 4.0.

A basic microstrip patch antenna consists of a metallic patch radiator on a thin dielectric substrate with the ground of metallic material generally copper. As stated earlier, these days the use of wireless communication is in demand [10, 11] and an antenna is an integral part of a communication system. Due to its smaller size and use at high frequency applications, the microstrip patch antenna has been a major attraction for researchers among various available antennas over the last few decades. The microstrip patch antenna is easy to fabricate and major research on microstrip patch antennas is to achieve small size, wide bandwidth, increased gain, and multiband operation [12, 13].

In today's daily life, we are surrounded by wireless devices that can communicate with each other through the internet. Electronic circuits, sensors, actuators, and software are embedded in these devices, which help them to communicate with each other. The system formed by the interconnection of these devices is called the Internet of Things (IoT). For these devices to communicate with each other through a wireless medium, an antenna is required. In order to decrease the size of these devices, a small antenna size is required. As these devices may be heterogeneous and may be working on different frequency bands, an antenna with multiband operation is required. The IoT network consists of devices called nodes which sense the information, collect it in reasonable form, and transmit this information to a gateway that provides the internet/cloud connectivity. The communication between these nodes and gateways should be done in such a way that power consumption is minimal. Thus, in order to deploy an IoT network, an antenna with multiple bands is required. The scope of the present work is to design a compact microstrip patch antenna which should work at multiple frequency bands.

21.2 IOT ARCHITECTURE

IoT is basically defined as the network devices which have the capability of sensing, accumulating, and transferring the data on the internet without any interference from humans. There are various functional blocks in the IoT system which facilitates different kinds of utilities. The first component of an IoT system is the device. The devices present in the IoT system can sense, actuate, control, and monitor the activities. IoT devices and other connected applications can exchange or collect data from one another through applications based on the cloud network [14]. Various interfaces are present in the IoT device so that it can communicate with other wired and wireless devices. These interfaces are for sensors, connectivity with the internet, and audio/video. All these interfaces are shown in Figure 21.1. The second functional block in the IoT system is communication [15]. The remote servers and the devices communicate in this block. The communication protocols in IoT work under the data link, network, application, and transport layer. The third functional block is service. Different functions are served by the IoT system in this block, such as modeling and controlling the device, publishing and analyzing the data, and discovering the device. The fourth functional block is management. Different functions are provided by the management block to govern the IoT system. The fifth functional block of the IoT system is security. Various functions such as authorization, integrity of message, authentication, privacy, and security of data are provided by this block to secure the IoT system. The last functional block of the IoT system is the application layer. It provides the module by the interface through which various aspects of the IoT system

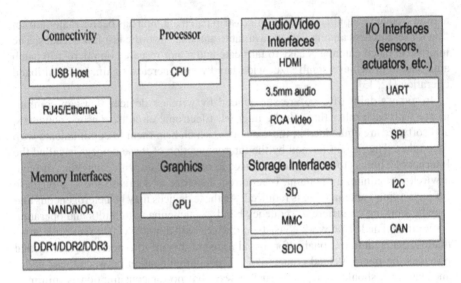

FIGURE 21.1 Various interfaces in device component of IoT.

are controlled and monitored. This also allows the visualization and analysis of the status of the system in its present stages and its future prospects [16, 17].

An object which is equipped with sensors is termed as a "thing." The data is gathered and transferred in the network with the help of sensors which can be connected with humans and these things are allowed to act with the help of actuators. Figure 21.2 shows an Internet of Thing /Internet of Everything architecture model. Many objects in daily use are included in this concept such as buildings, automobiles, fridges, street lamps, production machinery, and so on. It is not necessary that the sensors be attached physically to the object; sensors can also be used to monitor things. Gateways are used to move the data from things to the cloud and vice versa. The things and the cloud are connected through gateways. The preprocessing and filtering of the data are done in the gateways before the data is moved to the cloud. This is done for the volume reduction of the data so that it can be easily processed and stored further. Then the control commands from the cloud to the things are transmitted. The commands are then executed by things with the use of their actuators [18, 19]. The data compression is facilitated by cloud gateways. Cloud gateways also ensure the secure transmission of data between cloud IoT servers and field gateways. The various protocols need to be compatible so that the communication with the field gateways is supported by the protocols depending upon the gateways. The input data is effectively translated to the data lake through a streaming data processor. This also ensures the control of applications so that no data is lost or corrupted. The connected devices generate the data in its natural format which is stored in the data lake. The data is extracted from the data lake when it is needed for insights. The data is then loaded to the warehouse of big data (batches and streams). The big data warehouse stores the data that is extracted from the data lake for meaningful insights.

Multiband Antenna Design for Internet of Things (IoT) Applications 365

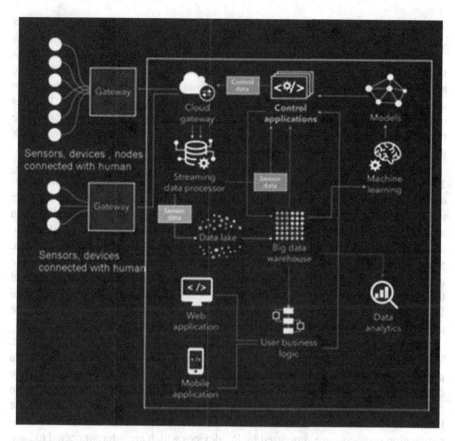

FIGURE 21.2 IoT/IoE architecture.

This data is preprocessed and filtered. Only the matched and structured data is present in the big data warehouse, unlike the data present in the data lake which contains all types of sensor generated data. The information related to the things and sensors is also stored in a big data warehouse. The application sent to the things is controlled by commands. The data from the big data warehouse can be used by data analytics, which use the data to analyze the performance of the devices which further helps in identifying the inefficiencies and then working out ways to improve the IoT system [20]. This makes the system reliable and customer-oriented. This can also contribute to control applications by creating algorithms. Machine learning helps create opportunities for efficient and precise models used in control applications. The new models are tested based on their applications and efficiencies and approved by data analysts. The commands to the actuators are sent through control applications. The big data warehouse stores the commands sent to actuators through control applications. The security of the data is also ensured through the storing commands that come in big amounts [21]. In rule-based control applications, the specialists decide some rules and the control apps work according to these rules. In machine learning control applications, the models are used by control apps which are updated regularly; this can be once a week or month.

The models are updated using the data stored in the big data warehouse. Users should always have an option to influence the nature of the application, although the better automation of IoT systems is ensured by control apps. The IoT systems and the users are connected through software components called user applications. The user can control and monitor the smart things through user applications. The state of the things can be monitored using the web or mobile app. Users can also send commands to control the applications and set their behavior.

Installing the app is not enough for the IoT devices to function sufficiently. The performance of the devices also needs to be managed through certain procedures. This includes the interaction between the devices and the data transmission to be secure. The identity of the device is established through device identification which ensures that the device has the software that transmits the data reliably. The devices are tuned through control and configuration according to the IoT system. Some parameters are written on the device after installation and some may require updates. The devices are diagnosed and monitored to ensure the performance of the device is smooth and secure. The risk of breakdowns is therefore reduced. The functionality and security vulnerabilities are added to the devices through software updates and maintenance [22].

There are various utilities of IoT. The IoT devices are self-adapting and dynamic. With the changing contexts, these devices adapt dynamically, and the actions are taken according to the environment being sensed. The IoT devices and systems have the capability to self-configure, which allows various devices to work together so that proper functionality can be attained. They have the capability to upgrade the latest software and networking setup with minimal user interference. Various communication protocols are supported by IoT devices which are interconnected to each other and with the infrastructure as well. A unique identity is processed by each IoT device [23]. The interface in the IoT device allows the user to do status monitoring, remote control, and infrastructure management. The integration of IoT devices with an information network allows them to communicate with other systems and exchange files with them. The knowledge of the surrounding nodes can be gained through the sensor nodes present in the IoT devices.

21.3 ANTENNA DESIGN EQUATIONS AND CONFIGURATION

Microstrip patch antennas (MPAs) are a type of planar antenna that has been widely investigated and developed over the last four decades. They have been popular among antenna designers and have been employed in a variety of applications in wireless communication systems, both in the military and commercial sectors. G.A. Deschamps suggested the microstrip antenna in 1953, but it did not become practical until the 1970s, when it was further refined by researchers such as Robert E. Munson. The frequency of a microstrip antenna is inversely proportionate to its size. Because of the sizes required, microstrip patches do not make sense at frequencies lower than microwave. At X-band, a microstrip antenna is around 1 centimeter long (easy to realize on soft-board technology). A microstrip antenna to receive FM radio at 100 MHz would be on the order of 1 meter long (which is a very huge circuit for any sort of substrate). The microstrip patch would be the size of a football field for AM radio at 1000 KHz, making it completely unfeasible. Satellite radio receivers are one common use for microstrip patches (XM and Sirius). A microstrip patch antenna has recently been widely used for IoT-based applications [24, 25].

The patch is often constructed of a conducting material such as copper or gold and can be any form such as rectangular, round, triangular, elliptical, or any other common shape. Typically, the radiating patch and feed lines are photo-etched on the dielectric substrate. The fringing fields between the patch edge and the ground plane are principally responsible for the radiation of microstrip patch antennas. A thick dielectric substrate with a low dielectric constant <6 is preferable for superior antenna performance because it gives more efficiency, a bigger bandwidth, and better radiation. However, this design results in a bigger antenna size. To create a small microstrip patch antenna, a substrate with a higher dielectric constant <12 must be utilized, resulting in reduced efficiency and a narrower bandwidth [26]. As a result, a balance must be struck between antenna size and antenna performance. Excitation directs the electromagnetic energy source to the patch, causing negative charges to form around the feed point and positive charges to form on the other side of the patch. This charge difference causes electric fields in the antenna, which are responsible for the patch antenna's radiations. Electromagnetic waves of three sorts are emitted. The first component is emitted into space and is referred to as "useful" radiation. The second component is diffracted waves, which are reflected back into space between the patch and the ground plane and contribute to power transmission [27–29]. Due to complete reflection at the air-dielectric separation surface, the last component of the wave stays trapped in the dielectric substrate. The trapped waves in the substrate are often unfavorable.

A rectangular patch that resembles a truncated microstrip transmission line is the most often used microstrip antenna. It is around one-half wavelength long. The length of the rectangular microstrip antenna is roughly one-half of a free-space wavelength when air is utilized as the dielectric substrate. The length of the antenna reduces as the relative dielectric constant of the substrate grows when the antenna is loaded with a dielectric as its substrate. Because of the prolonged electric "fringing fields," which increase the electrical length of the antenna significantly, the resonant length of the antenna is slightly shorter. A piece of microstrip transmission line with identical loads on either end to approximate radiation loss is an early model of the microstrip antenna.

There are several ways to feed microstrip patch antennas. These approaches can be both contacting and non-contacting. The radio frequency power is delivered directly to the radiating patch via a connecting device such as a microstrip wire in the contacting approach. Power is exchanged between the microstrip line and the radiating patch via electromagnetic coupling in the non-contacting approach. There are several feed ways, but the four most common are microstrip line, coaxial probe (both contacting systems), aperture coupling, and proximity coupling (both non-contacting schemes). In a microstrip line feed, a conducting strip is linked directly to the microstrip patch's edge. When compared to the patch, the conducting strip is narrower. This type of feed arrangement has the benefit of allowing the feed to be etched on the same substrate, resulting in a planar structure [30]. In coaxial feed, the coaxial connector's inner conductor extends through the dielectric and is attached to the radiating patch, while the outside conductor is linked to the ground plane. The main advantage of this is that the feed may be placed in any of the patch's desirable positions to match its input impedance. The ground plane separates the radiating patch and the microstrip feed line in the Aperture Coupled Feed method. The patch and feed line are connected via a slot in the ground plane. Due to the symmetry of the design, the coupling slot is centered below the patch, resulting in reduced cross-polarization. Spurious radiation is reduced because the ground plane separates the patch and the feed line. The primary downside of this feed technology is that it is difficult to manufacture because of the many layers, which also increases antenna thickness. In the Aperture Coupled Feed technique, two dielectric substrates are employed, with the feed line running between them. The radiating patch is positioned on top of the upper substrate. The fundamental benefit of this feed approach is that it avoids spurious feed radiation while still providing a very high bandwidth (as high as 13 percent). The main drawback of this feed method is that it is difficult to manufacture due to the two dielectric layers that must be properly aligned. Microstrip patch antennas are well known for their high performance and sturdy construction. Microstrip patch antennas are used in a variety of sectors, including medicine, satellites, and military equipment such as rockets, airplanes, and missiles, among others. They are now thriving commercially because of the cheap cost of the substrate material and production [31].

Figure 21.3 depicts the geometry of a rectangular microstrip antenna, excluding the ground plane and dielectric that would lie beneath it. The dimension L is often understood to be the long dimension, which induces resonance at its half-wavelength frequency. The radiating edges are at the extremities of the rectangle's L-dimension,

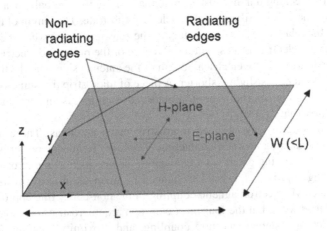

FIGURE 21.3 Geometry of microstrip patch antenna.

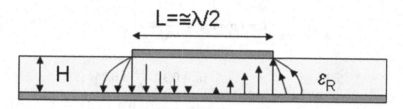

FIGURE 21.4 E-Field produced in microstrip patch antenna.

which creates a single polarization. Cross-polarization is the term used to describe radiation that happens at the extremes of the W-dimension.

Figure 21.4 depicts the E-field beneath the patch. The fields under the L-edges are of opposing polarity (owing to the patch's half-wave nature), and as the field lines curve out and eventually propagate out into the direction normal to the substrate, they are now in the same direction (both facing left). This is because the fields are in phase, the radiation from both sides adds up in the distant field perpendicular to the substrate, and this forms an antenna. It can be seen from the boresight directions that the intensity decreases as the fields of the two edges go further and further out of phase. The fields perfectly cancel at two angles. As a result, the strength of the microstrip patch radiation varies according to the direction from which it is perceived [32].

In order to design a multiband antenna, dimensions are chosen on the basis of frequency 5.775 GHz. FR-4 epoxy is used as substrate with a loss tangent of 0.02 and dielectric constant ε_r of 4.4 with the ground plane being 25.5873×21.5675mm². The thickness of copper is taken as 0.03 mm and the height of the substrate is taken as 1.63mm. Inset feed has been used for the proposed design to achieve an impedance of 50Ω. The Split Ring Resonator (SRR) was developed in [33] and its complementary structure, known as the complementary split ring resonator (CSRR), has found applications in designing planar microwave filters, phase shifters, microwave power dividers, etc. The SRR has also been used for designing high gain antennas in [34–38]. The circular SRR is the most common configuration used by researchers. In the proposed antenna two symmetrical circular SRRs are used for achieving resonance at multiple frequencies. Various parameters of the patch antenna are calculated using the following equations [39–44].

Patch Width, $W=Wp$ in m:

$$Wp = \frac{c}{2f\sqrt{\frac{\varepsilon_r + 1}{2}}} \tag{1}$$

Patch Length $L = Lp$, in m:

Also, $Leff = L + 2 \times \Delta L$ or $L = Leff - 2 \times \Delta L$

$$L = Lp = L_{eff} = \frac{c}{2f\sqrt{\varepsilon_{reff}}} \qquad (3)$$

$$\text{Where, } \Delta L = 0.412h \times \frac{(\varepsilon_r + 0.3)\left(\frac{Wp}{h} + 0.264\right)}{(\varepsilon_r - 0.258)\left(\frac{Wp}{h} + 0.8\right)} \qquad (4)$$

Where, c is velocity of light in m/sec
f is resonant frequency in Hz
ε_r is relative dielectric constant of substrate
ε_{reff} is effective dielectric constant of substrate given by

$$\varepsilon_{reff} = \frac{\varepsilon_r + 1}{2} + \frac{\varepsilon_r - 1}{2}\left[1 + \frac{12h}{Wp}\right]^{-1/2} \qquad (5)$$

h is height of substrate in m
Ground plane width and length are calculated as:

$$W_g = 2 \times W_p \text{ or } 6 \times h \times W_p \qquad (6)$$

$$L_g = 2 \times L_p \text{ or } 6 \times h \times L_p \qquad (7)$$

Depth of Notch

$$y_0 = \frac{Lp}{\pi}\cos^{-1}\left(\frac{Z_0}{R_{in}}\right) \text{ in mm} \qquad (8)$$

Where, $\eta_0 = 377\Omega$ intrinsic impedance of free space

$$\text{And } R_{in} = \frac{1}{2(G_1 + G_{12})} \qquad (9)$$

$$\text{Where } G_1 = \frac{1}{\pi\eta_0}\int_{\theta=0}^{\theta=\pi}\left[\frac{\sin\left(\frac{k_0 W_p}{2}\cos\theta\right)}{\cos\theta}\right]^2 \sin^3\theta\, d\theta \qquad (10)$$

$$\text{and } G_{12} = \frac{1}{\pi\eta_0}\int_{\theta=0}^{\theta=\pi}\left[\frac{\sin\left(\frac{k_0 W_p}{2}\cos\theta\right)}{\cos\theta}\right]^2 J_0(k_0 L_p \sin\theta)\sin^3\theta\, d\theta \qquad (11)$$

$$\text{where,}\ k_0 = \frac{2\pi}{\lambda_0} = \frac{2\pi f}{c} \qquad (12)$$

Gap of Notch is taken between

$$0.2 \times W_0 < x_0 < 0.5 \times W_0 \qquad (13)$$

Width of Strip line

$$W_0 = \begin{cases} h \times \dfrac{8 \times e^A}{e^{2A} - 2} & \dfrac{W_0}{h} \leq 2 \\ h \times \dfrac{2}{\pi}\left\{ (B-1) - \ln(2B-1) + \dfrac{\varepsilon_r - 1}{2\varepsilon_r}\left[\ln(B-1) + 0.39 - \dfrac{0.61}{\varepsilon_r}\right]\right\} & \dfrac{W_0}{h} > 2 \end{cases} \qquad (14)$$

Where

$$A = \frac{Z_0}{60}\sqrt{\frac{\varepsilon_r + 1}{2}} + \frac{\varepsilon_r - 1}{\varepsilon_r + 1}\left(0.23 + \frac{0.11}{\varepsilon_r}\right) \qquad (15)$$

$$B = \frac{\eta_0 \pi}{2Z_0 \sqrt{\varepsilon_r}} \qquad (16)$$

Length of Strip line

$$l_0 = \frac{c}{4 f_r \sqrt{\varepsilon_{effml}}}$$

where $\qquad (17)$

$$\varepsilon_{effml} = \frac{\varepsilon_r + 1}{2} + \frac{0.5(\varepsilon_r - 1)}{\sqrt{1 + 12(h/W_0)}}$$

21.4 RESULTS AND DISCUSSIONS

Figure 21.5 represents the schematic representation of the proposed multiband antenna for IoT applications

The various dimensions have been given in Table 21.1. Circular split ring resonators have been inserted on both sides of the feed line. The outer and inner radius of the outer circle are taken as 2 mm and 1.5 mm respectively while the outer and inner radius of the inner circle are taken as 1 mm and 0.5 mm respectively. Slot size for both outer and inner circles is taken as 0.4 mm. The simulation results in Figure 21.6 show that the proposed antenna resonates at six frequency bands centered around 5.5839 GHz, 8.5034 GHz, 9.651 GHz, 11.503 GHz, 15.268 GHz, and 17.886 GHz.

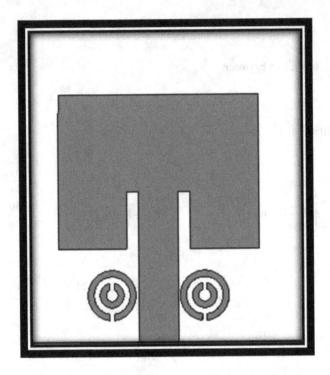

FIGURE 21.5 Schematic representation of the proposed multiband antenna for IoT applications.

TABLE 21.1
Dimensions of Proposed Antenna

Parameter	Value (mm)
Width of Patch, W	15.8073
Length of Patch, L	11.7875
Width of Patch Ground, Wg	25.5873
Length of Patch Ground, Lg	21.5675
Notch Length, yo	4.3813
Feed line Length, lo	7.1018
Feed line Notch Width, n	0.9836
Feed line Width, Wo	3.2787

Multiband Antenna Design for Internet of Things (IoT) Applications

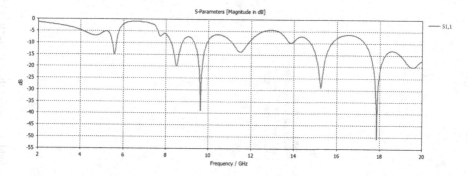

FIGURE 21.6 S-Parameter variation with frequency.

The S-parameter values at these frequencies are -14.1547 dB, -19.2765 dB, -31.6412 dB, -13.8393 dB, -28.2370 dB, and -31.8950 dB respectively.

The far-field radiation pattern of the proposed antenna at 5.5 GHz is shown in Figure 21.7. Absolute directivity, Far-field directivity theta (phi=0) and Far-field directivity phi (phi=0) are shown in Figure 21.8. It is observed that the proposed antenna shows good directivity characteristics.

Figure 21.8 shows the absolute far-field directivity pattern of the proposed antenna at different frequencies. Absolute far-field directivity abs (phi=0) and absolute far-field directivity abs (phi=90) are shown in Figures 21.8(a) and 21.8(b) respectively. From comparative analysis, it has been observed that the proposed antenna shows good directivity characteristics at 5.5839 GHz, 8.5034 GHz, 9.651 GHz, 11.503 GHz, 15.268 GHz, and 17.886 GHz.

Measured values of gain and front to back ratio are plotted in Figure 21.9 and Figure 21.10 respectively. It has been observed that the proposed antenna performed well in terms of gain and front to back ratio.

21.5 CONCLUSION

Microstrip patch antennas are used in a variety of sectors, including medicine, satellites, and military equipment such as rockets, airplanes, and missiles, among others. They are now thriving commercially because of the cheap cost of the substrate material and production [45–48]. In this paper, a novel rectangular patch antenna has been proposed with an inset feed. Design equations have been developed for the proposed antenna. The proposed antenna resonates at 5.5839 GHz, 8.5034 GHz, 9.651 GHz, 11.503 GHz, 15.268 GHz, and 17.886 GHz with reasonable directivity and gain characteristics. The proposed antenna shows wide frequency coverage ranging from around 5 GHz to 18 GHz. This antenna is quite suitable for IoT applications where different devices operate at different resonating frequencies.

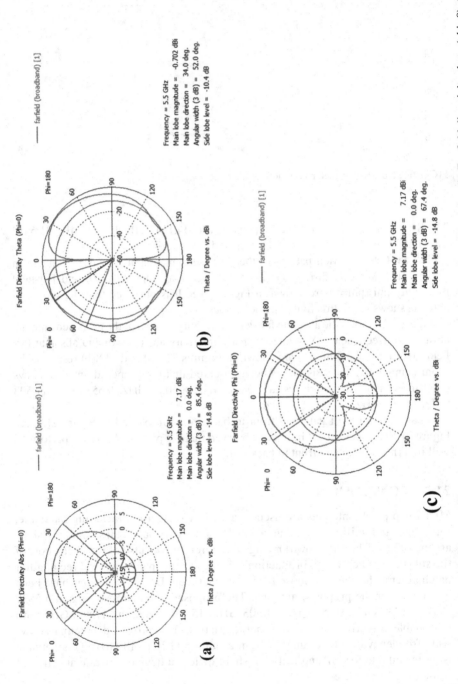

FIGURE 21.7 Far-field directivity pattern of the proposed antenna at 5.5 GHz; (a) Absolute directivity, (b) Far-field directivity theta (phi=0), (c)

Multiband Antenna Design for Internet of Things (IoT) Applications 375

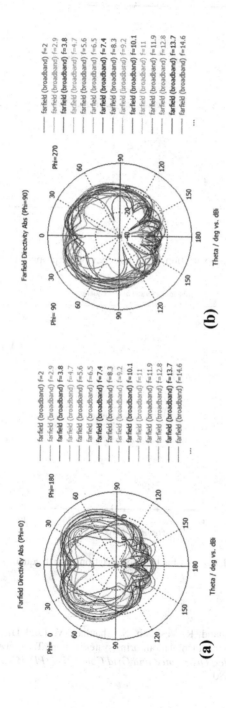

FIGURE 21.8 Absolute Far-field directivity pattern of the proposed antenna at different frequencies; (a) Far-field directivity abs (phi=0), (b) Far-field directivity abs (phi=90).

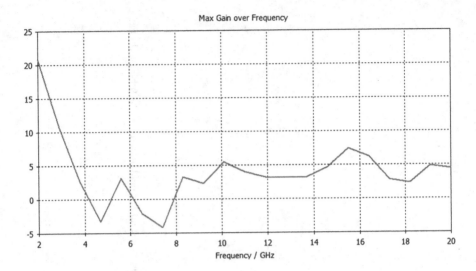

FIGURE 21.9 Measured gain variation with frequency of the proposed antenna.

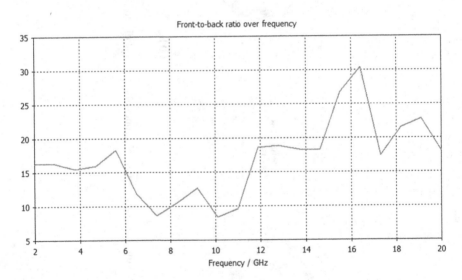

FIGURE 21.10 Measured front to back ratio variation with frequency of the proposed antenna.

REFERENCES

1. Sindhwani, N., R. Anand, R. Vashisth, S. Chauhan, V. Talukdar and D. Dhabliya. Thingspeak-based environmental monitoring system using IoT. In *Seventh International Conference on Parallel, Distributed and Grid Computing (PDGC) 2022*. IEEE, 2022, November, pp. 675–680.

2. Jain, N., A. Chaudhary, N. Sindhwani and A. Rana. Applications of Wearable devices in IoT. In *9th International Conference on Reliability, Infocom Technologies and Optimization (Trends and Future Directions)(ICRITO) 2021*. IEEE, 2021, September, pp. 1–4.
3. Gupta, A., A. Asad, L. Meena and R. Anand. IoT and RFID-based smart card system integrated with health care, electricity, QR and banking sectors. In *Artificial Intelligence on Medical Data. Proceedings of International Symposium, ISCMM 2021*. Springer Nature, 2022, July, pp. 253–265.
4. Gupta, A., A. Srivastava and R. Anand. Cost-effective smart home automation using internet of things. *Journal of Communication Engineering and Systems* 2019, 9(2), 1–6.
5. Kumar Pandey, B., D. Pandey, V. K. Nassa, T. Ahmad, C. Singh, A. S. George and M. A. Wakchaure. Encryption and steganography-based text extraction in IoT using the EWCTS optimizer. *The Imaging Science Journal* 2021, 69(1–4), 38–56.
6. Kumar, M. S., S. Sankar, V. K. Nassa, D. Pandey, B. K. Pandey and W. Enbeyle. Innovation and creativity for data mining using computational statistics. In *Methodologies and Applications of Computational Statistics for Machine Intelligence*. IGI Global, 2021, pp. 223–240.
7. Gupta, A., A. Srivastava, R. Anand and P. Chawla. Smart vehicle parking monitoring system using RFID. *International Journal of Innovative Technology and Exploring Engineering* 2019, 8(9S), 225–229.
8. Pandey, D., S. Wairya, M. Sharma, A. K. Gupta, R. Kakkar and B. K. Pandey. An approach for object tracking, categorization, and autopilot guidance for passive homing missiles. *Aerospace Systems* 2022, 5(4), 553–566
9. Badotra, S., S. Tanwar, A. Rana, N. Sindhwani and R. Kannan. *Handbook of Augmented and Virtual Reality* (Vol. 1). Walter de Gruyter GmbH & Co KG, 2023.
10. Daneshmandian, F., Dekhoda, P. and A.Tavakoli. A miniaturization circularly polarised microstrip antenna for GPS applications. *IEEE 22nd Iranian Conference on Electrical Engineering (ICEE)*. IEEE, 2014, pp. 1653–1656.
11. Behera, S. and D. Barad. A novel design of microstrip fractal antenna for wireless sensor network. *IEEE, InernationalConfrence of Power, Energy, Information and Communication* 2015, 0470–0474. https://doi.org/10.1109/ICCPEIC.2015.
12. Vaid, V. and S. Agarwal. Bandwidth optimization using fractal geometry on rectangular microstrip patch antenna with DGS for wireless applications. *International conference on medical Imaging, M-health and Emerging Communication Systems (MedCom)*, 2014, pp. 162–167.
13. Raj, V. D., A. M. Prasad, M. Satyanarayana and G. M. V. Prasad. Implementation of printed microstrip Apollonian gasket fracxtal antenna for multiband wireless applications. *IEEE, International Conference on SPACES*, 2015, pp. 200–204.
14. Prema, N. and A. Kumar. Design of multiband microstrip patch antenna for C and X. *Optik* 2016, 127(20), 8812–8818.
15. Wang, C., M. Daneshmand, X.M. MischaDohler, R.Q. Hu and H. Wang. Guest editorial-special issue on internet of things (IoT): Architecture, protocols and services. *IEEE Sensors Journal* 2013, 13(10), 3505–3510.
16. Kraijak, S. and P. Tuwanut. *A Survey on IoT Architectures, Protocols, Applications, Security, Privacy, Real-World Implementation and Future Trends*, 2015, pp. 6–6.
17. Desai, P., A. Sheth and P. Anantharam. Semantic gateway as a service architecture for iot interoperability. In *2015 IEEE International Conference on Mobile Services*. IEEE, 2015, pp. 313–319.

18. Castellani, A.P., N. Bui, P. Casari, M. Rossi, Z. Shelby and M. Zorzi. Architecture and protocols for the internet of things: A case study. In *2010 8th IEEE International Conference on Pervasive Computing and Communications Workshops (PERCOM Workshops)*. IEEE, 2010, pp. 678–683.
19. Karagiannis, V., F. Vazquez-Gallego, P. Chatzimisios and J. Alonso-Zarate. A survey on application layer protocols for the internet of things. *Transaction on IoT and Cloud Computing* 2015, 3(1), 11–17.
20. Lloret, J., J. Tomas, A. Canovas and L. Parra. An integrated IoT architecture for smart metering. *IEEE Communications Magazine* 2016, 54(12), 50–57.
21. Li, H., K. Ota and M. Dong. Learning IoT in edge: Deep learning for the Internet of Things with edge computing. *IEEE Network* 2018, 32(1), 96–101.
22. Tang, J., D. Sun, S. Liu and J.-L. Gaudiot. Enabling deep learning on IoT devices. *Computer* 2017, 50(10), 92–96.
23. Ziouvelou, X., C.M. Angelopoulos, P. Alexandrou, O. Evangelatos, J. Fernandes, N. Loumis, F. McGroarty, S. Nikoletseas, A. Rankov, T. Raptis, A. Ståhlbröst, S. Ziegler Fernandes, N. Loumis, F. McGroarty et al. Crowd-driven IoT/IoE ecosystems: A multidimensional approach. In *Beyond the Internet of Things*. Springer, 2017, pp. 341–375.
24. Wang, T., G. Zhang, A. Liu, Md Z.A. Bhuiyan and Q. Jin. A secure IoT service architecture with an efficient balance dynamics based on cloud and edge computing. *IEEE Internet of Things Journal* 2018, 6(3), 4831–4843.
25. Kaur, S.P. and M. Sharma. Radially optimized zone-divided energy-aware wireless sensor networks (WSN) protocol using BA (bat algorithm). *IETE Journal of Research* 2015, 61(2), 170–179.
26. Sharma, M. and H. Singh. SIW based leaky wave antenna with semi C-shaped slots and its modeling, design and parametric considerations for different materials of dielectric. In *Fifth International Conference on Parallel, Distributed and Grid Computing (PDGC)*. IEEE, 2018, pp. 252–258.
27. Sharma, M., B. Sharma, A.K. Gupta and B.S. Singla. Design of 7 GHz microstrip patch antenna for satellite IoT-and IoE-based devices. In *The International Conference on Recent Innovations in Computing*. Springer, 2020, pp. 627–637.
28. Sharma, B., M. Sharma, B.S. Singla and S. Goyal. Design and analysis of thin micromechanical suspended dielectric RF-MEMS switch for 5G and IoT applications. In *Soft Computing for Intelligent Systems*. Springer, 2021, pp. 133–146.
29. Singh, S., B.S. Singla, M. Sharma, S. Goyal and A. Sabo. Comprehensive study on Internet of things (IoT) and design considerations of various microstrip patch antennas for IoT applications. In *Mobile Radio Communications and 5G Networks*. Springer, 2021, pp. 19–30.
30. Sharma, M., D. Pandey, P. Palta and B.K. Pandey. Design and power dissipation consideration of PFAL CMOS V/S conventional CMOS based 2: 1 multiplexer and full adder. *Silicon*, 2021, 1–10.
31. Sharma, M., H. Singh and D. Pandey. Parametric considerations and dielectric materials impacts on the performance of 10 GHzSIW-LWA for respiration monitoring. *Journal of Electronic Materials* 2022, 51(5), 2131–2141.
32. Sharma, M., P. Palta, D. Pandey, S. Goyal, B.K. Pandey and V.K. Nassa. Modeling and analysis of positive feedback adiabatic logic CMOS-based 2: 1 mux and full adder and its power dissipation consideration. In *Mobile Radio Communications and 5G Networks*. Springer, 2022, pp. 281–295.
33. Sharma, M., S. Singh, D. Khosla, S. Goyal and A. Gupta. Waveguide diplexer: Design and analysis for 5G communication. In *2018 Fifth International Conference on Parallel, Distributed and Grid Computing (PDGC)*. IEEE, 2018, pp. 586–590.

34. Singh, N., S. Singh, A. Kumar and R. K. Sarin. A planar multiband antenna with enhanced bandwidth and reduced size. *IJEER, International Journal of Electronics Engineering Research* 2010, 2(3), 341–347.
35. Sharma, N., A. Kaur and V. Sharma. A novel design of circular fractal antenna using inset line feed for multi band application. *IEEE 2016 International Conference on Power Electronics, Intelligent Control and energy systems*. IEEE, 2016.
36. Ali, M. M. M., A. M. Azmy and O. M. Haraz. Design and implementation of reconfigurable quad-band microstrip antenna for MIMO wireless communication applications. *IEEE 31st National Radio Science Conference (NRSC)*. IEEE, 2014, pp. 27–34.
37. Sharma, M., Gupta, A. K., Arora, T., Pandey, D., and Vats, S. Comprehensive Analysis of Multiband Microstrip Patch Antennas used in IoT-based Networks. In *2023 10th International Conference on Computing for Sustainable Global Development (INDIACom)* IEEE, 2023, March, pp. 1424–1429.
38. Jose, S. K. and Dr. S. Suganthi. *Circular-Rectangular Microstrip Antenna for Wireless Applications* 4(1, January–February) (2015).
39. Patnaik, S. *Optimization of Z Shaped Microstrip Antenna with I- Slot Using Discrete Particle Swarm Optimization Algorithm* (2016), p. ICCC-2016.
40. Sharma, N. and V. Sharma. An optimal design of fractal antenna using modified Sierpinski carpet geometry for wireless applications. *International Conference on Smart Trends in Computer Communication and Information Technology (Springer, SmartCom 2016),CCIS 628*, August, pp. 400–407, Jaipur, 2016.
41. Singh, N., D. P. Yadav, S. Singh and R. K. Sarin. Compact corner truncated triangular patch antenna for WiMax application. *IEEE, Microwave Symposium (MMS), 2010 Mediterranean Conference, Cyprus*, 2010.
42. Kumar, G. and K. P. Ray. *Broadband Microstrip Antennas*. Artech House, 2003.
43. Balanis, C. A. *Antenna Theory: Analysis and Design*, 3rd edn. John Wiley, Hoboken, NJ, 2005.
44. Pendry, J. B., A. J. Holden, D. J. Robbins and W. J. Stewart. Magnetism from conductors and enhanced nonlinear phenomena. *IEEE Transactions on Microwave Theory and Techniques* 1999, 47(11), 2075–2084.
45. Anand, R. and P. Chawla. Bandwidth optimization of a novel slotted fractal antenna using modified lightning attachment procedure optimization. In *Smart Antennas: Latest Trends in Design and Application*. Springer International Publishing, 2022, pp. 379–392.
46. Chibber, A., R. Anand and S. Arora. A staircase microstrip patch antenna for UWB applications. In *2021 9th International Conference on Reliability, Infocom Technologies and Optimization (Trends and Future Directions)(ICRITO) 2021*. IEEE, 2021, September, pp. 1–5.
47. Sindhwani, N. and M. Singh. FFOAS: Antenna selection for MIMO wireless communication system using firefly optimisation algorithm and scheduling. *International Journal of Wireless and Mobile Computing* 2016, 10(1), 48–55.
48. Sindhwani, N. and M. Singh (2014). Transmit antenna subset selection in MIMO OFDM system using adaptive mutation Genetic algorithm. *arXiv Preprint ArXiv:1410.6795*.

22 A Counter-Propagation Based Neuro Solution Model for Categorization and Fee Fixation of Engineering Institutions

Krishna Kumar Nirala, Nikhil Kumar Singh, and Vinay Shivshanker Purani

22.1 INTRODUCTION

Every year, the authorities receive applications from across the state for the existing and new engineering institutes for the fixation of the fees. The process of engineering college fees fixation is being dealt with differently by different states. In the state of Gujarat, it is being done by a committee of academicians, chartered accountants, trustees nominated from self-financed institutions, and senior government officials headed by a retired high court judge. However, the fact that there has been involvement of vast human intervention at different levels may result in grievances and litigation on the decisions.

In this study, a counter-propagation neural network (CPN) [1, 2] is investigated for the distinct application of administrative issues which tried to obtain an AI based solution [3–5] with the aim of improving governance and transparency in the decision-making system. The study examines the accuracy and efficiency of the widely used multi perceptron approach of the standard error backpropagation (BPN) algorithm [6–9]. The BPN approach has attracted maximum attention in the literature due to its simplicity but has a slow rate of learning, while on the other hand, the CPN approach is rarely researched for modeling complex engineering problems [3]. The most untouched domain is public administration and governance, wherein the human intervention-free AI based transparent system, carrying huge potential for research, is explored intensively in this work.

For multiple networks with huge numbers of training patterns, CPN is found to be more efficient and much faster than BPN due to its interactive combination of supervised and unsupervised learning with a befitting transfer function. This work aims to develop a human intervention-free process of evaluation of the category of the institute and for prediction of the fees. In this context, the use of artificial neural networks (ANN) [1] is considered here as one of the most promising techniques, because ANN has the capability to directly use the available field/practical results for clustering, mapping, and classifying input and output in the absence of any mathematical relationship [10–15]. Among the available networks, the CPN algorithm is used in this work which is unique in its capabilities and could carry out such tasks rapidly and accurately.

This chapter is organized into six sections. Section 22.2 explains the counter-propagation neural network used in the proposed work. Section 22.3 contains some of the existing research related to the proposed work; Section 22.4 elaborates on the proposed CPN Application for the prediction of fees. Discussion of results for the proposed work is in Section 22.5 and the chapter is concluded in Section 22.6.

22.2 A COUNTER-PROPAGATION NEURAL NETWORK

The majority of the existing networks perform only supervised learning wherein what the network should learn from the data is well defined. Continuous feedback throughout the training and across the entire network is essential in this learning, which results in a huge period of network training. An unsupervised network (self-organizing) consists of hard coded organized information into its architecture. A counter-propagation neural network is a combination of the supervised and unsupervised mapping of neural networks. This network consists of two feed forward layers as shown in Figure 22.1. The hidden layer is a Kohonen layer with competitive units that do unsupervised learning. The output layer is a Grossberg layer, which is fully connected with the hidden layer and is not competitive. As counter-propagation uses a training methodology that combines supervised and competitive unsupervised learning techniques in an innovative way, and also there is no feedback across the network in the process or delay activated during the recall operation, its training occurs much faster than backpropagation. It has been used successfully for function

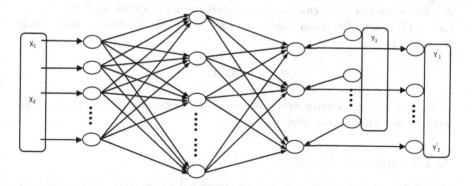

FIGURE 22.1 Counter-propagation network architecture.

approximation, pattern mapping and association, data compression, and classification. The objective of the counter-propagation network is to map input data vectors X_j into bipolar binary response Z_j, for $j = 1$ to P number of training patterns. The learning process of this network consists of the following two stages.

22.2.1 Unsupervised Competitive Learning

The unsupervised competitive learning network is allowed to develop its weight first. Counter-propagation trains the weights from input to the Kohonen layer nodes through the unsupervised winner-takes-all learning rule. In this rule, for each iteration, the network is presented with all the input samples of training data sets. Let W_j be the arbitrary weight vector assigned to the links between input nodes and jth node in the competition layer. For each input sample, the nearest node in the competition layer is obtained by means of Euclidean distance d_j with the help of

$$d_j = [\Sigma (X - W_j)]^{1/2} \quad (1)$$

The node having the minimum Euclidean distance wins the competition and its activation is set to 1 while all others are set to 0. Let jth neuron win the competition, hence output vector Z_j of the competition layer becomes,

$$Z_j = [\, y_1\ y_2\ y_3\ \ldots\ y_j\ \ldots\ y_p\,] = [0\ 0\ 0\ \ldots\ 1\ \ldots\ 0] \quad (2)$$

and weights corresponding to the winning node are the only weights adjusted in the given step using the following equation for (n+1)th iteration:

$$W_{ji}(n+1) = W_{ji}(n) + a\,[\,X_i - W_{ji}(n)\,]\,Z_j \quad (3)$$

where a = small positive learning coefficient and for one iteration the learning rate remains constant but as iteration progresses, the learning rate is shrunk. It is suggested that the learning rate must satisfy the inequality $0 < a < 1$. However, unsupervised learning is less sensitive to the learning rate, and shrinking in the learning rate as a function of iteration number as mentioned in Eq. (4) shows comparatively fast convergence:

$$a = 1/(n+1)^2 \quad (4)$$

In this learning, the error term is defined as the average Euclidean distance of all winning nodes for that iteration.

22.2.2 Supervised Learning:

Once the unsupervised learning phase is completed and weights are stabilized, the interpolation layer starts to learn the desired output. The input data set from the training pattern is presented to the competition layer as a fresh and the winning node

A Counter-Propagation Based Neuro Solution Model

j in the competition layer is selected. Then the supervised learning is performed on the connections from winning unit j to all the units in the output layer. Weights between winner node j and the node in the interpolation layer are adjusted according to the learning rule suggested by Grossberg as,

$$Vji(n+1) = Vji(n) + b[Yi - Vji(n)]Zj \qquad (5)$$

where b = small positive learning constant and is taken in the range of 0 < b < 1.

The interpolation layer uses a weighted summation function as a transfer function. Output of the network from the ith node of interpolation layer is determined by

$$Yi = \Sigma Vji.Zi \qquad (6)$$

This rule typically ensures that the output pattern becomes similar to the desired output after repetitive application of Eq. (5). The root mean square error is suggested as an error term for this learning.

The learning algorithm of counter-propagation is shown in Figure 22.2. During training, only one node of the competition layer can win the competition at both the learning stages. After connection weights are stabilized through winner-takes-all and outstar learning rules, the performance of the network is tested using untrained instances. During testing, the number of winning nodes in the competition layer may be set to a value of more than one depending on the type of problem. In the case of

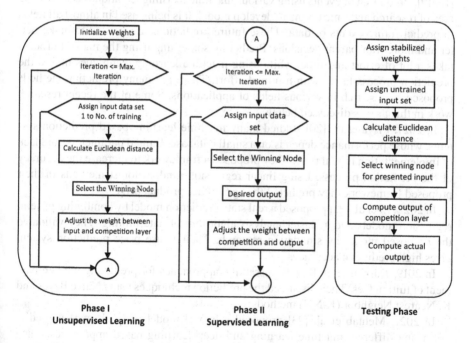

FIGURE 22.2 Counter-propagation training and testing algorithm.

the number of winning nodes during the verification stage, their nonzero outputs are set such that the winner node associated with the weight vector closest to the given untrained instance has the largest output. However, the sum of the outputs of all winning nodes in the competition layer must remain equal to 1.0. Actual output is then computed through interpolation of these winning nodes' output. If the number of winning nodes in testing is set to one, CPN behaves as a simple nearest neighbor classifier.

Authors felt that though it is a known fact that AI possesses noteworthy applications in almost all disciplines, however as a matter of fact, pure administrative and government departments have yet to embed AI in citizen-centric applications that can promote good governance [4, 16]. Keeping this in mind an attempt is made to showcase the potential use of the ANN in the unexplored area of administration. Graphical User Interface (GUI) based Windows application is developed under this work. With the available use of the open source modules and the basic algorithms for the ANN applications, innovative supervised and unsupervised learning in a CPN environment is assembled in a Windows 64-bit GUI application using Visual Studio. The training and verification patterns for various CPN architectures are developed arbitrarily. Each CPN contributes to deciding the institute category individually and the ANN based institute category is finally being predicted to decide the institute fees.

22.3 LITERATURE REVIEW

Machine Learning techniques are very useful in solving public administrative issues [17–19]. Prediction of events using various machine learning techniques is a growing area of research and time taking to develop model. It is being used in almost all fields to predict future events or data. These future predictions are needed to make better and more transparent decisions. Predicting something using the manual process takes a lot of effort and time, while using machine learning techniques makes the prediction process faster with minimal effort. There are many prediction methods proposed by research in various fields of applications. Some of the recent research work in this area is discussed below.

In 2021, Priya et al. [20] studied many machine learning based prediction systems, whose performance depends on a small database, limited parameters, and lack of implementation in real time. They proposed a framework to increase the accuracy of the prediction process. Using linear regression random forest methods in their proposed framework they predicted the agriculture product prize.

In 2022, Li et al. [21] proposed a real state prediction model by combining general regression models and a cluster analysis algorithm of data mining. They validated the proposed model by simulation research. Their house price prediction system gives high prediction accuracy.

In 2019, Kusrini et al. [22] discussed the approaches for predicting the late payment of tuition fees. They discussed the prediction techniques using Naive Bayes and K. Nearest Neighbor (K-NN) methods.

In 2021, Mehtab et al. [23] proposed a hybrid model of stock market prediction using different machine learning and deep learning based approaches. They used NIFTY 50 index values of the National Stock Exchange (NSE) of India from

December 29, 2014 to July 31, 2020 as training data to build eight regression models. Their long- and short-term memory (LSTM) based univariate model, which takes one week's prior data as input to predict next week's open value of NFTY 50 time series, is the most accurate of all the proposed models.

In 2019, Manjula et al. [24] found the relationship between gold price and the influencing parameters using three different machine learning algorithms. Out of three, random forest regression has better prediction accuracy for the entire period, which gradient boosting gives better accuracy when two periods are taken separately.

In 2021, Gajera et al. [25] proposed a machine learning based platform to predict the price of used cars. They used supervised learning algorithms on available consumer data and features.

In 2017, Ravikumar et al. [26] proposed a price prediction system for real estate markets and housing. They used advanced data mining algorithms over the existing approaches to get high accuracy.

22.4 PROPOSED CPN APPLICATION FOR FEE PREDICTION

Fixation of the fees for different engineering colleges is decided based on several independent parameters and submissions made by the institutions to the authorities. In this process, multiple human intervention is involved, and it requires sufficient care and accuracy to correlate different parameters influencing the fees' determination. The process itself is time-consuming, clumsy, and cannot be addressed concurrently for the sets of applications when processed manually. The verbal participation and silence of the committee members on a case-to-case basis in decision-making may not guarantee the uniform process for the fees' decision-making. Artificial Intelligence can be the right way if the appropriate techniques are used with customized parameters suitable for the problem type. Hence, in this study six fees influencing parameters: (i) Admission Merit (ii) University Results (iii) Faculty Position (iv) Quality of Education (v) Expenditure per student, and (vi) Placement, are identified. It is proposed in this work to employ the ANN [27] for each of these individual fees influencing parameters to predict the institute category out of six from (i) Low, (ii) Medium, (iii) Moderate, (iv) Good, (v) Very Good and (vi) Best. A mainframe program is assembled with all six individual CPNs with their weights, which on presenting the input, predicts the institute category from M1 to M6, and from all the outputs from these six CPNs, the final category of the institute is predicted. The fee for the institute is then decided by the authority for all these six categories. Table 22.1 depicts the fees influencing parameters, the category of the institute, and expenditure for the fees to be decided. Further, to substantiate the ANN capabilities, sets of suggestions are also framed for each individual CPN prediction and the same is being displayed along with the ANN output.

All these six fees influencing parameters, six classifications in each of these parameters and prediction of the institute category to decide the fees, are interactively used with an innovative multi-neuro approach. An attempt is made to explain the approach adopted to develop the ANN based model and flow of information among the CPNs, and its presence in the application is demonstrated in Figure 22.3.

TABLE 22.1
Fee Influencing Parameters

Level	Category	Extent of Merit in Admission	Extent of Quality of Education					ANN Output	
			Teachers	Results	Accreditation	Placement	Expenditure	Fees	Suggestions
1	Low	< 50%	Grade-1	Poor	0	< 10%	<50%	F1 (Minimum)	Set-1
2	Medium	< 60%	Grade-2	Low	1	< 20%	<60%	F2=1.2F1	Set-2
3	Moderate	< 70%	Grade-3	Moderate	2	< 30%	<80%	F3=1.5F1	Set-3
4	Good	< 80%	Grade-4	Good	3	< 40%	<90%	F4=2F1	Set-4
5	Very Good	< 90%	Grade-5	Very Good	4	< 50%	100%	F5=3F1	Set-5
6	Best	> 90%	Grade-6	Best	5	> 50%	>100%	F6=4F1	Set-6

A Counter-Propagation Based Neuro Solution Model

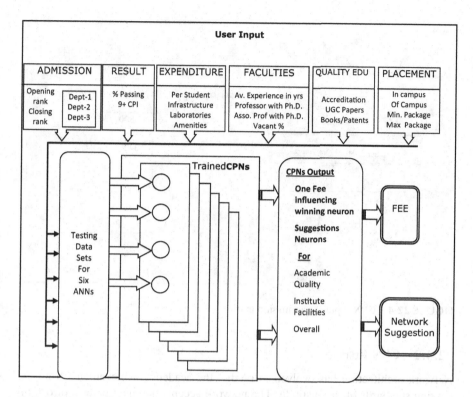

FIGURE 22.3 Flow of information for the CPN application for prediction of fee.

CPN architecture for each of these six fees influencing parameters is developed as follows:

22.4.1 CPN Admission

Based on the common merit system across the state, preferences in the choice of the institutes by the students are one of the parameters to be considered for deciding the category of the institute. The institute's classes are broadly grouped into six subcategories based on the admission of meritorious students. For the top three departments of the institute, the opening and closing ranks of the students are taken into consideration. The normalized data for three opening and three closing ranks as input, and the extent of merit of the institute as output is prepared for the 6-250-1 CPN Admission. Based on the training data sets, only one winner node is kept for this network during training. The trained network is then verified by 20 sample verification patterns which predict the admission grade of the institute from M1 to M6. The GUI facilitates the users to enter the data in a user-friendly environment as shown in Figure 22.4. Entered data is converted into the input pattern and on its submission, CPN Admission is activated and predicts the results. Using trial and error and the noise complexity of the training patterns, CPN Admission prediction is set with three winning nodes which brings about a 99% correct prediction.

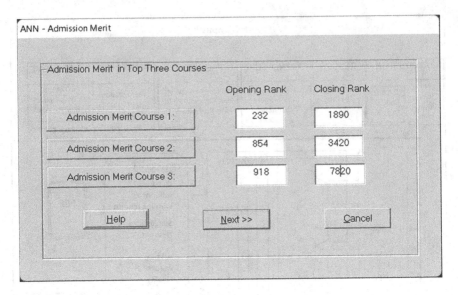

FIGURE 22.4 CPN Input for admission merit.

22.4.2 CPN Results

Network architecture has evolved to predict the academic achievements from the passing standards of the institute. The network accepts the input pattern of overall passing percentages of students from the top three performing branches and the number of students who have scored 9+ Cumulative Performance Index (CPI) in their passing results. To judge the student performance for both these parameters, network training patterns for the 2-180-1 CPN architecture are framed with non-dimensional input parameters. Samples for the testing were separately tested using the available CPN simulator and the output was found in close agreement with the actual [28–30]. The developed CPN structure is shown in Figure 22.5 and its GUI interface is shown in Figure 22.6. The samples of the training and verification patterns are depicted in Table 22.2.

22.4.3 CPN Faculties

4-180-1 Network is developed, trained, and enrolled in sequence among the CPNs for the proposed application for fee determination. It is considered that the faculty strength is one of the prime parameters in deciding the institute category. To distinguish between the best to poor practicing institutes, four different parameters are selected for the CPN input. Since the faculty position, its endorsement by the affiliation university, qualification, average experience, and regular appointment of the institute heads is always non-debatable and public domain information for all the institutes, the same is used for the prediction of the level of a faculty position. The CPN is trained with the four-input pattern, (i) Average teaching experience (ii)

A Counter-Propagation Based Neuro Solution Model

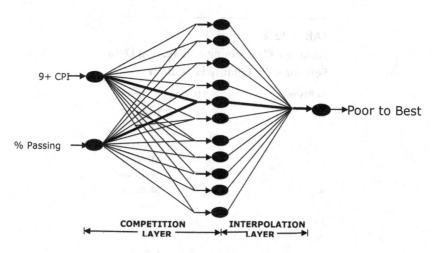

FIGURE 22.5 CPN Result architecture 2-180-1.

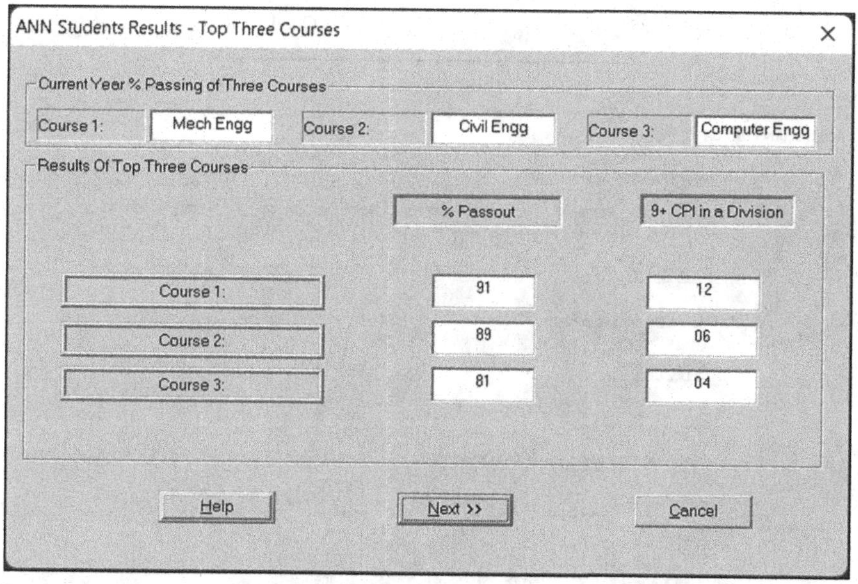

FIGURE 22.6 CPN input for institute result.

percentage of Ph.D. holders from among the filled positions (iii) Principal appointment status, and (iv) Feeder cadre faculties with Ph.D. degrees. All these parameters are clubbed together and, based on the overall faculty strength of the institute, the CPN Faculty will predict the category of the institute. On successful training of the network, verification patterns were presented to the network, and with two winning nodes, CPN fairly predicts the institute category based on the faculty position. The network input interface is developed as shown in Figure 22.7.

TABLE 22.2
Sample CPN Training and Testing Data Sets for CPN Quality Education

% Passing	9+CPI	Level
100	30	Best (10)
90	2	Low
80	20	Good
70	08	Moderate
80	25	Very Good
50	10	Moderate
40	12	Good
30	10	Good
100	05	Poor
50	05	Low
Verification Pattern		
75	25	Very Good
85	15	Good

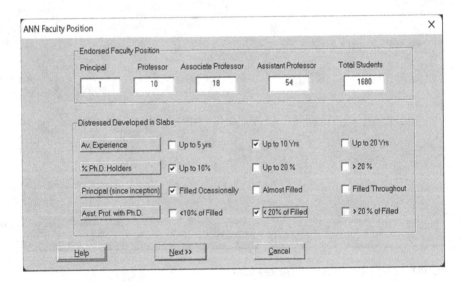

FIGURE 22.7 CPN faculty network input interface.

22.4.4 CPN Expenditure

Most often, fee fixation is predominantly sought high by self-financed institutes based on the expenditure incurred during the subsequent financial years. Sometimes it is becoming very difficult to derive a decision based on the expenditure reflected in the audited report. However, the institute development plan is mainly dependent upon

A Counter-Propagation Based Neuro Solution Model

the expenditure incurred per student per year. The network 1-20-1 CPN Expenditure takes input for the average expenditure of the last three years incurred against the salary of faculties, laboratory development, and infrastructure development. A similar GUI is developed to facilitate user input for this network.

22.4.5 CPN Quality Education

The quality of education at engineering colleges is mainly decided on by the institute ranking, accreditation of programs running at the institute by the National Board of Accreditation, research and innovation paper publications to journals of merit by faculties and students, patents filed and obtained, etc. However, there is no system to take into account the contribution of faculties and students in the promotion of quality standards of the institute. CPN Quality Edu, which is the network with 3-220-1, is taking cognizance of the three foremost quality parameters: (i) number of programs accredited out of total in percentage, (ii) University Grant Commission (UGC) recognized research papers/books published by the institute in the last three years, and (iii) patents applied for/granted during the last three years from the institute by faculties and staff lead innovations. A total of 220 databases were prepared for the training of the network and 20 data sets were kept for the verification of the network performance. It is observed that CPN is capable of predicting the same category as per the output patterns of the verification sets.

22.4.6 CPN Placement

Employability of the students and packages both on-campus and off-campus are vital parameters for the institute's performance. Categorization of the institute based on the placement and the packages are embedded into CPN Placement with 4-60-1 network architecture. The network is enough to predict the accurate category in a range from one to six using 60 training patterns prepared to train the network. Training patterns consist of four input parameters: (i) on-campus placement in percentage, (ii) off-campus placement in percentage, (iii) maximum package in Indian national rupee (INR) per annum, and (iv) Minimum package in INR per annum; and one output parameter that is a category of the institute from the range one to six as poor to best respectively. Arbitrarily framed training patterns with input and output are put in the main program after the successful supervised and unsupervised learning with an appropriate learning rate [31–33]. Two winning nodes for this network are derived as optimal for the accurate prediction of the institute category.

22.5 RESULTS

All the six trained networks are put in sequence to predict the institute category in the range of poor to best with a non-dimensional integer from one to six. The post-processor of the Windows application receives the output of the institute category and supplies it to the CPN Summary, a 6-60-1 CPN architecture. This network can perform the regression of the input information and finally predicts the overall

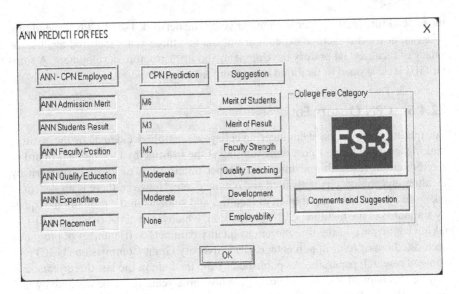

FIGURE 22.8 The output of the network.

category of the institute and hence the fee structure category from FS-1 as minimum to FS-6 as maximum. Further, the weightage of the individual category of all six CPNs' output can be decided by the authority in the final prediction of the institute category, if so required. The output of the network is processed by the post-processor and gives a user-friendly output as shown in Figure 22.8.

The suggestion button included in the CPN Summary dialogue box contains the pre-defined information associated with the individual categorization and the final suggestion/remarks are also produced by the application by clicking the button. The role of the CPN Summary is outlined with sequentially enrolled CPNs in the form of Table 22.3 as shown below.

Based on the network prediction, fees for the institute can be decided using empirical the formula suggested in this paper. The administrative decision is simply to decide upon the minimum fee level as the benchmark value and for the rest, all category fees can be decided without any human intervention and interpretation.

22.6 CONCLUSION

Prediction of the behavior of complex administrative problems using neural networks exhibits promising potential because its modeling process is more direct since there is no necessity to specify a mathematical relationship between input and output variables. CPN shows an excellent convergence with no trial-and-error procedure required for selecting the best architecture and the verification instances proved that the CPN can produce results within a few percentages of the exact or desirable values. The use of AI and its application is rare in public administration, especially when human intervention for case to case is streamlined using ANN. Fee deciding

TABLE 22.3
Different CPNs Prediction for CPN Summary

Sr No	CPN	Training Sets Generated from	Prediction	No. of Winner Nodes
1	CPN_ADM	Top 3 Course Opening Rank, Closing rank	Merit of Students M1 to M6	3
2	CPN_RES	Top 3 Course % Passing, 9+ CPI	Poor to Best	2
3	CPN_FT	Av. Experience, Professor, Asso. Prof., % Vacant	Grade-1 to Grade-6 Grade-1 to Grade-6	2
4	CPN_QT	Accreditation, UGC Papers Av. API, Books/Patents	0 to 6 0 to 6	2
5	CPN_EXP	Per Student, On Infrastructure, On Laboratories, On Amenities	50% to >100%	1
6	CPN_PL	On-campus, Off-campus Min. Package, Max. Package	Poor to Best Poor to Best	2
7	CPN_FEE	Output of above 6 CPNs	Fees Suggestions	1 3

procedure for the engineering institutes using the most befitting ANN techniques circumvents all human intervention and provides a transparent, fast, reliable, and acceptable solution in comparison to the classic mode of approach. Pre- and Post-processor with user-friendly GUI gives a professional touch to the AI application and promises to handle the citizen-centric activity of the public administration. Even if a smaller number of training patterns and smaller CPN architecture are used for the classification of the problem type, trained CPN with up to three winning nodes is able to predict the true category of the institute. A multiple network approach in complex administrative problems offers a wide scope to build a dedicated neural network based practical application. The concept of pre- and post-processor in artificial neural networks makes applications more user-friendly and extends new hopes for framing complicated and practical application models in the field of ANN.

REFERENCES

1. Hecht-Nielsen, R. Counterpropagation networks. In *Proceeding of 1st International Conference on Neural Networks*. IEEE Press, 1987, 2, 19–32.
2. Rothman, D. *Artificial Intelligence by Example*. Ingram Publications, 2018.
3. Rich, E., Knight, K., and Nair, S. *Artificial Intelligence*. Tata McGraw Hill, 2009.
4. Hecht-Nielsen, R. Applications of counterpropagation networks. *Neural Networks*, 1988, 2(2), 131–139.
5. Zurada, J. M. *Introduction to Artificial Neural Networks*. Jaico Publishing House, 1994.

6. Sindhwani, N., Anand, R., Meivel, S., Shukla, R., Yadav, M. P., and Yadav, V. Performance analysis of deep neural networks using computer vision. *EAI Endorsed Transactions on Industrial Networks and Intelligent Systems*, 2021, 8(29), e3–e3.
7. Sindhwani, N., Rana, A., and Chaudhary, A. Breast cancer detection using machine learning algorithms. In *2021 9th International Conference on Reliability, Infocom Technologies and Optimization (Trends and Future Directions)(ICRITO)*. IEEE, 2021, September, pp. 1–5.
8. Verma, S., Bajaj, T., Sindhwani, N., and Kumar, A. Design and development of a driving assistance and safety system using deep learning. In *Advances in Data Science and Computing Technology*. Apple Academic Press, 2022, pp. 35–45.
9. Chawla, P., Sindhwani, N., and Anand, R. Smart coal payload delivery system using pic microcontroller. *International Journal of Advanced Science and Technology*, 2020, 29(10s), 1485–1490.
10. Pandey, B. K., Pandey, D., Wariya, S., Aggarwal, G., and Rastogi, R. Deep learning and particle swarm optimisation-based techniques for visually impaired humans' text recognition and identification. *Augmented Human Research* 2021, 6, 1–14.
11. Kumar, M. S., Sankar, S., Nassa, V. K., Pandey, D., Pandey, B. K., and Enbeyle, W. Innovation and creativity for data mining using computational statistics. In *Methodologies and Applications of Computational Statistics for Machine Intelligence*. IGI Global, 2021, pp. 223–240.
12. Tiwari, I., Juneja, S., Juneja, A., and Anand, R. A statistical-oriented comparative analysis of various machine learning classifier algorithms. *Journal of Natural Remedies* 2020, 21(3), 139–144.
13. Singh, H., Pandey, B. K., George, S., Pandey, D., Anand, R., Sindhwani, N., and Dadheech, P. Effective overview of different ML models used for prediction of COVID-19 patients. In *Artificial Intelligence on Medical Data: Proceedings of International Symposium, ISCMM 2021*. Springer Nature, 2022, July, pp. 185–192.
14. Pandey, D., Pandey, B. K., Sindhwani, N., Anand, R., Nassa, V. K., and Dadheech, P. An Interdisciplinary Approach in the Post-COVID-19 Pandemic Era 2022, pp. 1–290.
15. David, S., Duraipandian, K., Chandrasekaran, D., Pandey, D., Sindhwani, N., and Pandey, B. K. Impact of blockchain in healthcare system. In *Unleashing the Potentials of Blockchain Technology for Healthcare Industries* Academic Press, 2023, pp. 37–57.
16. AI and Its Impact on Public Administration. National Academy of Public Administration, 2019. https://doi.org/10.1016/j.ijleo.2022.168789
17. Purani, V. S., and Patodi, S. C. Neuro solution for damage detection and categorisation of earthquake-affected buildings. *Indian Concrete Journal*, 2002, 76(8), 504–509.
18. Nirala, K.K., Singh, N.K., and Purani, V.S. A survey on providing customer and public administration based services using AI: Chatbot. *Multimedia Tools and Applications*, 2022, 81, 1–32.
19. AI-Chatbot to determine citizen centric solution of Aadhaar Card services in Public Administration: Aadhaarbot. Bangal, Past and Present, 2022, 140(VII), SSN 0005-8807.
20. Priya, S., Terence, S., and Immaculate, J. A novel framework to detect effective prediction using machine learning security issues and privacy concerns in Industry 4.0 applications, 2021, 179–194. onlinelibrary.wiley.com/doi/10.1002/9781119776529.ch9
21. Li, X. Prediction and analysis of housing price based on the generalized linear regression model. *Computational Intelligence and Neuroscience*, 2022, 2022, 1–9.
22. Kusrini, K., Luthfi, E.T., Muqorobin, M., and Abdullah, R.W. Comparison of naive Bayes and K-NN method on tuition fee payment overdue prediction. In *4th International Conference on Information Technology, Information Systems and Electrical Engineering (ICITISEE)*. IEEE, 2019, pp. 125–130.

23. Mehtab, S., Sen, J., and Dutta, A. Stock price prediction using machine learning and LSTM-based deep learning models. In *Symposium on Machine Learning and Metaheuristics Algorithms, and Applications*. Springer, 2021, pp. 88–106.
24. Manjula, K. A., and Karthikeyan, P. Gold price prediction using ensemble based machine learning techniques. In *3rd International Conference on Trends in Electronics and Informatics (ICOEI)*. IEEE, 2019, pp. 1360–1364.
25. Gajera, P., Gondaliya, A., and Kavathiya, J. Old car price prediction with machine learning. *International Research Journal of Modernization in Engineering Technology and Science*, 2021, 3, 284–290.
26. Ravikumar, A. S. Real estate price prediction using machine learning. PhD diss., Dublin, National College of Ireland, 2017.
27. Artificial neural networks in public policy: Towards an analytical framework, Joshua A. Lee, George, Ph.D. Dissertation, Mason University Fairfax, VA, 2020.
28. Srivastava, A., Gupta, A., and Anand, R. Optimized smart system for transportation using RFID technology. *Mathematics in Engineering, Science and Aerospace (MESA)*, 2021, 12(4), 953–965.
29. Anand, R., Nirmal, V., Chauhan, Y., and Sharma, T. An image-based deep learning approach for personalized outfit selection. In *2023 10th International Conference on Computing for Sustainable Global Development (INDIACom)*. IEEE, 2023, March, pp. 1050–1054.
30. Jain, S., Kumar, M., Sindhwani, N., and Singh, P. SARS-Cov-2 detection using Deep Learning Techniques on the basis of Clinical Reports. In *2021 9th International Conference on Reliability, Infocom Technologies and Optimization (Trends and Future Directions)(ICRITO)*. IEEE, 2021, September, pp. 1–5.
31. Badotra, S., Tanwar, S., Rana, A., Sindhwani, N., and Kannan, R. Handbook of augmented and virtual reality. In *Handbook of Augmented and Virtual Reality*. De Gruyter, 2023.
32. Meslie, Y., Enbeyle, W., Pandey, B. K., Pramanik, S., Pandey, D., Dadeech, P., Belay, A., and Saini, A. Machine intelligence-based trend analysis of COVID-19 for total daily confirmed cases in Asia and Africa. In *Methodologies and Applications of Computational Statistics for Machine Intelligence*. IGI Global, 2021, pp. 164–185.
33. Lelisho, M. E., Pandey, D., Alemu, B. D., Pandey, B. K., and Tareke, S. A. The negative impact of social media during COVID-19 pandemic. *Trends in Psychology*, 2022, 31, 1–20.

23 Industry 4.0 in the Nutraceutical Industry and Public Health Nutrition

Swapan Banerjee, Damanjeet Kaur, Digvijay Pandey, Sulagna Ray Pal, Ahamefula Anselm Ahuchaogu, Muhammad Omer Iqbal, and Binay Kumar Pandey

23.1 INTRODUCTION

Nutraceutical cum functional food production and technology are always significantly correlated. The food, beverage, dietary, and other supplements industry is one of the broader industries globally that needs various raw materials, skilled staff, product experts, and a well-planned logistics supply chain system. Consumers get tasty and nutritious functional foods and supplements duly meeting with compliance by various regulatory bodies and customers [1]. People gradually become busier with their professional, educational, and other activities. Hence, there is a demand for ready-to-eat foods, also called convenience foods, to replace cooked food. There is a trend of using supplements on many days apart from functional foods. Besides clinical or therapeutic nutrition products, muscle or bodybuilding, probiotics, ayurvedic, multivitamins, minerals, and saturated fatty acids have been becoming more popular globally for decades. There are habits and choices of the buyers as per current trends, necessity, and urgent demand. Dieticians also recommend some nutraceuticals to get better results considering that food is not entirely functioning; hence, there is a need for various supplements. However, the unnecessary supplement consumption element is a type of eating disorder unwanted at the physiological level [2].

23.1.1 INDUSTRIAL TRANSFORMATION

The fourth industrial revolution, or Industry 4.0, has drastically progressed the manufacturing and operational process for all industries, including the nutraceuticals industry. The course has significantly changed the various sectors with more significant revenue globally through the latest computerized technology. More flexibility, skills, tools, and techniques are integral to this digitalization by which employees also work smartly and comfortably [3]. A better version of automation, computerization, and interconnection in Industry 4.0 promises to deliver better products and services to consumers through improved manufacturing quality and operational performance [4]. Experts predicted that Industry 4.0 should reach an investment of US$500 billion in 2021 in the economy [5]. Some countries like Cambodia, the Philippines, Indonesia, Thailand, and Vietnam may be affected by around 55% of actual employment before 2040, mainly in the food cum beverages industry and retail and wholesale markets. The manufacturing and operational sectors are more concerned [6].

23.1.2 Progression of Industry 4.0

In the eighteenth century, the first industrial revolution (industry 1.0) started with the mechanization of production, and steam-generated power, giving people a great advantage. The food and nutraceutical industry was not created at that time by using steam power or mechanization. People were mainly dependent on actual production and crop cultivation. Industry 2.0, or the second industrial revolution, was observed early in the twentieth century, especially in using electrical energy. Fredrick Taylor, an American mechanical engineer, came into the industry with his concept to optimize workers, workplace techniques, and the best utilization of resources. Industry 3.0 went into the sector with electronics products and their hazards at the industry and domestic levels. Programmable Logic Controller (PLC), introduced in the 1960s, was a landmark invention with a remarkable automation application using electronics as well as possible. In the public cum private sectors, many enterprises started with versatile management systems. These are enterprise resource planning (ERP), inventory or stock management, logistics channels, product flow scheduling (PFS), and total quality management at the factory level [7, 8]. Finally, Industry 4.0, or the fourth industrial revolution, came into the world of sectors with intelligent decision-making processes and had already impacted the industrial Internet of Things (IIoT). The Cyber-Physical Systems in Industry 4.0 can analyze, assess, and guide innovative actions through intelligent technologies. They can show and apply preventative measures and remedial action as necessary, including shipping cum logistics, manufacturing, production scheduling, quality control and assurance, capacity building, and efficiency developments [9].

Table 23.1 shows the various stages of the progression of Industry 4.0, starting from mechanical production (Industry 1.0) to mass production (Industry 2.0), from the display (Industry 3.0) to digital-physical production (Industry 4.0). The progression has been significantly observed in the food industry globally.

23.2 LITERATURE REVIEW

23.2.1 Global Nutraceuticals Industry

Nutraceuticals are recognized as having a vital role in public health nutrition and promotion. Nutraceuticals are food or parts of food that give therapeutic benefits in

TABLE 23.1
Step-by-step Progression of Industry 4.0 [10]

Industry 1.0-Mechanization	Industry 2.0-Electrification	Industry 3.0-Automation	Industry 4.0-Cyber-Physical systems
Hydraulic and steam-powered mechanical production	Conveyor belt system and use of electricity for the labor work	Utilizing information technology and electronics to elevate production	Artificial intelligence applications, Machine Learning, Big Data, Robotics, Cybersecurity, Augmented reality, etc.
18th Century (Mid-1780s)	19th Century (Mid-1870s)	20th Century (Mid-1960s)	2020 onwards (2011 – ongoing)

prevention and treatment. There is much scope for growth in nutraceutical products. Nutraceuticals are considered functional foods, dietary supplements, and medical/pharmaceutical foods [11]. There is a difference between pharmaceuticals and nutraceuticals. Pharmaceuticals are products that mitigate and treat diseases, while nutraceuticals are intended to prevent infections [12].

23.2.2 PRECISION MEDICINE

Clinical practice has gained momentum, enabling the disease prevention and treatment concept. Precision nutrition provides efficient intervention for disease treatment and management with various methods like genetics, lifestyles, physiology, and the gut microbiome [13]. The near future will tell if and how these new AI-based digital technologies will uncover new targets and pathogenic disease profiles, optimize clinical trial designs, and impact medication development in the pharmaceutical industry. The coming together of patient-centric real-world evidence (RWE) tools, electronic health records (EHRs), multi-omics profiling, digital biomarkers, and AI-based data analysis will pave the way for biomarker-enabled algorithm-based precision research and development [14].

23.2.3 MOLECULAR FINGERPRINTING

In this new era, molecular technology and fingerprinting are considered dynamic processes. The impact of nutrition among specific populations is related to molecular fingerprinting. This molecular fingerprinting should provide further information on the potential biomarkers of nutritional status and the natural history of diseases [15]. The combination of capillary electrophoresis and microsatellite markers was quite amenable to solving the adulteration problem out of the several technologies applied to identify the adulteration. Furthermore, DNA-based approaches detect contaminants in hybrid seed manufacturing lines and processed foods [16].

23.2.4 HUMAN GENOME AND NUTRIGENOMICS

The human genomic project provides information on the genome's various structural and functional properties. These functions are related to nutrients and infections [17]. Multiple studies have found that several nutrients and bioactive phytochemicals act as signaling molecules in the genes and expression changeability function. Nutrigenomics addresses the changes in the transcriptome, proteome, and metabolome. Nutrigenetics deals with the effect of genetic disposition and epigenetic modifications on nutritional biology, characterized by a mechanism other than an alteration in DNA sequencing [18]. Nutrigenomics can be fruitful in assessing the interindividual variability of nutrient absorption and nutrient metabolism and facilitating personalized dietary recommendations for health outcomes. Potential implications of nutrigenomics on public health are a few examples of setting the safe upper limits for the subgroups, better understanding of data from epidemiological and interventional to identify the health impacts of dietary factors, optimizing the intervention strategies and appropriate diagnostic tools to assess and monitor micronutrient status and their response to designed, optimized interventions [19].

23.2.5 INDUSTRY 4.0 AND QUALITY ASSURANCE

To provide the intended results, the software-enabled procedures that lead to this stage of industrialization require extensive testing and quality assurance. There is a pressing need to incorporate quality into every service and product offered to clients. This is critical to provide them with a consistent experience and ensure optimum retention [10–12].

In Table 23.2, we have tried to show an overview of all the standard digital applications cum software used in various industries, including food and nutraceutical companies. These tools are helpful not only in the food industry but in every sector worldwide for better quality products and services.

23.3 METHODS

The most popular form of publication in the medical literature is narrative reviews. There currently needs to be a method available to assess the quality of narrative reviews, unlike systematic reviews and randomized controlled trials (RCT) papers. SANRA is an acronym for the Scale for the Assessment of Narrative Review Articles [23]. More field testing, especially in terms of validity, is desirable. The feasibility, inter-rater reliability, item homogeneity, and internal consistency of SANRA are sufficient for a six-item measure. Based on the "explanations and instructions" shared with SANRA, we recommend special training on rating. Using SANRA, we conducted a narrative assessment of critically relevant material available in open-access databases. Step by step, we went through the process of reviewing: 1) Justification of the article's importance, 2) specific objectives, 3) a review of pertinent literature, and 4) Study assumption based on open-access data. Scientific reasoning, on the other hand, was not included due to the lack of an online survey of randomized

TABLE 23.2
Industry 4.0 with Digital Applications in the Food and Nutraceutical Industry [20–22]

Additive Manufacturing	Additive Manufacturing can help to identify greasy food and other nutritional issues. The meals can be customized according to a dietitian's advice based on this. Data-driven meal plans allow for customized texture and healthy interest. Printing systems will enable the production of specific flavors and mesoscale porosity. It's developing a novel utility in 3D food printing.
Artificial Intelligence	Artificial intelligence assures improvement in the quality of cafés, restaurants, delivery food chains online, hotels, and food outlets. AI with data science can hike the total production based on algorithms for sales prediction. AI can significantly upgrade the packaging, shelf life, and menu using algorithms and food safety by making a more transparent supply chain management system.
Autonomous Robot	Autonomous Systems and Robotics are a great boon in the food and nutraceutical industry 4.0 to improve productivity and enhance shipping, operation, quality compliance, and feedback systems. These two assure food security for all and proper resource applications at every management level.
Augmented Reality	AR technologies can alter existing technology by altering sensory science's pattern or shape. The five principal areas' five critical innovations are food consumption patterns, biometrics, food structure of food and taste cum texture, sensory marketing management, and augmenting sensory perception.
Big Data	Big Data helps analyze and review all the data used through current sources or traditional processes. The analysis can help give a competitive edge to the global food and beverage industry.
Cloud Computing	With instant data sharing, the cloud makes collaboration across an organization that could be easily reachable at any point in time. Cloud computing can boost productivity and also assures a higher degree of accuracy. Hence, Industry 4.0 in the food and beverage industry can grow with digitalization and interaction with other sectors.
Cybersecurity	Cybersecurity has become an important issue, and food manufacturers are careful to prevent cyberattacks from retaining all their data. Cloud-based enterprise resource planning (ERP) providers such as QAD have a solid commitment to security programs, expertise, and processes.
Digital Twin Process	The digital twin process puts forward the digital models, such as production models, that help to interact with the physical system. It can be utilized to design, optimize, and monitor its overall performance.
Edge Computing	Smart agriculture with the latest technologies is often considered one of the greatest innovations in the industry that can help diminish world hunger and undernourishment. Edge computing provides a long-term trackable model for intelligent agriculture in current agribusiness. Undoubtedly, agriculture is the nation's backbone.
Internet of Things	A high-population country like India procures versatile foods for various states. Food wastage has continuously been critical in lower-middle-class countries where more than 60% do not get daily food. However, technological advances can detect and distribute excess food to non-governmental houses or street habitats, mostly in democratic countries.

(Continued)

TABLE 23.2 (CONTINUED)
Industry 4.0 with Digital Applications in the Food and Nutraceutical Industry [20–22]

Machine Learning	Machine learning cum Deep learning is the latest advanced technology for extensive data analysis for speech recognition, image processing, object detection, etc. They have automatic feature learning to start various applications in food science such as food category recognition, quality detection of fruit and vegetables, estimation of food calories, etc.
Simulation	Food processing has highly relevant modeling and simulation as a part of Industry 4.0. The technologies can balance knowledge and other hybrid approaches. These approaches are state-of-the-art in various fields of processing and ultra-high-pressure processing.
Vertical and Horizontal System Integration	The intelligent factory horizontal integration assures IoT and other mechanical devices and engineering methodologies to work seamlessly. Vertical integration ensures the generation of data used at top management levels. The systems are related to staffing, shipping, operating, marketing, and other decisions. A report showed that networking and integration 30 years later (from 1998) would be more critical factors to increase.

control trials (RCT) research [23]. In our study, we have considered and discussed some parts of digital tools of 4.0 in the Indian and Australian food and nutraceuticals industry perspective

23.4 DISCUSSION

23.4.1 FOOD AND NUTRACEUTICALS CLASSIFICATION:

Nutraceuticals are foods or parts of foods that substantially change and sustain proper human physiological functions. The present demographic and health trends are the primary drivers of the nutraceutical market's global expansion. Dietary fiber, prebiotics, probiotics, polyunsaturated fatty acids, antioxidants, and other herbal foods are utilized as nutraceuticals. These nutraceuticals aid in treating some of the century's most pressing health issues. The health issues include obesity, cardiovascular disease, cancer, osteoporosis, arthritis, diabetes, and dyslipidemia [24]. While discussing various lifestyle disorders, some supplements are always mentioned for general or therapeutic use. Some supplements are for bodybuilding, protein supplementation, functional foods, and functional band beverages. Similarly, dieticians recommend medical nutrition therapy for hospital and institutional purposes, adding therapeutic products for all types, including the critical care unit.

Nutraceuticals are foods or components that provide medicinal or health advantages, such as illness prevention and treatment. They are natural compounds, including certain plants used as nutritional supplements and controlled as foods in a broader sense. Plant-derived nutraceuticals have gained much attention because of their putative safety and numerous dietary and therapeutic properties. They are widely regarded as a powerful tool for preventing and treating nutritionally induced acute and chronic disorders and increasing overall health and longevity [25].

23.4.2 THE FOOD AND NUTRITION INDUSTRY AND USES OF SUPPLEMENTS: INDUSTRY 4.0 IN THE INDIAN SCENARIO

India has much room to expand from a small foundation, which is estimated to be 1–2% of worldwide nutraceutical sales. On the other hand, food and pharmaceutical firms must solve various legislative, cultural, economic, and structural hurdles in the Indian market to realize that proprietarily primary drivers of growth are already in place. A trend toward health maintenance, early intervention, and disease risk reduction – significant pillars of the nutraceuticals market – is favored by increased disposable income, rising chronic disease prevalence, and willingness to pay for personal healthcare [26]. The food processing industry is projected to adopt Industry 4.0 gradually. Big Data and analytics, autonomous robots, simulation, horizontal and vertical integration, cybersecurity, the Industrial Internet of Things (IIoT), the cloud, augmented reality, and additive manufacturing are among the nine technological developments that drive Industry 4.0. In addition, the research looks at how they can be used in the food industry. Examples are intelligent manufacturing, food safety, training, marketing, and other functions expected in the food sector [27].

Figure 23 demonstrates the three major segments of food or dietary supplements that are the most in-demand in India. In our chapter, we highlight food manufacturing through its proper quality processing. Food processing enhances the shelf life of food goods by adding value to agricultural products through various procedures such as grading, sorting, packaging, etc. An active and robust food processing sector would accelerate a country's overall economic setup. In addition, the food processing sector provides critical linkages and synergies between industry and agriculture and immediate opportunities for growth and employment. Our Indian food processing industry has grown significantly over the previous five decades, beginning with a small number of plants primarily serving household or cottage markets [29].

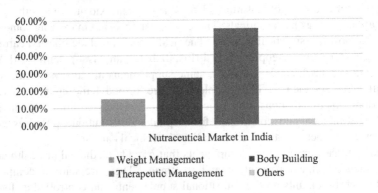

FIGURE 23.1 Category-wise nutraceutical market as Indian people purchase. *Source*: Nutraceuticals Market in India (Deloitte, 2018) [28]

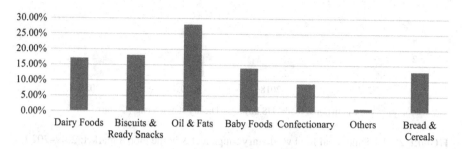

FIGURE 23.2 Category-wise functional food market in India based on consumers' purchase. *Source*: Nutraceuticals Market in India (Deloitte, 2018) [28]

Figure 23.2 classifies the six core sectors of essential food markets in India in daily need. There is a priority to consume daily food that offers almost all the vital nutrients, including multivitamins and minerals. However, some persons might have therapeutic needs also.

23.4.3 Food and Nutrition Industry and Uses of Supplements: 4.0 in the International Scenario

International dietary supplements market status: In this chapter, we have separately discussed and depicted the core digitalization of various tools using Australia as a boon of Industry 4.0. As the market for dietary supplements becomes more worldwide, many scientific and regulatory difficulties in studies on their safety, quality, and efficacy are universal to all countries. This article also describes some of the challenges in supplement science. In 2021, the worldwide dietary supplements market was worth USD 96 billion, and it is predicted to rise at a CAGR of 6.9% over the forecast period. The need for nutritional supplements is being propelled by a paradigm change toward preventative health management methods, pushed by rising healthcare expenditures and the rising burden of lifestyle diseases. The future expansion of dietary supplements will be aided by retail digitization. Furthermore, the popularity of herbal and probiotic supplements is projected to fuel future market expansion [30–32].

International Regulatory issues: in the international market, some of the regulatory problems are encountered herewith to understand the rules and regulations. In regulatory situations, science is critical, and there's no reason why science and law can't coexist. The issues in supplement science and its rules show the new opportunities for scientists and regulators to collaborate and unify approaches to promote public health on a national and international level [30]. In regulatory situations, science is critical, and there's no reason why science and regulation can't coexist. The issues in supplement science and law present new opportunities for scientists and regulators to collaborate and unify approaches to promote public health on a national and international level (30)0-32.

Figure 23.3 shows the international market for dietary and functional food supplements from 2017 to 2020. It compares both, where nutritional supplements are increasing over the same period as functional food groups moving from 2.5 USD

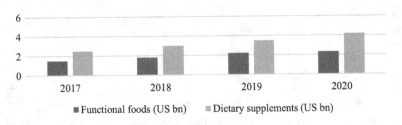

FIGURE 23.3 Functional food vs. dietary supplements in the global market: 2017–2020.

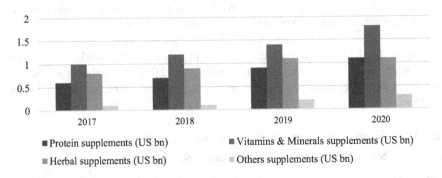

FIGURE 23.4 Protein vs. vitamins/minerals and herbal supplements global market: 2017–2020.

billion to 4.2 USD billion within three years [31]. Similarly, Figure 23.4 compares three major groups with a minimum of 1 USD bn to a maximum of 1.8 USD bn for the vitamins cum minerals group due to various levels of demand worldwide [32].

23.4.4 Quality Control and Quality Assurance

Business 4.0 is swiftly becoming a reality, thanks to tangible advances in the Internet of Things (IoT) and Wireless Sensor Networks (WSN), which profoundly impact every manufacturing industry, from logistics to quality control. Quality control measurements will no longer be performed in a separate metrology unit but on the ion line in real-time. Intelligent sensors may record and send data, but there is no meaningful added value if the data is not used to optimize a process [33]. Figure 23.5 signifies the importance of maintaining the core factors while discussing food quality and customer satisfaction.

23.4.5 Industry 4.0 and Society 5.0

As the food business gradually moves with Industry 4.0, the transition to 5.0 is planned. Japan's government was the first to offer this new paradigm, dubbed "Society 5.0." Unlike Industry 4.0, Society 5.0 is not limited to manufacturing but uses technology to address societal concerns focusing on human factors. The

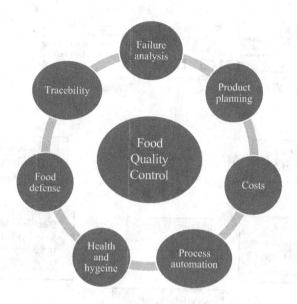

FIGURE 23.5 The relationship among the significant factors in quality control and quality assurance.

industry is at a crossroads, with serious environmental challenges and human components, to consider the future of production within the forthcoming fifth industrial paradigm, which promises technology for more sustainable, resilient, and egalitarian cyber-physical systems [34].

The 5th Science and Technology Basic Plan, issued on December 18, 2015, introduces us to Society 5.0 first. It is intended that many issues, such as unemployment, poverty, and air pollution, will be resolved with the help of artificial intelligence. Similar technologies include artificial intelligence, cyber-physical systems, Big Data, the Internet of Things, robots, augmented reality, and the cloud. As a necessary practice to reduce the gap and ensure long-term economic stability, Society 5.0 confronts us. Numerous nations, including the USA, France, China, South Korea, and Italy, have developed their own Industry 4.0-inspired application plans. In Japan, there are currently 26.7% of persons over the age of 65, and that percentage is predicted to rise to 40% by 2050.

In Figure 23.6, we have tried to show the impact of the Industry 4.0 digital revolution on personal health in the community. These are all together in a chain that helps the nation's prosperity.

23.4.6 Industry 4.0 Application and Influence in Industries in Australia

The Australian industries are transitioning into and within the "fourth industrial revolution" or Industry 4.0, with considerable progress restricted among prominent local and multinational organizations. Australian CEOs consider a lack of customer demand (30%) followed by skill shortage (21%) as the primary barrier to embracing

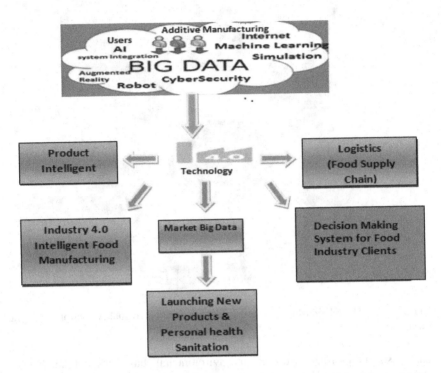

FIGURE 23.6 Big data and other digital tools impact personal health, society, and community.

Industry 4.0 Technology (I4T); however, investment in digitization has been plummeting since 2013. The Australian Bureau of Statistics (ABS) depicts that larger businesses (77%) utilize the Internet more proactively as compared to medium (69%) or smaller (56%) firms. Internet for marketing and advertising is being used majorly by the service sector (87%), manufacturers (64%), and constructors (55%), which validates significant capability for growth in this area, yet to be explored by small-medium enterprises (SMEs) and manufacturing sectors. Additionally, a large proportion of Australian businesses underestimate the value of the Internet of Things (IoT) (60%) and Radio-frequency Identification Devices (RFID) (80%). The Australian mining, transport, retail, warehousing, information media, and telecommunication sectors value such novel technologies. Australian firms that have successfully embarked on the I4T journey are Australian energy retailers (AGL). Some more companies are Brisbane-based manufacturer B&R Enclosures, Watkin's Steel, construction and engineering firm Laing O'Rourke, Adelaide-based manufacturer REDARC, and company Weir [35, 36].

In 2016, Australia officially announced its 4IR intention, indicating a significant economic transformation. The Prime Minister's Industry 4.0 Task Force strategically initiated Industry 4.0 Test Labs with the primary objective of improving the competitiveness of Australian manufacturing industries by adopting 4IR technologies, especially supporting SMEs and commencing workforce formation. These test labs are innovative factory prototypes based in leading research and educational organizations across Australia, designed, developed, and trialed in partnership with industry and other stakeholders. They will support any company in Australia, especially SMEs embarking on the Industry 4.0 journey. They will also help companies like Boeing, Imagine Intelligent Materials, Thales, Quickstep, Marand, Bosch, Austin 'Carbon Revolution, and CNand C Design. The pilot program has been established in six Australian Universities with funding of $5 million [35, 37]. In 2018, a further $2.4 billion was invested in Australia's mounting research, science, and technology.

23.4.7 INDUSTRY 4.0 IN THE FOOD AND AGRICULTURE INDUSTRY IN AUSTRALIA: DATA EXCHANGE AND DIGITAL TOOLS

Australia has been extensively known to adopt technologies early compared to other countries [38, 39]. The government is more proactive in investing and more likely to invest in novel I4T disruptive technologies [40]. The conjunction of automation and data exchange technologies in manufacturing makes Industry 4.0 possible. These intelligent factories are characterized by wireless connectivity and self-regulation, with the possibility of end-to-end production without direct human intervention. The Australian government has listed intelligent sensors, the Internet of Things (IoT), 3D printing, location detection technologies, advanced robotics, cloud, and computing as being technologies relevant to Industry 4.0 [41].

Industry 4.0 Technologies (I4Ts) have recently emerged as a novel frontier in transmuting conventional business operations and effectively counteracting unanticipated supply chain disruptions. These can transform logistics and supply chain practices, deemed more challenging in agri-food. A descriptive online survey to collect information from the Australian supply chain (n=188) was conducted with the Australian Supply Chain and Logistics Association (SCLAA) [42]. IoT technology was the most widely adopted (48%) I4T among Australian supply chain organizations. At the same time, Big Data Analytics (BDA) is estimated to have a considerable impact shortly, including predictive analytics and cost reduction. Fourteen technologies adopted by various Australian sectors are anticipated to sign an impression, as shown in Table 23.3.

Due to the highly perishable nature of fresh foods, tracking food products, maintaining quality regulations, and timely actions indicate that BDA will have the most significant impact throughout the FPI supply chain [43, 44]. Additionally, larger Australian firms are more digitally prepared and more likely to have adopted I4Ts than smaller firms, indicating that the firm's size is critical in adopting this emerging technology, except for blockchain technology. Smaller firms have commonly adopted such technology (12%) compared to larger firms (6%). It is estimated that larger Australian firms will appoint more new skilled employees, conduct external research, invest internally, and source BDA and IoT resources.

The shorter shelf life and seasonal production system of the Food Processing Industry (FPI) pose several threats, such as product demand forecasting errors, leading to risks of over or undersupply, temporary disruption in the process operations

TABLE 23.3
The Rate of Adoption, Estimated Impact, and Estimated Expenditure on Various Industry 4.0 Technologies by Varied Australian Industrial Sectors [42]

Consulting	Distribution	Govt.	Import Export	IT	Manufacture	Material Handling	Resource	Logistic	Retail	Wholesale
Highest Recent Adoption(%)										
BDA/IoT (42)	BDA/IoT (60)	Drones (60)	IoT (67)	BDA (100)	Robotics (69)	BDA (100)	Drones/IoT (67)	IoT (60)	Robotics (55)	IoT (60)
Highest Expected Impact (%)										
BDA(55)	AVs(71)	BDA (44)	BDA (50)	BDA (80)	BDA(53)	Robotics/IoT (50)	BDA (83)	AVs (69)	BDA(92)	IoT (33)
Highest Expected Expenditure (%)										
BDA (64)	BDA/Blockchain (80)	IoT (60)	IoT (83)	BDA (80)	Robotics (73)	IoT (100)	BDA/Drones (83)	BDA (76)	IoT (73)	IoT (80)

Abbreviations: Big Data Analytics (BDA); Internet of Things (IoT); Autonomous Vehicles (AVs) (42) tarnishing customer demands and firm profitability, lack of coordination in refrigerated transportation, temperature maintenance, and unexpected delays. Moreover, the recovery of supply chains from the global COVID-19 pandemic and the adoption of I4Ts has significantly improved overall extended performance and flexibility in FPIs to combat such unforeseen shocks. These supply-demand mismatches (SDM), process risks (PR), and transportation risks (TR) are the main jeopardies that cause SDM leading to severe disruptions in the food industry [35, 45]. Empirical research focusing on managers of Australian FPI, which has a market size of US$100 billion globally, revealed that I4Ts significantly alleviate these threats. However, the impact on TR was positive but not significant [35]. The frequent use of RFID and electronic data interchange (EDI) technological resources enhances the exchange of information on the production, distribution, and sale of goods, integrating the supply chain process. Proficiencies in product distribution and improved competence in estimating future customer demands will help control discrepancies in the supply-demand of FPIs by using 4ITs like big analytics, cloud computing, and IoT as critical resources [46].

In labour-intensive FPIs, possessing robotics and drones will assist in minimizing threats related to labour scarcities, high labour costs, and input losses. The short shelf life of perishable foods in FPIs causes considerable food wastage and generates revenue loss due to unpredicted machine breakdowns. IoT-connected smart sensors in FPI processes can generate quick alerts, diminishing the effects of significant machine breakdowns [47]. Financial losses can be abbreviated by using sensor-supported cameras, which scan hidden objects and help build pliability against accidents [47]. Geographical Information systems, EDI, and sensors provide real-time information on the delivery of goods due to unexpected delays or intensive weather patterns and RFID resources which would be beneficial in keeping effective transportation functioning [48]. Soosay and Kannusamy's article showed that transformation in the organization and culture of manufacturing industries goes beyond ICT integration. Adopting I4T based on the Industry 4.0 Maturity Index Model outlines four consecutive I4T maturity stages. The first stage is computerization, followed by visibility and transparency. The third stage is Predictive Capability, followed by the fourth, Adaptability. These stages aid in recognizing the current stage the firm is in and the consecutive developments required for the complete establishment of Industry 4.0.

Based on preliminary findings of an online cross-sectional survey (n=360), mostly in SMEs using a 7-point Likert scale, the current end-to-end autonomous supply chain in the Australian agri-food industry is far from reaching the Industry 4.0 milestone. Traditional labor-intensive practices are still standard, especially concerning perishable food commodities and the fragmented adoption of advanced technologies in various supply chain stages. Table 23.4 clearly shows the I4T adoption in the Australian agri-food industry [43].

It is evident from the mean scores in Table 23.3 that overall, the downstream players, like food retailers, manufacturers, and logistic providers, are utilizing novel I4Ts vigorously compared to the downstream participants of the supply chain, including farmers, producers, and growers. All supply chain stages utilized

TABLE 23.4
Industry 4.0 Stages of Adoption by the Agri-food Industry in Australia [43]

Factors involved	Input Suppliers	Producers	Packaging Providers	Food Manufactures	Logistics suppliers	Wholesale dealers	Retailers
Computerization & Connectivity							
Barcode	53.3	29.4	80	24.1	91	76.2	100
CRM	20	2.9	60	64.8	34.3	17.5	75.9
E-Business	83.3	86.7	45	77.8	65.7	12.7	70.7
EDI	0	4.4	0	79.6	28.4	30.2	81
EPOS	30	0	0	0	0	3.2	100
E-Procurement	43.3	41.1	80	90.7	80.6	15.9	89.7
Mean	38.3	27.4	44.2	56.2	50.0	26	86.2
Visibility and Transparency							
GPS	0	0	0	0	80.6	49.2	3.4
TTI	0	7.4	0	0	44.8	46.0	12
Data Loggers	0	16.2	0	13	52.2	30.2	3.4
TMS	0	0	15	9.3	80.6	7.9	0
WMS	13.3	0	70	9.3	10.4	39.7	75.9
Mean	2.7	4.7	17	6.32	53.7	34.6	18.9
Predictive Capability							
ERP	0	1.5	55	79.6	70.1	66.7	75.9
MES	0	0	75	77.8	0	0	0
RFID	10.0	0	40	27.8	80.6	36.5	79.3
Mean	3.3	0.5	56.7	61.7	50.2	34.4	51.7
Adaptability and Self-Learning							
CPFR	10	0	0	44.4	28.4	3.2	58.6

(*Continued*)

TABLE 23.4 (CONTINUED)
Industry 4.0 Stages of Adoption by the Agri-food Industry in Australia [43]

Factors involved	Input Suppliers	Producers	Packaging Providers	Food Manufactures	Logistics suppliers	Wholesale dealers	Retailers
ECR	10	0	0	29.6	0	6.3	60.3
VMI	0	0	10	44.4	4.5	0	29.3
Mean	6.7	0	3.3	39.5	11	3.2	49.4

Abbreviations: Customer Relationship Management (CRM); Electronic Data Interchange (EDI); Electronic Point of Sale (EPOS); Global Positioning System (GPS); Time Temperature Integrators (TTI); Transport Management System (TMS); Warehouse Management System (WMS); Enterprise Resource Planning System (ERP); Manufacturing Execution Systems (MES); Radio-Frequency Identification Systems (RFID); Collaborative Planning, Forecasting and Replenishment (CPFR); Efficient Consumer Response (ECR); Vendor Managed Inventory (VMI)

Barcoding Identification Technology (BIT); however, farmers and producers of fresh and perishable fruits, vegetables, and animal protein have less utilization. BIT adoption aids in controlling inventory and better communication with suppliers using Electronic Data Interchange (EDI) [49]. Electronic Point of Sale (EPOS) was used by input suppliers, wholesalers, and retailers, permitting them to develop novel products and enable cooperative purchase patterns [35, 50]. Most agri-food producers, farmers, and cultivators use the E-marketplace as a business model to expand and diversify their business opportunities and augment profitability by surpassing geographical boundaries. E-procurement like e-sourcing, e-auctions, and EDI are being used considerably among food manufacturers, logistic suppliers, and packaging suppliers essential for trading with suppliers online, thus meeting demand requirements by alleviating delivery timings. Therefore, Australian agri-food businesses have adopted the first stage of computerization and connectivity of I4R [50].

Global Positioning System (GPS), Transport Management System (TMS), Warehouse Management System (WMS), and Time Temperature Integrators (TTI) are the leading technologies utilized in the Australian food supply chain, especially by logistic providers, as part of the second stage of maturity in I4T. Using GPS and TMS, efficient vehicle scheduling and routing and managing transportation activities and distribution are possible [51]. WMS is a critical stage in managing perishable food products that farmers and producers have less widely adopted, affecting inventory management [44, 52]. To maintain quality control, prevent deterioration and instability of food, and avoid loopholes in the food supply chain TTI is essential, surprisingly used by logistic providers and wholesalers only, as shown in Table 23.3 [53]. Integrating information is necessary for efficient traceability and ensuring discernibility by different supply chain stakeholders to meet food safety and regulatory requirements. Although technological solutions are present, cultural barriers could be a reason for poor I4T adoption by upstream participants like growers, farmers, and raw material suppliers compared to logistic providers in Australia.

Predictive capability, the third maturity stage of embracing 4IT, minimizes unforeseen circumstances like production delays, bullwhip effect, and forecasting errors, which would alleviate supply chain inefficiencies like food wastage and extra cost [49, 54]. Table 23.3 shows a clear difference between upstream agri-farm producers and downstream retailers and food manufacturers in adopting this stage. Compared to upstream food providers, SC downstream activities are performed by more prominent firms that have adopted Enterprise Resource Planning (ERP) solutions and Radio-Frequency Identification Technology (RFID). Hence the Australian food industry is at an early stage concerning predictive capability. A reassessment of the powers with individual orientation is quintessential for I4T progression in the supply chain [52, 54].

The fourth stage of I4T adoption is the adaptability with self-learning characterized by a self-sufficient SC, based on the development and implementation of the three initial steps. Australian food manufacturers and retailers have majorly reached this stage of I4T adoption, inclusive of Vendor Management Inventory (VMI), Collaborative Planning Forecasting and Replenishment (CPFR), and Efficient Consumer Response (ECR). These are essential for transparency, visibility of inventory, point of sale data and demand forecasting, inventory management to alleviate supply chain efficiencies, and order fulfillment critical for shelf life, perishability, and time sensitivity [52, 55].

23.4.8 Industry 4.0 in Australian Perspective: Food Manufacturing and Quality Assurance

As manufacturing is critical for the modern Australian economy, the Australian government's vision by 2030 is to increase the value of Australian Food and Beverage (F&B) Manufacturing two-fold. This will be achievable by focusing on intelligent food and beverage manufacturing techniques, food safety and traceability, and innovative foods to create a prompt, vigorous, and responsive F&B manufacturing industry. The Australian government announced an investment of $1.5 billion in the Modern Management Strategy (MMS) on October 1, 2020, to up-scaling the manufacturing industry, enabling more competition, and creating robust supply chains. Intelligent technologies for customer-driven products will amplify productivity and competitiveness and modify the farm-to-fork process. Implementation of various I4T in the F&B manufacturing industry of consumer-driven products will be a primary goal of the country, which will include [49–54]:

1. Automatic continuous and batch control systems, line control, and inspection systems augment production and minimize errors through minimum human intervention.
2. Investment in robotics includes high-speed robot palletizing systems and soft robots to improvise food sorting and handling practices.
3. Advanced monitoring systems or sensors for temperature, humidity, and pressure verifications. This will indicate areas of productivity improvements and ensure effectiveness.
4. Alleviating food wastage by digitalized forecasting and sale predictions.
5. Optimization of supply chain management and tracking of food products.
6. Transforming packaging lines and equipment to meet national packaging targets.

According to the Data 61 report by CSIRO, Industry 4.0 can provide $315 billion to the Australian economy in the forthcoming decade. However, industry stakeholders consider the high cost of adopting I4T a barrier. Nevertheless, digitization will provide instantaneous discernments of supply chains and support decision-making through Artificial Intelligence (AI), data analytics-enabled proactive decisions, and IoT to make adjustments in real-time manufacturing. These technologies will equip and help manage the manufacturing sector during unforeseen shocks like Covid-19, which increases the vulnerability and volatility of this sector [56–58].

A robust F&B industry is not achievable in seclusion; hence, leveraging agricultural production is essential, which will simultaneously enable the growth of the F&B sector. Australia has envisioned its agricultural sector into a $100 billion worth industry by 2030. This is achievable by converting traditional farming into intelligent farming, also called Agriculture 4.0. The Australian government has allocated the Smart Farms Program $136 million. Farmers and producers will embrace sustainable approaches using I4T to monitor and manage various levels, enhancing productivity and alleviating the environmental impact by minimizing waste, using intelligent sensing and analysis systems automation in the agriculture sector. A few technologies adopted in Australia are autonomous farm vehicles, timely detection of crop disease, bee tracking, animal physiology/behavior tracking, quality inspection and yield estimates of crops, pest dissuasion, and 3D canopy condition monitoring [57, 58].

Vertebrate pest detection and deterrence is a recent project that uses sensors to detect agricultural pests and deter them by light emission and sound signals for crop protection. More than 500 sensing devices are involved in trials in Australia and Africa with financial support from OLAM's (OLAM is a leader in food, feed and fibre in high-growth emerging markets) giant agribusiness worldwide. AgScan3D+ is another sensor technology that detects the nutritional and disease status of the whole orchard, assisting growers in canopy management and ensuring high-quality fruits in vineyards [55–57].

The Australian F&B manufacturing sector has an international reputation as a clean, safe, high-quality product with trusted regulatory and quality assurance systems. This sector has been in lower gear concerning digital transformation. Automating data collection, barcoding, and RFID technology in the production line will enable efficient customer connectivity. Access to unlimited data with information integration and customer feedback will improve food quality assurance. Standardized digital information throughout the supply chain will be possible by implementing digital product management systems. Food safety during the handling, preparation, and storage of value-added foods can be done by utilizing blockchain technologies that steadily capture and store information about food origin [59].

Digital labeling can be provided to retailers and customers, such as QR codes, to identify and confirm the food products' origin, traits, and quality. A digitalized food system will build on the National Traceability Framework, which will be a tool to guide Australian food producers, agri-food industries, and governments. This will augment the accuracy of traceability systems internationally, improving quality control and other assurances to customers and trading partners. Australia's recent novel challenge project investigates technologies for monitoring human food handling and preparing methods in commercial kitchens and restaurants to ensure good food quality [59, 60].

23.4.9 Impact of Industry 4.0 on Public Health

A patriarchal medical model gives a co-managed and integrated approach to Western industrialized health and care policy. Meanwhile, the fourth industrial revolution (Industry 4.0) changes production with the digital consumer revolution. Some of the same capabilities are being used by digital health and care projects to optimize healthcare provision. In an organization-centric delivery paradigm, this is mainly limited to self-management. True co-management and integration with other organizations and people are challenging since formal care organizations need to

Industry 4.0 in the Nutraceutical Industry

share control and extend confidence. This chapter discusses a more person-centered deployment of Industry 4.0 capabilities for care through co-design. It introduces "Care 4.0 linking with society 5.0," a new paradigm focusing on trustworthy, interconnected networks of organizations, people, and technologies. It could transform how people construct digital health and care services [61]. The current status of progress, future challenges, and proposed remedies were examined. The research gaps highlighted the system's inadequacies while also offering knowledge that can be utilized to improve the technology and make it more valuable for other uses, not just healthcare [62]. In addition to general healthcare, nutrition for the public has also taken place in the Industry 4.0 platform [63–68]. In Figure 23.7, we have tried to show the intelligent, eco-friendly food industry, also called food industry 4.0. Food industry 4.0 aims to improve social well-being with better technologies.

A new industrial paradigm known as "Industrial 4.0" encompasses technologies including robotics, big data, cloud manufacturing, augmented reality, and cyber-physical systems (CPS). It significantly affects the industrial sector because it brings about advances pertinent to innovative and futuristic manufacturing. For the challenges caused by virus, primary Industry 4.0 technologies are needed. Daily reports on an infected patient's condition are helpful, broken down by state, age group, and geographic location. These technologies would benefit from proper adoption to improve public health education and communication. With the help of

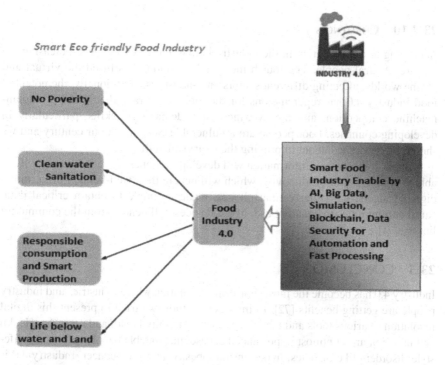

FIGURE 23.7 Smart, eco-friendly food industry with version 4.0 for social causes.

Industry 4.0, businesses may provide more efficient and effective production and service at a lower cost. The Internet of Health Things, Medical Cyber-Physical Systems, Machine Learning, and Big Data are just a few technologies that have helped the healthcare industry realize the importance of these data sets. Organizations will take preventative measures to ensure the well-being of their staff as Industry 4.0 moves toward greater automation and a marked decrease in menial tasks. To safeguard their reputations, businesses will work to prevent problems rather than address them once they have arisen. Furthermore, workers will have more safety nets because they will be harder to replace.

In this article, we have pointed out some measures communities and governments have taken to reduce excess calorie intake and enhance the nutritional status and some of the conflicts of interest and implementation issues they've encountered. A comprehensive approach involving governments and communities working in genuine partnership is essential to create meaningful change in nutritional intake quality and lower the long-term prevalence of obesity. Taking no or little action today will result in a lower life expectancy for many of today's youth in the future. Technologies can indirectly produce the best nutritious products to help our daily nutrition. Nowadays, dieticians and doctors spread much awareness online and physically through various camps and campaigns. Hence, proper dietary habits and daily exercise practice not only keep a person fit but also improve all types of communities and the entire nation [69, 70].

23.4.10 CHALLENGES

According to this research, in the industrial era 4.0, literacy impacts increasing the welfare of rural people. The fourth industrial revolution has fused the virtual and offline worlds, affecting different sectors and industries. Additionally, the potential food industry 4.0 has repercussions for the joblessness rate, the advent of human-machine competition, and the ever-increasing demand for skills, particularly in developing countries. Poor people are a vulnerable category in our century and we should be concerned about improving their survival ability.

Village areas that get information will develop awareness and, as a result, will be able to increase their productivity, which will benefit their well-being. Furthermore, digital gadgets, particularly smartphones, made it simple to obtain critical data. Rural communities that cannot use digital devices will benefit from the community library [71].

23.5 CONCLUSION

Industry 4.0 has become the most significant revolution where industries and industry people are getting benefits [72]. In this review study, we tried to present this digital revolution's various tools and techniques for community health and nutrition. Health and nutrition are of utmost importance because uncountable people suffer from lifestyle disorders like diabetes, hypertension, obesity, and even cancer. Industry 4.0 is indirectly helping to produce better quality products with better packaging, labeling, and logistics: these altogether provide affordable and low-calorie foods and nutritious food for all people globally. Currently, food is not available openly in groceries but as packet food with safety-assured labeling as ready-to-eat or functional food. Altogether, Industry 4.0 can provide us with more food and nutraceuticals with affordable pricing and safety through better quality control and quality-assured techniques.

REFERENCES

1. Demir, Y., & Dincer, F. (2020). The effects of Industry 4.0 on the food and beverage industry. *Journal of Tourismology, 6.* https://doi.org/10.26650/jot.2020.6.1.0006.
2. Banerjee, S. (2020). Food companies' role in supply nutritious foods as per buyers changing lifestyles, buying habits and the recent trends. *International Journal of Innovative Research in Science, Engineering and Technology, 9*(3), 1062–1067. https://doi.org/10.15680/IJIRSET.2020.0903127.
3. Leso, V., Fontana, L., & Iavicoli, I. (2018). The occupational health and safety dimension of Industry 4.0: Industry 4.0 and occupational health. *La Medicina del Lavoro | Work, Environment and Health, 109*(5), 327–338. https://doi.org/10.23749/mdl.v110i5.7282.
4. Cooper, J., & James, A. (2009). Challenges for database management in the Internet of things. *IETE Technical Review, 26*(5). https://doi.org/10.4103/0256-4602.55275.
5. ERPS-European Parliamentary Research Service. (2015). Industry 4.0 - Digitalisation for productivity and growth. http://www.europarl.europa.eu/RegData/etudes/BRIE/2015/568337/EPRS_BRI(2015)568337_EN.pdf. Accessed on 15th August 2021.
6. International Labour Office. (2016). Asian in transformation. The future of Jobs is at risk of automation. http://www.ilo.org/public/english/dialogue/actemp/downloads/publications/2016/asean_in_transf_2016_r2_future.pdf. Accessed on 15th August 2021.
7. Jin, C., Bouzembrak, Y., Zhou, J., Liang, Q., van den Bulk, L. M., Gavai, A., Liu, N., van den Heuvel, L. J., Hoenderdaal, W., & Marvin, H. J. P. (2020). Big Data in food safety- A review. *Current Opinion in Food Science, 36,* 24–32. https://doi.org/10.1016/j.cofs.2020.11.006.
8. KPMG. Neutraceuticals: The future of intelligent food. 2015. http://www.kpmg.com Accessed on 15th August 2021.
9. Landherr, M., Schneider, U., & Bauernhansl, T. (2016). The application center industries 4.0 - Industry-driven manufacturing, Research, and Development. *Procedia CIRP, 57,* 26–31. https://doi.org/10.1016/j.procir.2016.11.006.
10. Financial Express. (2021). Industry 4.0 technology: The key game-changer for the Indian manufacturing sector. https://www.financialexpress.com/industry/industry-4-0-technology-the-key-game-changer-for-indian-manufacturing-sector/2199098/ Accessed on 15th August 2021.
11. Ganesh, G. N. K., Ramachandran, A., Senthil, V., Suresh, V., & Baviya, P. R. (2018). Nutraceuticals - A regulatory review. *International Journal of Drug Regulatory Affairs, 3*(2), 22–29. https://doi.org/10.22270/ijdra.v3i2.165.
12. Patil, J. S. (2016). Nutraceuticals: Emerging trend in public health promotion. *Advances in Pharmacoepidemiology and Drug Safety, 5*(2), 4172. https://doi.org/10.4172/2167-1052.1000e142.
13. Picó, C., Serra, F., Rodríguez, A. M., Keijer, J., & Palou, A. (2019). Biomarkers of nutrition and health: New tools for new approaches. *Nutrients, 11*(5), 1092. https://doi.org/10.3390/nu11051092.
14. Hartl, D., de Luca, V., Kostikova, A., Laramie, J., Kennedy, S., Ferrero, E., Siegel, R., Fink, M., Ahmed, S., Millholland, J., Schuhmacher, A., Hinder, M., Piali, L., &

Roth, A. (2021). Translational precision medicine: An industry perspective. *Journal of Translational Medicine, 19*(1), 245. https://doi.org/10.1186/s12967-021-02910-6.

15. Reddy, V. S., Palika, R., Ismail, A., Pullakhandam, R., & Reddy, G. B. (2018). Nutrigenomics: Opportunities & challenges for public health nutrition. *The Indian Journal of Medical Research, 148*(5), 632–641. https://doi.org/10.4103/ijmr.IJMR_1738_18.
16. Vemireddy, L. R., Satyavathi, V. V., Siddiq, E. A., & Nagaraju, J. (2015). Review of methods for the detection and quantification of adulteration of rice: Basmati as a case study. *Journal of Food Science and Technology, 52*(6), 3187–3202. https://doi.org/10.1007/s13197-014-1579-0.
17. Dauncey, M. (2013). Genomic and epigenomic insights into nutrition and brain disorders. *Nutrients, 5*(3), 887–914. https://doi.org/10.3390/nu5030887.
18. Müller, M., & Kersten, S. (2003). Nutrigenomics: Goals and strategies. *Nature Reviews. Genetics, 4*(4), 315–322. https://doi.org/10.1038/nrg1047.
19. Moore, J. B. (2020). From personalised nutrition to precision medicine: The rise of consumer genomics and digital health. *The Proceedings of the Nutrition Society, 79*(3), 300–310. https://doi.org/10.1017/S0029665120006977.
20. Lipton, Z., Kale, D., Elkan, C., & Wetzel, R. (2015). Learning to diagnose with LSTM recurrent neural networks. https://arxiv.org/abs/1511.03677.
21. Stock, T., & Seliger, G. (2016). Opportunities of sustainable manufacturing in Industry 4.0. *Procedia CIRP, 40*, 536–541. https://doi.org/https. https://doi.org/10.1016/j.procir.2016.01.129.
22. Vaidya, S., Ambad, P., & Bhosle, S. (2018). Industry 4.0 – A glimpse. *Procedia Manufacturing, 20*, 233–238. https://doi.org/10.1016/j.promfg.2018.02.034.
23. Baethge, C., Goldbeck-Wood, S., & Mertens, S. (2019). Sandra—A scale for the quality assessment of narrative review articles. *Research Integrity and Peer Review, 4*(1), 5. https://doi.org/10.1186/s41073-019-0064-8.
24. Das, L., Bhaumik, E., Raychaudhuri, U., & Chakraborty, R. (2012). Role of nutraceuticals in human health. *Journal of Food Science and Technology, 49*(2), 173–183. https://doi.org/10.1007/s13197-011-0269-4.
25. Nwosu, O., & Ubaoji, K. (2020). Nutraceuticals: History, classification and market demand. https://doi.org/10.1007/978-3-030-42319-3_2.
26. Reuters Report. https://www.reutersevents.com/pharma/column/nutraceuticals-india-challenging-opportunity. Accessed on 15th August 2021.
27. Hasnan, N., Zafira, N., & Yusoff, Y. (2018). Short review: Application areas of Industry 4.0 technologies in food processing sector. 1–6. https://doi.org/10.1109/SCORED.2018.8711184.
28. Chatterjee, R. (2018). Nutraceuticals market overview. *Allied Market Research*. Accessed on 15th August 2021.
29. K, S., Roy, A. & Ajmal, H. (2012). A changing scenario of food processing technology in India. *Journal of Biological and Information Sciences, 1*, 17–20.
30. Dwyer, J. T., Coates, P. M., & Smith, M. J. (2018). Dietary supplements: Regulatory challenges and research resources. *Nutrients, 10*(1), 41. https://doi.org/10.3390/nu10010041.
31. Melkozernova, A. (2019). *2019 Global Sustainability Studies Programs*. 2–3. Accessed on 15th August 2021.
32. Frost&Sullivan. (2010). *Global Nutraceutical Industry: Investing in Healthy Living*. 50. Accessed on 15th August 2021.
33. Godina, R., & Matias, J. (2019). Quality Control in the Context of Industry 4.0, 177–187. https://doi.org/10.1007/978-3-030-14973-4_17.
34. Boz, Z. Moving food processing to Industry 4.0 and beyond. https://www.ift.org/news-and-publications/food-technology-magazine/issues/2021/july/columns/processing-food-processing-industry. Accessed on 20th August 2021.

35. Ali, I., Arslan, A., Khan, Z., & Tarba, S. Y. (2021). The role of Industry 4.0 technologies in mitigating supply chain disruption: Empirical evidence from the Australian food processing industry. *IEEE Transactions on Engineering Management*, 1–11. https://doi.org/10.1109/TEM.2021.3088518.
36. Yang, F., & Gu, S. (2021). Industry 4.0, a revolution that requires technology and national strategies. *Complex and Intelligent Systems*, 7(3), 1311–1325. https://doi.org/10.1007/s40747-020-00267-9.
37. Denny, L. (2019). Heigh-ho, heigh-ho, it's off to work we go – The Fourth Industrial Revolution and thoughts on the future of work in Australia. *Australian Journal of Labour Economics*, 22(2), 95–120.
38. Fallis, A. (2013). A portrait of global patterns and several testable hypotheses. *Journal of Chemical Information and Modeling*, 53(9), 1689–1699.
39. Fayezi, S., Zutshi, A., & O'Loughlin, A. (2015). How Australian manufacturing firms perceive and understand the concepts of agility and flexibility in the supply chain. *International Journal of Operations and Production Management*, 35(2), 246–281. https://doi.org/10.1108/IJOPM-12-2012-0546.
40. Power, D., & Sohal, A. (2002). Implementation and usage of electronic commerce in managing the supply chain: A comparative study of ten Australian companies. *Benchmarking: An International Journal*, 9(2), 190–208. https://doi.org/10.1108/14635770210421854.
41. Gill, A. Q., Phennel, N., Lane, D., & Phung, V. L. (2016). IoT-enabled emergency information supply chain architecture for elderly people: The Australian context. *Information Systems*, 58, 75–86. https://doi.org/10.1016/j.is.2016.02.004.
42. Hopkins, J. L. (2021). An investigation into emerging industry 4.0 technologies as drivers of supply chain innovation in Australia. *Computers in Industry*, 125, 103323. https://doi.org/10.1016/j.compind.2020.103323.
43. Soosay, C., & Kannusamy, R. (2018). Scope for industry 4.0 in agri-food supply chain. In T. R. C. M. Kersten & W. Blecker (Eds.), *The road to a digitalized supply chain management: Smart and digital solutions for supply chain management. Proceedings of the Hamburg international conference of logistics (HICL)*, Vol. 25 (pp. 37–56). epubli GmbH. https://doi.org/10.15480/882.1784.
44. Tamplin, M. L. (2018). Integrating predictive models and sensors to manage food stability in supply chains. *Food Microbiology*, 75, 90–94. https://doi.org/https://. https://doi.org/10.1016/j.fm.2017.12.001.
45. Wagner, S. M., & Bode, C. (2008). An empirical examination of supply chain performance along several dimensions of risk. *Journal of Business Logistics*, 29(1), 307–325. https://doi.org/10.1002/j.2158-1592.2008.tb00081.x.
46. Wamba, S. F., Gunasekaran, A., Akter, S., Ren, S. J., Dubey, R., & Childe, S. J. (2017). Big data analytics and firm performance: Effects of dynamic capabilities. *Journal of Business Research*, 70, 356–365. https://doi.org/10.1016/j.jbusres.2016.08.009.
47. Winkelhaus, S., & Grosse, E. H. (2020). Logistics 4.0: A systematic review towards a new logistics system. *International Journal of Production Research*, 58(1), 18–43. https://doi.org/10.1080/00207543.2019.1612964.
48. Ali, I., & Gölgeci, I. (2019). Where is supply chain resilience research heading? A systematic and co-occurrence analysis. *International Journal of Physical Distribution and Logistics Management*, 49(8). https://doi.org/10.1108/IJPDLM-02-2019-0038.
49. Kumar, P., Reinitz, H. W., Simunovic, J., Sandeep, K. P., & Franzon, P. D. (2009). Overview of RFID technology and its applications in the food industry. *Journal of Food Science*, 74(8). https://doi.org/10.1111/j.1750-3841.2009.01323.x.
50. Cox, H., & Mowatt, S. (2004). Consumer-driven innovation networks and e-business management systems. *Qualitative Market Research: an International Journal*, 7(1), 9–19. https://doi.org/10.1108/13522750410512840.

51. Aung, M. M., & Chang, Y. (2014). Traceability in a food supply chain: Safety and quality perspectives. *Food Control, 39*, 172–184.
52. Mason, S., Ribera, P., Cross, J., & Kirk, R. (2003). Integrating the warehousing and transportation functions of the supply chain. *Transportation Research Part E: Logistics and Transportation Review, 39*(2), 141–159. https://doi.org/10.1016/S1366-5545(02)00043-1.
53. Koutsoumanis, K., Taoukis, P. S., & Nychas, G. J. (2005). Development of a Safety Monitoring and Assurance System for chilled food products. *International Journal of Food Microbiology, 100*(1–3), 253–260. https://doi.org/10.1016/j.ijfoodmicro.2004.10.024.
54. Wu, N. C., Nystrom, M. A., Lin, T. R., & Yu, H. C. (2006). Challenges to global RFID adoption. *Technovation, 26*(12), 1317–1323. https://doi.org/10.1016/j.technovation.2005.08.012.
55. White, A., Daniel, E., Ward, J., & Wilson, H. (2007). The adoption of consortium B2B e-marketplaces: An exploratory study. *Journal of Strategic Information Systems, 16*(1), 71–103. https://doi.org/10.1016/j.jsis.2007.01.004.
56. Chibani, A., Delorme, X., Dolgui, A., & Pierreval, H. (2018). Dynamic optimisation for highly agile supply chains in e-procurement context. *International Journal of Production Research, 56*(17), 5904–5929. https://doi.org/10.1080/00207543.2018.1458164.
57. Food and Beverage National Manufacturing Priority Road Map. Growth opportunities | Department of industry, science, energy, and resources. Accessed 20th August 2021.
58. Prime Minister's Industry 4.0 Taskforce. (2017). *Industry 4.0 Testlabs in Australia Preparing for the Future.* https://www.industry.gov.au/sites/g/files/net3906/f/July 2018/document/pdf/industry-4.0-testlabs-report.pdf. Accessed 20th August 2021.
59. Rejeb, A., Keogh, J. G., Zailani, S., Treiblmaier, H., & Rejeb, K. (2020). Blockchain technology in the food industry: A review of potentials, challenges, and future research directions. *Logistics, 4*(4), 27. https://doi.org/10.3390/logistics4040027.
60. Bablani, L., Ni Mhurchu, C., Neal, B., Skeels, C. L., Staub, K. E., & Blakely, T. (2020). The impact of voluntary front-of-pack nutrition labeling on packaged food reformulation: A difference-in-differences analysis of the Australasian Health Star Rating scheme. *PLOS Medicine, 17*(11), e1003427. https://doi.org/10.1371/journal.pmed.1003427.
61. Chute, C., & French, T. (2019). Introducing care 4.0: An integrated care paradigm built on Industry 4.0 capabilities. *International Journal of Environmental Research and Public Health, 16*(12), 2247. https://doi.org/10.3390/ijerph16122247.
62. Shahid, S. M., & Bishop, K. S. (2019). Comprehensive approaches to improving nutrition: Future prospects. *Nutrients, 11*(8), 1760. https://doi.org/10.3390/nu11081760.
63. Gupta, A., Anand, R., Pandey, D., Sindhwani, N., Wairya, S., Pandey, B. K., & Sharma, M. (2021). Prediction of breast cancer using extremely randomized clustering forests (ERCF) technique: Prediction of breast cancer. *International Journal of Distributed Systems and Technologies (IJDST), 12*(4), 1–15.
64. Pandey, D., Pandey, B. K., Sindhwani, N., Anand, R., Nassa, V. K., & Dadheech, P. (2022). An interdisciplinary approach in the post-COVID-19 pandemic era. *An Interdisciplinary Approach in the Post-COVID-19 Pandemic Era*, NOVA Science Publishers, 1–290.
65. Pandey, B. K., Pandey, D., Anand, R., Singh, H., Sindhwani, N., & Sharma, Y. (2022). The impact of digital change on student learning and mental anguish in the COVID era. *An Interdisciplinary Approach in the Post-COVID-19 Pandemic Era*, 197–206.

taylorfrancis.com/chapters/edit/10.1201/9781003319238-13/building-integrated-systems-healthcare-considering-mobile-computing-iot-rohit-anand-ashy-daniel-lenin-fred-tarun-jaiswal-sapna-juneja-abhinav-juneja-ankur-gupta

66. Singh, H., Pandey, B. K., George, S., Pandey, D., Anand, R., Sindhwani, N., & Dadheech, P. (2022, July). Effective overview of different ML models used for prediction of COVID-19 patients. In *Artificial Intelligence on Medical Data. Proceedings of the International Symposium, ISCMM 2021*. Singapore: Springer Nature Singapore, pp. 185–192.

67. Kaur, J., Sindhwani, N., Anand, R., & Pandey, D. (2022). Implementation of IoT in various domains. In *IoT Based Smart Applications*. Cham: Springer International Publishing, pp. 165–178.

68. Jain, S., Sindhwani, N., Anand, R., & Kannan, R. (2022, March). COVID detection using chest X-ray and transfer learning. In *Intelligent Systems Design and Applications: 21st International Conference on Intelligent Systems Design and Applications (ISDA 2021) Held During December 13–15, 2021*. Cham: Springer International Publishing, pp. 933–943.

69. Banerjee, S. (2020). Uses of technologies & social media for diet and exercise awareness among obese, hypothyroid, and pre-diabetic women – A case study in West Bengal. *Journal of Xi'an University of Architecture and Technology*, *12*(3), 4682–4688. https://doi.org/10.37896/JXAT12.03/426.

70. Shobaruddin, M. (n.d.). Industry 4.0: Welfare literacy to face the challenges of the rural community. *Proceedings of the Annual International Conference of Business and Public Administration (AICoBPA 2018)*, pp. 124–127. https://doi.org/https. https://doi.org/10.2991/aicobpa-18.2019.30.

71. Banerjee, S. (2021). A comprehensive review on the economic status of the global convenience food industry. *International Journal of Business, Management, and Economics*, *2*(1), 43–52. https://doi.org/10.47747/ijbme.v2i1.236.

72. Lelisho, M. E., Pandey, D., Alemu, B. D., Pandey, B. K., & Tareke, S. A. (2022). The negative impact of social media during COVID-19 pandemic. *Trends in Psychology*, *31*, 1–20.

Index

A

Acceleration sensors, 48
ACO, *see* Ant colony optimization
Adaptability, new jobs, 234
Agent-based theory, 161
Agricultural management; *see also* Crop yields
    data-driven insights, 260
        applications, 263, 264
        benefits, 262–264
        challenges, 261
        crop yields, 265–268
        data collection, 261
        definition, 262
        predictive modeling, 261
        resource optimization, 268–270
    I4T
        benefits, 265
        blockchain technology, 264
        data analysis, 263
        industrial production, 265
        limitations, 271
AI, *see* Artificial Intelligence
Alerts wearable robotics, 159
Algorithm-centered analysis, 15
Amazon Cloud Platform (AWS), 8–9
Amazon Forecast, 9
Amazon Machine Images (AMIs), 9
Amazon Polly, 9
Amazon SageMaker, 8
Amazon's Augmented AI, 8
Amazon's "Personalize" tool, 9
Amazon Translate, 9
ANNs, *see* Artificial neural networks
Anomaly detection, 179
Ant colony optimization (ACO)
    accuracy, 164, 166, 167
    confusion matrix, 163–165
    F1-score, 168, 169
    precision, 166, 167
    process flow, 163, 164
    recall value, 166, 168
Apache Hadoop, 77
Area monitoring technology, 4
Articulated robot, 49, 51
Artificial Intelligence (AI), 2, 3, 22, 65, 69
    algorithmic black box, 162
    applications, 152, 175–177
    automation, 74–75, 116–118, 123

    concept of, 111
    confusion matrix, 124
    cultural anthropomorphism, 138
    features, 114
    gaming, 175
    handwriting recognition, 176
    health care, 180, 190, 191; *see also* Healthcare
    human-level cognition, 112
    industrial automation, 110
    industrial robots, 188
    intelligent robots, 176
    IoRT, 188
    limited memory, 153
    ML
        advantages, 178
        algorithms, 176–177
        BDA, 177, 186–187
        data mining, 177
        deep learning, 179–180, 184
        supervised, 178, 184
        unsupervised, 178–179, 184
        usage, 179
        value stream mapping, 186
        VANET, 188
    MVO, 123, 124
        accuracy, 124, 125
        error rate, 124, 125
        time consumption, 124, 126
    need of, 152–153
    NLP, 175, 183–184
    personalized medical treatment, 161
    reactive machines, 153
    research methodology, 123
    robots, 96, 101, 110, 111
    self-aware, 153
    soft computing, 110; *see also* Soft computing
    speech recognition, 176
    systems of expert, 176
    theory of mind, 153
    types of, 153
    vision systems, 176
    voice recognition, 110–111
    WK CNN model, 188
Association mining, 179
Augmented Reality (AR) technology, 352
Australian Food and Beverage (F&B)
        Manufacturing, 413–414
Author-required key transfer mechanism, 34

Automated employee data management, 249–250
Automated order picking systems, 52
Automated performance management systems, 249
Automatic identification and data capture (AIDC) sensors, 303
Automation, 80, 116, 123; *see also individual entries*
   industrial, 75, 135–136; *see also* Industrial automation
   optimization
      benefits, 209–210
      challenges in, 210, 211
      continuous improvement, 208
      data analysis, 207
      goal of, 206
      ML, 208
      process mapping, 207
      return on investment, 206
      role, 207
      simulation, 208
      statistical process control, 209
      techniques for, 208
Autonomous vehicles, 333
Aviation industry, automation, 118
AWS, *see* Amazon Cloud Platform
Azure Data Bricks, 9

B

Backpropagation (BPN) algorithm, 380–381
Barcoding Identification Technology (BIT), 412
Big data, 2, 4, 90, 122, 138
   IoT, 364–365
   RFID system, 305–306
Big Data Analytics (BDA), 177, 186–187, 407
Binary classification, 178
Biogeography-Based Optimization-Fitness Function-Proportional–integral–derivative (BBO-FF-PID) Controller Hybrid, 101
Bionics
   Grippy, 156
   limbs, 155–156
   mechanical foot, 156
   Tentacle Gripper, 156
BI tools, *see* Business intelligence tools
Blind and visually impaired, wearable robotics, 158
Blockchain-based DL as a Service (BDS), 186
Blockchain-based traceability (BT) systems
   automated systems
      challenges, 337, 342–343
      cyber-attacks, 341–342, 345, 346
      decentralized data management, 340–341
      enhanced security, 339
      financial systems, 345

      fraud prevention, 345, 346
      high computational power requirement, 343
      immutable and transparent ledger, 341
      integration, 343
      opportunities of, 339, 340
      regulatory frameworks, 344
      role of, 337, 338
      scalability, 344
      secure data sharing, 342
      security risks, 338
      tamper-proof record, 338
   benefits, 325–326
   cryptographic algorithms, 327
   SC, 327–328
   SCM
      advanced analytics, 332
      AI algorithms, 332
      autonomous vehicles, 333
      benefits and applications, 328–329
      challenges, 331–332
      consumer demand, 330
      decentralized and tamper-proof ledger records, 332
      drones, 333
      IBM Food Trust, 329
      Maersk and IBM's Trade Lens, 329
      provenance, 330
      real-time tracking, 332
      RFID, 333
      solutions, 329
      stages, 329, 330
      Walmart's Blockchain Traceability System, 329
Bluetooth-enabled devices, 54
BPN algorithm, *see* Backpropagation algorithm
Brute force attacks, 219
Business Intelligence (BI), 205
   advantages, 288
   automated systems
      advantages, 291–292
      benefits, 289, 290, 300
      challenges, 293–295
      dashboards, 296
      data collection and analysis, 290
      data visualization systems, 296
      informed decision-making, 297
      limitations, 292–293
      ML, 297–298
      NLP, 298
      predictive analytics systems, 296, 297
      process automation systems, 296
      reporting systems, 296
      RPA, 298–299
      success factor, 295–296
      types of, 296

Index

definition, 289
features, 289, 291
process, 289
Business Process Automation (BPA), 116
Business Process Management (BPM), 117

C

Camshft tracking technique, 52
Carbon emissions, 79–80
Cartesian robot, 49
Center for Internet Security (CIS) Controls, 221–223
Change management, 294
Character text generation, 180
Clinical decision support systems (CDSSs), 15
Cloud computing (CC), 1, 12
   advantages, 30
   challenges, 32–33
   cloud services, 7
   energy-efficient cluster-based routing, 35
   fog computing, 33
   infrastructure investment, 7
   need for, 30, 31
   platforms, 8
      AWS, 8–9
      Google's Cloud Platform, 9
      IBM's Cloud Platform, 9–10
      Microsoft Azure, 9
   precision agriculture, 30–32, 40
      accuracy, 36–38
      confusion matrix, 36, 37
      data transmission, 35, 36
      F1-score, 38, 40
      precision, 38, 39
      recall value, 38, 39
      research methodology, 36
      sensor data, 35, 38
   primer of, 7, 8
   resource utilization, 33–34
   services, 29
Cloud security risks, 218
Clustering, 178
Cognitive automation, 202
Collision avoidance, swarm robotics, 97–98
Communication
   skills, new jobs, 234
   soft computing, 69
Compliance, HR professionals, 247
Confusion matrix
   ACO, 163–165
   AI, 124
   industrial automation, 16–18, 81–82
   precision agriculture, 36, 37
   PSO, 101
Convenience foods, 396

Convolutional neural network (CNN), 140, 162
Cost savings, HR professionals, 251–252
Counter-propagation neural network (CPN), 380
   architecture, 380–381
   fee prediction
      2-180-1 architecture, 388, 389
      admission, 387, 388
      ANN, 385
      expenditure, 390–391
      faculty network input interface, 388–390
      information flow, 385, 387
      multiple human intervention, 385
      output, 392, 393
      parameters, 385, 386
      placement, 391
      quality education, 391
      result, 388, 389
      training and testing data, 388, 390
   prediction methods, 384–385
   supervised learning, 382–384
   training and testing algorithm, 383
   unsupervised competitive learning network, 382
Creativity, new jobs, 234
Critical thinking, new jobs, 234
Crop yields
   data-driven insights, 267–268
   factors affecting, 266, 267
   food security, 266
   predictive modeling, 265
   resistant crop varieties, 267
Cryptographic algorithms, 327
Customer Relationship Management (CRM) systems, 205
Cutting-edge computational intelligence, 13–14
Cyber-attacks, 341–342, 345, 346
Cyber-physical system, 28
Cybersecurity risks
   assessment, 216–217
   cloud security risks, 218
   definition, 216
   denial-of-service attacks, 217
   emerging technologies and trends, 225–226
   endpoint security risks, 218
   frameworks
      application, 222–223
      CIS Controls, 221–223
      evaluation, 222
      FAIR, 221, 223, 226
      PCI DSS, 221, 222
   human error, 219
   insecure communications, 220
   insider threats, 219, 220
   IoT devices, 217, 218, 220, 225
   management, 216
      strategies for, 223–224

network security risks, 219–220
physical attacks, 220
physical security risks, 218
Ransomware, 220
remote exploits, 220
threats and vulnerabilities, 217, 218
types of, 215, 218–219
Cybersecurity skills, new jobs, 234
Cylindrical robot, 49

D

Dashboards, 296
Data analysis, agricultural management
    I4T, 263
    resource optimization, data-driven insights, 268
Data analytics (DA)
    HR professionals, 247
    Industry 4.0, 275
        applications, 278
        benefits, 283
        challenges, 279–280, 283–284
        customer analytics, 279
        energy management, 278
        predictive analytics, 278
        process optimization, 278
        production process, 276
        product quality, 276
        quality control, 278
        supply chain optimization, 278
        techniques and tools, 277
Data compression, soft computing, 70
Data-driven statistical patterns, 80
Data mining, 2, 385
Data plumbing technologies, 6–7
Data predictive analytics, health care, 185
Data security and privacy, HR professionals, 254
Data visualization systems, 277, 296
Decentralized data management, 340–341
Decision support system, 30–31
Deep learning (DL), 2, 153, 154, 384
    automatic machine translation, 179–180
    character text generation, 180
    face recognition, 179
    image colorization, 179
    object classification, photographs, 180
    pre-training method, 184
Deep neural networks (DNN)/PID controller, 99
Defense Research and Development Organization (DRDO), 134
Defense sectors, robotics, 95–96
Delta robot, 49
Denial-of-service attacks, 217
Development programs, HR professionals, 250–251

Dielectric resonator antenna (DRA)
    curved dual band, 314
    dual band integrated, 317–318
    layers, 314
    L-shaped, 315, 316
    multilayer, 311
    octahedron, 315, 316
    rectangular, 316, 317
    Y-shaped, 316–317
Digital worker, 74, 117, 135
Dimensionality reduction, data collection, 179
Distinct texture-based features, 79
Distributed denial-of-service attacks, 219
Dual band integrated dielectric resonator antenna (DRA), 317–318
Dynamic waste management system, 13

E

Edge computing, 277
Education systems
    automation, workforce challenges
        collaboration and teamwork, 235
        creativity and innovation, 235
        diverse skill set, 235
        effectiveness, 237–239
        on employment, 233
        extent of, 232
        global education system, 237
        high-skilled jobs, 232
        income inequality, 231
        international cooperation, 239
        job displacement, 230–231
        low-skilled jobs, 232
        policies and programs, 236
        response to, 235–236
        skills and knowledge, 229, 231, 233–234, 237, 240
        social impact, 235
        upskilling and reskilling, 235
        vocational education, 236
    labor market, 229–230
Electronic data interchange (EDI), 412
Electronic health records (EHRs), 183, 185–186
Electronic Point of Sale (EPOS), 412
E-marketplace, 412
Embedded subscriber identity module (ESIM), 34
Emotional intelligence, new jobs, 234
Employment
    engagement, HR professionals, 247, 250
    negative effects, 233, 234
    positive effects, 233, 234
Endpoint security risks, 218
Enhanced training, HR professionals, 250–251
Ensem Convolutional Network, 185

Index

Enterprise resource planning (ERP), 90, 205, 412
Entertainment, robotics, 96
Environment boundary, swarm robotics, 98
E-procurement, 412
Evolutionary computing, 69–70
Exoskeleton legs, wearable robotics, 159

F

Face recognition, 47, 179
Factor Analysis of Information Risk (FAIR) framework, 221, 223, 226
Fee prediction, engineering institutions, 380; *see also* Counter-propagation neural network
Financial data leverage
  accounting software, 205
  automation
    benefits, 201–202
    definition, 201
    optimization of, 206–210
    production process, 201
    smart factory, 200
    types, 202, 203
  benefits, 204
  BI tools, 205
  bottlenecks in, 199–200
  challenges, 198–199
  CRM software, 205
  definition, 204
  ERP systems, 205
  manual data entry, 205
  POS systems, 205
  predictive analytics, 203
  profitability and competitiveness, 200
  sources of, 198, 204–205
Finite Element Method (FEM), 315
Finite Integration Technique (FIT), 315
Fixed automation, 202
Flexible automation, 202
Fog computing, 33
Food and nutraceutical industry
  in Australia, 405–407
    BDA, 407, 408
    F&B Manufacturing, 413–414
    FPI, 407, 409
    I4T disruptive technologies, 407, 409
    supply chain stages, 409–412
  challenges, 416
  classification, 401
  digital applications, 399–401
  human genomic project, 399
  India
    dietary supplements, 402
    food markets, 403
  international scenario
    dietary supplements market status, 403
    functional food *vs.* dietary supplements, 403, 404
    protein *vs.* vitamins/minerals and herbal supplements, 403–404
    regulatory issues, 403
  molecular fingerprinting, 398
  nutrigenomics, 399
  precision nutrition, 398
  public health, 414–416
  quality assurance, 399
    vs. quality control, 404
  SANRA, 399
Food Processing Industry (FPI), 407
Fuzzy analytic hierarchy process, 14
Fuzzy logic (FL), 66, 67, 69, 113–116

G

Genetic algorithm (GA), 66, 68, 98, 114
Gesture recognition, 53
Global education system, 237; *see also* Education systems
Global navigation satellite system, 31
Global nutraceuticals industry, 397–398
Global positioning system (GPS), 31, 412
Graphical structuring notation (GSN), 78
Grippy, 156

H

Handwriting recognition, 116, 176
  soft computing, 70
Hard computing, 112, 113
  vs. soft computing, 68, 69, 115
Healthcare
  applications, 183
  BDA, 352
  BinDaaS, 185, 186
  blockchain technology, 181, 183
  caregiving, definition, 180
  database, 183
  data predictive analytics, 185
  DeTrAs project, 185
  DL, 185, 186
  EHRs, 183, 185–186
  Ensem Convolutional Network, 185
  error rate, 192, 194
  federated learning, 186
  functions of, 350
  health data ecosystem, 181, 182
  image processing, 187
  Industry 4.0, 350–351
    3D-printed prosthetics, 353, 354
    AR technology, 352–354
    challenges and limitations, 356–357

continuous monitoring and feedback, 353–354
enhanced diagnosis and treatment, 354
nanotechnology, 353, 354
personalized medicine, 354–355
preventive healthcare, 356
RPM, 355
telemedicine, 355–356
VR technology, 352–354
IoT, 187, 351
LRF-PCA, 190
management system, 181
ML
  advantages, 178
  algorithms, 176–177
  BDA, 177, 186–187
  data mining, 177
  deep learning, 179–180, 184
  supervised, 178, 184
  unsupervised, 178–179, 184
  usage, 179
  value stream mapping, 186
  VANET, 188
neurosurgery, 184
OPF classifier, 187
process flow, 191, 192
research, 183
RFID system, 312
robotics, 183
shared private key, health information, 183
SM on MH
  impacts, 357
  outcomes, digital age, 359
  positive and negative effects, 357–358
  time consumption, 191, 193
Healthcare 4.0 paradigm, 185
Hearing aid, wearable robotics, 158–159
Home appliances, soft computing, 69
Home automation, 116
Human arm movement-detecting module, 139
Human error, cybersecurity risks, 219
Human resource information system (HRIS), 73, 179
Human resource management (HRM)
  benefits, 244–245
  challenges, 252–253
    automation balance, human touch, 255–256
    data security and privacy, 254
    job displacement, 253
    personal connections, 254–255
    resistance to change, 256
    retraining programs, 253–254
  definition, 245
  ethical implications, 245, 256, 257
  functions of, 245, 246
  human interaction and empathy, 245
  importance of, 245–247
  opportunities, 247–248
    automated employee data management, 249–250
    cost savings, 251–252
    development programs, 250–251
    employee engagement, 250
    enhanced training, 250–251
    streamlined recruitment processes, 248–249
Human-Robot Interaction (HRI)
  acceleration sensors, 48
  accuracy, 57, 58
  adaptability, 55–56
  Bluetooth-enabled devices, 54
  cooperative methods, 55
  error rate, 58, 59
  industrial robot, 46
    Camshft tracking technique, 52
    gesture recognition, 53
    temporal dynamics, 53
    types of, 49
  light sensors, 48
  navigation sensor, 49
  NLP skills, 47
  nonverbal behavioral indicators, 54
  proximity sensors, 48
  remote interaction, 46
  research methodology, 57
  safety
    characteristics, 49
    control and sensing systems, 50
    inline palletizing, 51
    layer depalletizing, 51–52
    mixed case palletizing, 52, 57
    palletizing, 50–51
    social cues, 50
  Smart Pads, 54
  sound sensors, 48
  temperature sensors, 48
  Theatrical Robot paradigm, 48
  types of, 46, 47

I

IBM Food Trust, 329
IBM's Cloud Platform, 9–10
IBM Watson Studio, 10
ILE, *see* Innovative learning environment
Image colorization, 179
Image processing, soft computing, 70
Industrial automation, 110, 117–118
  accuracy, 19, 82
  autopilot controls, 75
  benefits, 118, 119

Index

confusion matrix, 16–18, 81–82
deep learning approach, 22
distribution business, 75
efficiency, 135
error rate, 82, 83
F1-score, 20, 21
flow chart, 16, 17
IoRT, 132
IoT, 137
lower costs, 77, 119, 136
monitoring assets and equipment, 76–77
monitoring reduction, 119, 136
oil and gas business, 75
paper mills, 75
pneumatic controls, 79
precision, 19, 20
quality production, 76, 119, 136
recall value, 19–21
research methodology, 16
safety, 76, 136
steel mills, 75
time consumption, 83, 84
time saving, 76, 119, 136
workplace safety, 118
Industrial Internet of Things (IIoT), 397
Industrial robot, 46, 96, 159, 188
    Camshft tracking technique, 52
    gesture recognition, 53
    temporal dynamics, 53
    types of, 49
Industry 4.0, 99–100, 104–105; *see also individual entries*
    data-driven approach, 284
    definition, 199, 326
    digital transition, 160
    evolution of, 304, 305
    Maturity Index Model, 409
    progression of, 397, 398
    revolution, 138
Industry 4.0 technologies (I4T), 407, 409
    benefits, 265
    blockchain technology, 264
    data analysis, 263
    industrial production, 265
    limitations, 271
Information technology (IT) automation, 116
Innovative learning environment (ILE), 162
Insecure communications, cybersecurity risks, 220
Insider threats, cybersecurity risks, 219, 220
Instructional robots, 162
Integrated automation, 202
Integrated pest management (IPM), 269
Intelligent robots, 139, 161, 176
Intelligent Transportation Systems, 188
Intelligent wearable assistant, 159
Internet-connected gadgets, 12

Internet of Everything (IoE), 364, 365
Internet of Robotic Things (IoRT), 132, 140, 188
    accuracy, 144, 145
    error rate, 144, 146
    process flow, 142, 143
    research methodology, 142, 143
    time taken (sec), 144, 145
Internet of Things (IoT), 2–3, 15, 28, 275; *see also* Internet of Robotic Things (IoRT); Radio frequency identification (RFID) system
    agricultural equipment manufacturers, 33
    applications, 14
        big data, 4
        environmental monitoring, 5
        instruments and sensors, 5
        manufacturing, 3, 5
        marketing, 3
        personalized advertising, 5
        personalized experience, individual needs, 4
        physiology and pharmacology, 4
        product use and protection, 5–6
        public administration and public safety, 4
        residential and commercial needs, 6
        service delivery/public safety, 3
    architecture, 364, 365
    big data, 364–365
    components, 133
    cutting-edge computational intelligence, 13–14
    cyber-physical system, 28
    cybersecurity risks, 217, 218, 220, 225
    data processing, 133
    detection and control, 132–133
    device identification, 366
    dispersed processing, 14
    domain application, 28
    elements, 26–27
    environmental changes, 28
    health care, 187
    industrial automation, 16, 137
        accuracy, 19
        confusion matrix, 16–18
        DL approach, 22
        F1-score, 20, 21
        flow chart, 16, 17
        precision, 19, 20
        recall value, 19–21
        research methodology, 16
    interfaces, 363–364
    vs. Internet, 3
    layers, 133
    M2M capabilities, 362
    motto of, 27
    multiband antenna design, 371, 372; *see also* Microstrip patch antennas

optical sensors, 362
physical components, 29
precision agriculture, 40
   accuracy, 36–38
   confusion matrix, 36, 37
   data transmission, 35, 36
   F1-score, 38, 40
   precision, 38, 39
   recall value, 38, 39
   research methodology, 36
   sensor data, 35, 38
real-time sensors, 6
rule-based control applications, 365–366
servo motors, 6–7
smart homes, 14
structure of, 29
telemetry, 134
user-friendly web interface, 27, 41
IPM, *see* Integrated pest management

J

Job displacement, HR professionals, 253

K

K-Nearest Neighbor (K-NN) method, 384

L

Layered defense approach, cybersecurity risks, 223
Layer-picking systems, 51–52
Light sensors, 48
Linear Discriminant Analysis (LDA), 79
Linear regression random forest methods, 384, 385
Load balancing, 30
Logistic regression (LR), 184
Low-rank factorization with principal component analysis (LRF-PCA) technique, 190
L-shaped dielectric resonator antenna (DRA), 315, 316

M

Machine learning (ML), 127; *see also* Cloud computing
   actuators, 6–7
   algorithm-centered analysis, 15, 153
   application of
      analytics business, 73
      automated vehicles, 73
      customer's internet service, 73
      HRIS, 73
      smart assistants, 73
   automated BI systems, 297–298
   benefit of, 70–71
   controller, 6
   credit card fraud, 12
   cutting-edge computational intelligence, 13–14
   dispersed processing, 14
   DL, 2
   dynamic waste management system, 13
   e-health services, 12
   health care; *see also* Healthcare
   human-centered assessment, 15
   Industry 4.0, 275
      benefits, 280, 283
      challenges, 282–284
      limitations, 282
      predictive models, 284
      process optimization, 280
      techniques and tools, 280–281
      use cases, 281–282
   intelligent cloud, 1
   IoT, 2–3; *see also* Internet of Things
   NIC, 10–11
   real-time sensors, 6
   soft computing, 65–70, 80
   TNA, 13
   types, 71
      reinforcement learning, 71, 73
      semi-supervised learning, 71, 72
      supervised and unsupervised learning, 71, 72
Machine-to-machine (M2M) capabilities, 362
Maersk and IBM's Trade Lens, 329
Malware, 220
Man-in-the-middle attacks, 220
Medical sectors, robotics, 96
Mental Health (MH), *see* Social media
Meta-heuristic algorithms, 92
Microsoft Azure Cognitive Search, 9
Microstrip patch antennas (MPAs), 362–363
   absolute far-field directivity pattern, 373, 375
   Aperture Coupled Feed method, 368
   dimensions of, 371, 372
   E-field, 369
   electromagnetic energy source, 367
   far-field radiation pattern, 373, 374
   frequency of, 366–367
   gain and front to back ratio, 373, 376
   parameters of, 369–371
   rectangular, 368
   S-parameter, 371, 373
Military robots, 159
Minimum viable option (MVO), 123, 124
Mining robot, 96, 155
Mixed-type package palletizing, 52
Modern Management Strategy (MMS), 413
Molecular fingerprinting, 398

Index

Moth-flame optimization, 35
Multi-class classification, 178
Multi-layer perceptron (MLP), 15
Multispectral photography, 31
Multivariate analysis, 121

N

Naive Bayes, 384
Nanorobots, 353
Natural computing techniques, 11, 92
Natural language processing (NLP), 9, 175, 183–184, 239
    automated BI systems, 298
    skills, 47
Nature-inspired based methods, 12
Navigation sensor, 49
Network security risks, 219–220
Neural Network (NNs), 115, 184
NIST Cybersecurity Framework, 222, 226
Nonlinear optimization model, 122
Nonverbal behavioral indicators, 54
Nonverbal coordination technique, 50
Nursing Interventions Classification (NIC)
    autonomous entities, 11
    Industry 4.0, 91–92
    meta-heuristic algorithms, 11
    natural computing techniques, 11
    problem-solving techniques, 10
Nutraceutical industry, *see* Food and nutraceutical industry
Nutrigenomics, 399

O

Object classification, photographs, 180
Octahedron dielectric resonator antenna (DRA), 315, 316
Oil and gas industry, automation, 118
On-demand self-services, 30
Optimum-path forest (OPF) classifiers, 187

P

Paper industry, automation, 118
Parallel Distributed processing, 67, 114
Particle swarm optimization (PSO), 15, 35, 81–82
    accuracy, 103, 124, 125
    advantages, 94–95
    confusion matrix, 101
    disadvantages, 95
    error rate, 103, 124–126
    flow chart, 101, 102
    gradient descent approach, 94
    hospitable zone, 93
    Hybrid GA-PSO Algorithm, 100

Industry 4.0, 99–100, 104–105
    with digital technology, 89
    NIC, 91–92
    Smart Factory, 90–91
IoT services, 101
micro-PSO, 99
mobility and intelligence, 89
MVO, 123, 124
profit maximization, 93
quantum, 100
quasi-Newton approach, 94
research methodology, 101, 102
robotics
    collision avoidance, 97–98
    controller, 97
    discrete *vs.* continuous movement, 97–98
    low dimensionality, 98
    real-world noise, 98
    source localization activities, 97
stochastic optimization technique, 94
time consumption, 124, 126
Pattern recognition, 11, 92
Payment Card Industry Data Security Standard (PCI DSS), 221, 222
Performance management, HR professionals, 247
Personal connections, HR professionals, 254–255
Personal digital assistant, 153
Physical attacks, cybersecurity risks, 220
Physical security risks, 218
Phytogeomorphology, 31
Plant-derived nutraceuticals, 401–402
Point of sale systems (POS) systems, 205
Poker and Checker games, soft computing, 70, 116
Polar robot, 49
Precision agriculture, 30–32, 40
    accuracy, 36–38
    confusion matrix, 36, 37
    data transmission, 35, 36
    F1-score, 38, 40
    precision, 38, 39
    recall value, 38, 39
    research methodology, 36
    sensor data, 35, 38
Precision nutrition, 398
Predictive analytics systems, 296, 297
Process automation systems, 296
Programmable automation, 202
Proximity sensors, 48
PSO, *see* Particle swarm optimization

Q

Quantum computing, 225
Quantum particle swarm optimization algorithm (QPSO), 100
Quasi-newton approach, 94

R

Radio frequency identification (RFID) system, 303, 304, 412
  active tag, 308–309
  applications
    animal identification, 313
    anti-theft system, 313
    asset tracking, 313
    baggage, 313
    healthcare, 312
    libraries, 313
    national identification, 314
    security and control, 312
    security guard patrol, 312
    toll road, 313
    waste management, 314
  big data, 305–306
  BT systems, 333
  components, 307–308
  DRA
    curved dual band, 314
    dual band integrated, 317–318
    layers, 314
    L-shaped, 315, 316
    octahedron, 315, 316
    rectangular, 316, 317
    Y-shaped, 316–317
  frequency, 309–310
  impact of, 304–305
  manufacturing, 306
  operating principle of, 310
    backscattering, 312
    inductive coupling, 311–312
    multilayer DRA, 311
  passive tag, 308–309
  pillars of, 305, 306
  pre- and post-sale, 307
  production process, 304
  protocols control the communication, 307
  reader, 310–311
  semi-active tags, 308–309
  supply chain, 307
  tag detector, 310
  tag standards, 311
Ransomware, 220
Raspberry Pi, 137–138
Real state prediction model, 384
Real-time sensors, 6, 31
Rectangular dielectric resonator antenna (DRA), 316, 317
Regression modeling, 178
Reinforcement learning, 71, 73
Remote exploits, cybersecurity risks, 220
Remote heart monitoring, wearable robotics, 158
Remote patient monitoring (RPM), 355
Reporting systems, 296
Resource optimization, data-driven insights
  challenges, 269–271
  conservation agriculture, 268
  data analysis, 268
  IPM, 269
  predictive analytics, 268
  role of, 269–270
Retraining programs, HR professionals, 253–254
RFID system, *see* Radio frequency identification system
Risk management, *see* Cybersecurity risks
Robot aesthetics, 47
Robot-based inline palletizing system, 51
Robotic palletizing, 50–51
Robotic process automation (RPA), 117
Robotics; *see also individual entries*
  applications, 95
  Daksh, 134, 154
  social engineering, 159
  transhumanism, 160
Robotics 4.0, 189
Robotics Process Automation (RPA), 298–299

S

Scale for the Assessment of Narrative Review Articles (SANRA), 399
Search target, swarm robotics, 98
Search visual analytics tool, 15
Second-level cluster-to-node authentication matrix, 34
Secure Storage Management Expenditure, 30
Selective Compliance Assembly Robot Arm (SCARA) robot, 49
Self-driving cars, 179
Semi-supervised learning, 71, 72
Sensor networks, 79, 120
Sex robots, 159
Smart Contracts (SC), 326
Smart Factory, 137, 326
  characteristics, 90
  custom manufacturing, 91
  data analysis, optimal decision-making, 90
  IT-OT integration, 90
  multi-cloud IT infrastructure, 91
  supply chain, 91
Smart Healthcare and Agriculture Solutions, 35–36, 40
Smart health watches, 157
Smart homes, 14
Smart manufacturing, 137
Smart Pads, 54
Smartwatch, 156–157

Index

Social media (SM), 349–350
  on MH
    impacts, 357
    outcomes, digital age, 359
    positive and negative effects, 357–358
  ML, 178
  virtual assistants, 298
Society 5.0, 404–405
Soft computing, 83–84, 110
  advantages, 70, 116
  applications, 69–70, 115–116
  communication, 115–116
  computational intelligence, 65, 112
  data compression, 116
  disadvantages, 70, 116
  DM and DL analytics, 120
  electrical spinning equipment, 121
  elements, 66–67, 113–114
    ANN, 67–68
    FL, 67, 113, 114
    GA, 68, 114
  features, 66, 112
  handwriting recognition, 116
  vs. hard computing, 68, 69, 115
  home appliances, 115
  image processing, 116
  natural selection, 65, 68
  need for, 66, 113
  NNs, 113
  Poker and Checker game, 116
  problem-solving, 65
  real-world applications, 66
  research methodology, 80, 81
  robotics, 115
  transportation, 116
Sound sensors, 48
Speech recognition, 176
Strategic planning, HR professionals, 247
Streamlined recruitment processes, HR professionals, 248–249
Supervised learning, 71, 72, 178, 184, 385
  CPN, 382–384
Supply chain management (SCM); *see also* Blockchain-based traceability systems
  stakeholders, 325
  and traceability, 326
    challenges, 327
    transparency and accountability, 326
Support vector machine (SVM), 184
Swarm robotics
  collision avoidance, 97–98
  controller, 97
  discrete *vs.* continuous movement, 97–98
  low dimensionality, 98
  real-world noise, 98
  source localization activities, 97
Systems of expert, 176

T

Talent acquisition, HR professionals, 247
Technical skills, new jobs, 233
Telerehabilitation, 139
Temperature sensors, 48
Temporal dynamics, 53
Temporal network analysis (TNA), 13
Tentacle Gripper, 156
Theatrical Robot paradigm, 48
Time sensitive networking (TSN) capabilities, 120
Time Temperature Integrators (TTI), 412
Transhumanism, 160
Transportation, soft computing, 69–70
Transport Management System (TMS), 412

U

Unsupervised learning, 67, 114, 178–179, 184
  CPN, 382

V

Value stream mapping, 186
Vehicular Ad hoc Network (VANET), 79, 188
Virtual assistants, 179
Virtual Reality (VR) technology, 352
Vision systems, 176
Vocal coordination technique, 50
Vocational education, 236
Voice recognition, 110–111

W

Walmart's Blockchain Traceability System, 329
Warehouse Management System (WMS), 412
Water-stressed crops, 266
Wearable robotics, 155, 169
  AI components
    for blind and visually impaired, 158
    exoskeleton legs, 159
    hearing aid, 158–159
    intelligent wearable assistant, 159
    providing alerts, 159
    remote heart monitoring, 158
  bionics
    Grippy, 156
    limbs, 155–156
    mechanical foot, 156

Tentacle Gripper, 156
biosensors, 157
chargers, 157
classification of, 155, 156
CNN, 162
disadvantages, 157–158
healthcare, 157
research methodology, 163
Smartwatch, 156–157

Web of things (WoT), 137
Weilding robot (WR) development, 140
Wide kernel convolutional neural network (WK CNN) model, 188

Y

Y-shaped dielectric resonator antenna (DRA), 316–317

Printed in the United States
by Baker & Taylor Publisher Services

Printed in the United States
by Baker & Taylor Publisher Services